Impact of Climate Change on Plantation Crops

THE EDITORS

Dr. K.B. Hebbar, Currently Head of Plant Physiology, Biochemistry and Post Harvest technology at ICAR-CPCRI, Dr. K.B.Hebbar has made significant contribution to the studies on Climate change and abiotic stress tolerance of various field and plantation crops. Earlier he was involved in calibration and validation of a simulation model Infcrop-cotton, the same was later used to predict the production under future climate. His work on water use efficiency of groundnut was well recognized which led to the development of drought tolerant groundnut varieties. Similarly he has worked on other abiotic stresses like salinity and flooding tolerance. In recognition of his contribution on climate change and abiotic stress he has been awarded Borlaug Fellow under Global Research Alliance program on climate change. He has published his research findings in reputed national and international journals.

Dr. Soora Naresh Kumar is Principal Scientist at the Centre for Environment Science and Climate Resilient Agriculture, Indian Agricultural Research Institute, New Delhi, 110012, INDIA. His major research contributions include i) Assessment of regional impacts of climate change on wheat, rice, maize, sorghum, soybean, mustard and coconut; adaptation gains and vulnerable regions for these crops, reported in India's National communication to UNFCCC, IPCC reports, for policy support ii) Socio-economic impacts of climate change, implemented community level participatory adaptation iii) Lead scientist of InfoCrop models, compared models in global group of modellers under AgMIP iv) Developed simulation models for coconut (InfoCrop-COCONUT-CocoSim), green gram, spinach and grapes (VitisMod) and released InfoCrop version 2. and 2.1 being used in 44 countries. v) Quantified carbon sequestration potential,responses of coconut, areca nut and cocoa to elevated temperature and CO_2 vi)Characterized drought impacts on coconut, identified *in situ* tolerant palms, developed drought management strategies, characterized coconut germplasm for fatty acid profile. Current research interests include developmentand implementation of adaptation framework in climatically challenged areas; developing integrated assessment methodologies for managing agriculture in extreme weather events and climate change scenarios and developing processed based simulation models for various crops. Published about 100 high impact journal papers. A globally travelled scientist, winner of several awards and is a fellow of scientific societies. Carried out national assignments by the Govt. of India; IPCC-AR5 report expert-reviewer; member in global group of crop modelers; Network coordinator, ICAR-Network project on climate change (2010-2013); Invited by Parliamentary Forum on Global Warming and Climate Change to deliver a lecture. Member, Wheat Expert Group-Modelling, Wheat Initiative.

Dr. Pallem Chowdappa received M.Sc. in 1980 from Sri Venkateswara University, Tirupathi, Ph.D in 1985 from Mangalore University, Mangalore, Karnataka and post doctoral research at CABI Bioscience, U.K. He joined as Scientist-SI in 1985 at ICAR-Central Plantation Crops Research Institute, Kasaragod, Kerala and was elevated to Principal Scientist in 2006 at Indian Institute of Horticultural Research, Bangalore. Dr. Chowdappa served as Scientist-in-Charge, Central Plantation Crops Research Institute Research Centre, Hirehalli and Head, Central Horticultural Experimental Station, Hirehalli from December, 2000 till April, 2006. He became Director, Central Plantation Crops Research Institute, Kasaragod in September, 2014. Dr. Chowdappa is specialized in molecular plant pathology and has over 30 years of research experience in molecular characterization and management of *Alternaria*, *Colletotrichum* and *Phytophthora* associated with diseases of horticultural crops. He attended international training program on '*Oomycetes* bioinformatics' at Virginia Tech, USA in 2014. Dr. Chowdappa was awarded DFID fellowship for Post-Doctoral research at CABI Bioscience, UK in 1998 . Dr. Chowdappa has published more than 120 research papers in leading national and international journals, 12 books, 35 technical bulletins, 42 book chapters and 65 experimental manuals. He is a fellow of Scientific Academia and has won several awards of repute. He is also president of many scientific societies in India.

Impact of Climate Change on Plantation Crops

— Editors —

K.B. Hebbar

S. Naresh Kumar

P. Chowdappa

2017

Daya Publishing House®

A Division of

Astral International Pvt. Ltd.

New Delhi – 110 002

ISBN: 978-93-86071-71-2 (International Edition)

Publisher's Note:

Every possible effort has been made to ensure that the information contained in this book is accurate at the time of going to press, and the publisher and author cannot accept responsibility for any errors or omissions, however caused. No responsibility for loss or damage occasioned to any person acting, or refraining from action, as a result of the material in this publication can be accepted by the editor, the publisher or the author. The Publisher is not associated with any product or vendor mentioned in the book. The contents of this work are intended to further general scientific research, understanding and discussion only. Readers should consult with a specialist where appropriate.

Every effort has been made to trace the owners of copyright material used in this book, if any. The author and the publisher will be grateful for any omission brought to their notice for acknowledgement in the future editions of the book.

Published by : **Daya Publishing House®**
A Division of
Astral International Pvt. Ltd.
– ISO 9001:2015 Certified Company –
4736/23, Ansari Road, Darya Ganj
New Delhi-110 002
Ph. 011-43549197, 23278134
E-mail: info@astralint.com
Website: www.astralint.com

त्रिलोचन महापात्र, पीएच.डी.
एफ एन ए, एफ एन ए एस सी, एफ एन ए ए एस
सचिव एवं महानिदेशक

TRILOCHAN MOHAPATRA, Ph.D.
FNA, FNASc, FNAAS
SECRETARY & DIRECTOR GENERAL

भारत सरकार
कृषि अनुसंधान और शिक्षा विभाग एवं
भारतीय कृषि अनुसंधान परिषद
कृषि एवं किसान कल्याण मंत्रालय, कृषि भवन, नई दिल्ली 110 001

GOVERNMENT OF INDIA
DEPARTMENT OF AGRICULTURAL RESEARCH & EDUCATION
AND
INDIAN COUNCIL OF AGRICULTURAL RESEARCH
MINISTRY OF AGRICULTURE AND FARMERS WELFARE
KRISHI BHAVAN, NEW DELHI 110 001
Tel.: 23382629; 23386711 Fax: 91-11-23384773
E-mail: dg.icar@nic.in

Foreword

Plantation crops, an integral part of the horticulture, provide livelihood security to millions of people. They blend effectively with environment, contributing to sustaintability, conservation of biodiversity and stable ecosystem. Plantation sector plays vital role in employment generation and poverty allievation in rural area, contributing 12.72 per cent to total exports and 75 per cent of the agricultural produce although they occupy 2 per cent of the cultivable area.

Plantation crops are mainly confined in the economically and ecologically vulnerable regions, plays a crucial role as far as the issue of sustainability is concerned. General circulation models project increases in the earth's surface air temperatures and other climate change in the middle or later part of the 21st century, and therefore crops such as coconut, arecanut, cocoa, cashew, oil palm, rubber, coffee and tea will be grown in a much different environment than today. The 2nd National Communication to the United Nations Framework Convention on Climate Change projects an all-round warming over the Indian subcontinent of 1-4°C towards 2050s. The monsoon rainfall in some parts of the southern peninsula, where plantation crops are largely cultivated, is predicted to decrease by 10-20 per cent, along with a decrease in the number of rainy days. Climate change will affect plantation crops through higher temperatures, elevated CO_2 concentration, precipitation changes, increased weeds, pest, and disease pressure, and increased vulnerability of organic carbon pools.

The ecosystem services rendered by plantation crops need to be understood, assessed and realized for resilient plantation systems in India. New crop varieties, cropping systems, and agricultural management strategies are needed for long-term adaptation of plantation crops to climate change. I hope that book on "**Impact of Climate Change on Plantation Crops"** would be of immense help to plantation researchers, extension personnel, policy planners, students and plantation managers, who are involved in plantation sector development and need to address issues related to climate change. I congratulate the editors for this timely compilation for the benefit of all the stakeholders.

Dr. T. Mohapatra

Secretary DARE and DG, ICAR

Preface

The plantation industry is important in many parts of the world, not so much interms of area occupied but because of its contribution to the agricultural grossproduct to the economy. In India plantation crops occupy less than 2% of total cultivable areabut generate over Rs16,000 million contributing about 12.72% of export of all commodities. India is leading in production of coconut, tea, cashewand arecanut and offers direct and indirect employment to millions of people.Plantations crops are perennial in nature, grown for quality and high value produce. Owing to their perennial nature, they also experience changes in climatic conditions during their life cycle. Since most of the plantations crops are grown in climatically sensitive areas, they are prone to climatic risks. Climate change is projected to increase the frequency and intensity of the climatic stresses, potentially threatening the productivity of plantation sector. Slow recovery of plantation during post-stress period causes perennial economic loss to the farmer. Thus, plantation farmer is more vulnerable to climate change.Therefore, assessing the effect of climate change on plantation crops isimportant. However, this task is difficult due toi) their size ii) slow growth ratesiii) slow response to external factors and iv) perennial nature.

Most of the plantations have very narrow range of climatic requirements for their optimal growth and yield, in terms of quantity and quality. The plantation crops have immense potential for carbon sequestration. However, research studies and hence the information on climate change impacts on plantation crops is relatively very less as compared to that on food crops. Nonetheless, several efforts by scientists in India have led to generation of quite a good amount of information on climate change impacts on plantation crops. The studies used innovative and integrated approaches as evident from different chapters. This book gives a summarizedinformation on the basics of climate change; climate change projections for India and Indian agriculture; climatic requirement, sensitive stages, impacts of climate change, adaptive strategies toovercome climate change effects on plantation crops viz., coconut, arecanut, cocoa, black pepper and cardamom, cashew, oilpalm, tea and rubber. The approaches used in studies to quantify theeffect of climate change variables under controlled conditions and simulation studies on coconut may also be applicableto other crops. The ways to make the plantation industry more sustainable under changing climates including the genetic and management options available for each crop are provided. Further, carbon sequestration potential of plantation crops is also presented.

The publication of the book has become possible with the whole hearted supportand cooperation from all the contributors. It is hoped that this book will be usefulto all stakeholders involved in plantation research,extension, development and policy planning besides students and all those involved in assessing the impacts, and developing adaptationand mitigation strategies for improving plantations productivity in changing climates.

K. B. Hebbar

S. Naresh Kumar

P. Chowdappa

Contents

Foreword *v*

Preface *vii*

List of Contributors *xv*

1. Climate Change and Plantation Crops: An Overview **1**

1. Introduction; 2. Defining Climate Change; 2.1. Past Changes in Climate; 2.2. Future Projections on Change in Climate at Global Level and for South Asia; 2.3. Climate Change Seasonal Projections for India; 3. Challenges of Climate Change to Plantation Crops; 3.1. Current Sensitivity of Plantation Crops to Climate Change; 4. Future Thrust; 4.1. Adaptation Strategies in Plantations to Face Climate Change Impacts; References.

S. Naresh Kumar and P.K. Aggarwal

2. Coconut **15**

1. Causes of Climate change; 2. Past Trends in Global and Indian Climate; 3. Projected Climate Change Scenarios; 4. Climatic Change and Agriculture; 5. Observed Impacts on Agriculture: Some Examples; 6. Projected Impacts on Agriculture and Adaptation Gains; 7. Effects of Climatic Changes on Coconut; 7.1. Approaches for Climate Change Studies; 7.2. Impact of Climate Change on Coconut; 7.3. Adaptation Strategies to Climate Challenges on Coconut Plantations; Conclusions; Acknowledgements; References.

S. Naresh Kumar, V. John Sunoj, K.S. Muralikrishna, K.B. Hebbar, V. Rajagopal, K.V. Kasturi Bai and P. Chowdappa

3. **Physiological and Biochemical Response of Coconut to Climate Change Variables** 45

1. Introduction; 2. Responses to Moisture Deficit; 3. Responses to Temperature Variations; 4. Reponses to CO_2 Concentrations; 5. Anatomical and Morphological Traits; 6. Leaf Photosynthesis; 7. Biochemical Responses and Osmotic Adjustment; 8. Biomass Production and Water Use Efficiency; 9. Interaction Effect of CO_2, High Temperature and Moisture Deficits; 10. Strategies Adopted for Improving Existing Cultivars and Developing New Varieties; 10.1. Integration of Beneficial Traits into Existing Crops through Use of Germplasm Accessions; 10.2. Identified Germplasm/Variety and Traits that Tolerate Drought and Heat; 11. Field-level Evaluations of Crop Germplasm/Varieties; 12. Devised Cropping/Farming Systems to Alleviate the Effect of Climate Change; 12.1. Cultural Practices, Soil Conservation and Water Management Techniques are Evolved to Manage the Drought; 12.1.1. Optimize Land Use; 12.1.2. Optimize Water-use Efficiency; 12.1.3. Use Crop Models in Decision-Making; 13. Coconut is an Excellent Tree Crop for Climate Change Mitigation; 13.1. Carbon Sequestration and Carbon Stocks in Coconut; 13.2. Coconut can Check Erosion and Wind Speed; 14. Strategies for the Future; References.

K.B. Hebbar, Mukesh Kumar Berwal, M. Arivalagan and V.K. Chaturvedi

4. **Arecanut and Cocoa** 61

1. Introduction; 2. Climatic Conditions in Plantation Belt; 2.1. Suitable Climate for Arecanut; 2.2. Suitable Climate for Cocoa; 2.3. Data Acquisition; 3. Adaptation Strategies; 4. Arecanut and Cocoa for Carbon Sequestration; 5. Conclusions; References.

S. Sujatha, Ravi Bhat and P. Chowdappa

5. **Black Pepper and Cardamom** 75

1. Introduction; 2. Black Pepper (Piper nigrum L.); 2.1. Climate Change and Productivity; 2.2. Climate Change and Pathogens and Pests; 2.3. Climate Change and Quality; 3. Small Cardamom (Eletteria cardamomum Maton); 3.1. Climatic Influence on Cardamom Production; 4. Conclusions; References.

K.S. Krishnamurthy, K. Kandiannan, S.J. Ankegowda and M. Anandaraj

6. **Cashew** 87

1. Introduction; 2. Climatic Requirement; 3. Climate Change Impacts; 4. Extreme Events of Weather Impacts; 5. Adaptation and Mitigation Strategies; 5.1. Soil and Water Conservation Practices; 5.2. Mulching; 5.3. Green Manuring; 5.4. Site Specific Nutrient Management (SSNM); 5.5. Intercropping; 5.6. Contingency Plan for Rainfall Deficit Management for Cashew; 5.7. Pest and Disease Management; 5.8. Carbon Sequestration; 6. Future Line of Work; References.

T.R. Rupa

7. **Oil Palm** 101

1. Introduction; 2. Climate Change and Oil Palm; 2.1. Temperature; 2.2. Precipitation; 2.3. Atmospheric Carbon Dioxide Concentration; 2.4. Variations in Climate; 2.5. Pests and Diseases; 2.6. Greenhouse Gas (GHG) Emissions; 3. Impacts of Climate Change: Case Studies; 3.1. India; 3.2. Indonesia; 3.3. Malaysia; 3.4. Columbia; 4. Adaptation Strategies; 4.1. Timely Weather Forecasts; 4.2. Improved Water Use Efficiency (WUE); 4.3. Development of Drought Tolerant Oil Palm Hybrids; 4.4. Soil and Land Management; 5. Mitigation Strategies; 5.1. Carbon Sequestration; 5.2. Zero Burning; 5.3. Water Management; 5.4. Reduced Usage of Chemical Inputs; 5.5. Composting; 5.6. Biogas Production from POME; 5.7. Conversion of Biomass to Energy; 6. Future Research Strategies; 6.1. Quantification of CO_2 Flux, Energy Budget and Water Transfer in Oil Palm; 6.2. Assessment and Quantification of Impact of CO_2 and Tmperature in Oil Palm using CO_2 Enrichment Studies; 6.3. Understanding of Seasonal Variations in Vegetative and Reproductive Growth, Phenology and Yield Components to Climatic Variables; 6.4. Phenotyping of Drought, Salinity and High Temperature Tolerance Traits in Oil Palm; 7. Conclusions; References.

K. Suresh, S.K. Behera, K. Manorama and B.N. Rao

8. **Tea** 123

1. Introduction; 2. Tea Production; 2.1. Global Tea Production; 2.2. Indian Tea Production; 2.3. Tea Producing Regions of India; 3. Favourable Growth Conditions for Tea; 3.1. Climatic Requirements of Tea; 4. Relationship between Yield and Climate; 4.1. Yield; 4.2. Temperature; 4.3. Relative Humidity; 4.4. Rainfall; 5. Impact of Carbon Dioxide and Technological Advancements on Tea Production; 5.1. Impact of Carbon Dioxide on Tea Production; 5.2. Impact of Technological Advancements on Tea Production; 5.3. Interactive Effects of CO_2 and Technology on Tea Production; 6. Impact of ENSO and SST on Crop Yield; 6.1. El Niño Southern Oscillation; 6.2. Impact of ENSO on Crop Yield; 6.3. Impact of ENSO on Rainfall; 7. Future Climate Projections using GCM Models; 7.1. Downscaling of Temperature; 7.2. Downscaling of Rainfall; 7.3. Land Suitability of the Tea Planting Regions under Future Climate Scenario; 8. Conclusions; Acknowledgements; References.

E.Edwin Raj, K.V. Ramesh, B. Radhakrishnan and R. Raj Kumar

9. **Rubber** 145

1. Introduction; 2. Climate Change and Growth and Establishment of Rubber in the Field; 3. Climate Change and Diseases of Rubber; 4. Climate Warming and Rubber Yield; 5. Climate Change, Soil Organic Matter Content and Floral Diversity; 6. Carbon Sequestration by Rubber Plantations; 7. Natural Rubber and Synthetic Rubbers; References.

R. Krishnakumar, James Jacob, K.K. Jayasooryan and P.R. Satheesh

10. Carbon Sequestration in Plantation Crops 157

1. Introduction; 2. Carbon Sequestration of Plantation Crops; 2.1. Arecanut and Cocoa; 2.2. Coconut; 2.3. Rubber; 2.4. Tea; 2.5. Coffee; 2.6. Cashew; 2.7. Oil Palm; 3. Conclusions; References.

K.B. Hebbar, D. Balasimha and S. Naresh Kumar

11. Phenotyping Tools to Understand Effects of Climate Change 169

1. Plantation Crops in India; 2. Screening Tools for Temperature and Moisture Deficit Tolerane; 2.1. Vegetative Stage; 2.2. Reproductive Stage; 2. Conclusions; References.

V.S. John Sunoj, K.B. Hebbar and P.V. Vara Prasad

12. Biotechnological Tools to Mitigate Climate Change 189

1. Climate Change and its Impact on the Productivity of Perennial Crops; 2. Biotechnological Tools for Trait Specific Gene Discovery and Functional Validation; 2.1. Targeted Gene Expression Studies to Identify Candidate Genes; 2.2. Microarrays to Analyze Expression Pattern of known Genes; 2.3. Global Transcriptome Analysis to Discover Novel Genes; 2.4. Functional Validation of Genes Linked to Specific Traits; 2.5. Targeted Manipulation of Specific Traits; 3. Intragenesis and Cisgenesis; 4. Genome Editing; 5. Conclusions; References.

K.N. Nataraja, R.S. Sajeevan and K.H. Dhanyalakshmi

13. Farming System Approach to Reduce Impacts of Climatic Change 209

1. Introduction; 2. Climate Change Impact; 2.1. Moisture Deficiency; 3. Coconut Based Farming System; 3.1. Soil and Water Conservation; 3.2. Moderating Temperature and Moisture Content in Soil; 4. Conclusions; References.

K.B. Hebbar, M. Arivalagan and Ravi Bhat

14. Climate Resilience Agriculture: Experiences from Kerala 219

1. Introduction; 2. Climate Change/Variability over Kerala; 2.1. Onset of Monsoon; 2.2. Rainfall; 2.3. Temperature Projections; 2.4 Climate Change and Rice; 2.5. Climate Change and Coconut; 2.6. Cashew; 2.7. Coffee; 2.8. Black Pepper; 2.9. Cocoa; 2.10. Cardamom; 2.11. Rubber; 3. Climate Resilience Agriculture; References.

G.S.L.H.V. Prasada Rao

15. Climate Change in Plantation Crops: A Socio-Economic Reflection on Vulnerability, Risks and Way Forward 237

1. Introduction; 2. Assessment of Vulnerability to Climate Change; 3. Conceptual Approaches to Vulnerability Assessment; 3.1. Socio Economic Approach; 3.2. Biophysical or Impact Assessment Approach; 3.3. Integrated Approach; 4.

Adaptive Capacity and Vulnerability to Climate Change; 5. Climate Change and Plantations; 6. The Research Gaps and Way Forward; 7. Conclusions; References.

S. Jayashekar

Index **247**

Colour Plates **253**

List of Contributors

Aggarwal, P.K
Climate Change and Food Security Programme, CIMMYT Office, DPS Marg, New Delhi, 110012, India

Ankegowda, S.J
ICAR-Indian Institute of Spices Research, Kozhikode 673012, Kerala, India

Anandaraj, M
ICAR-Indian Institute of Spices Research, Kozhikode 673012, Kerala, India

Arivalagan, M
ICAR-Central Plantation Crops Research Institute, Kasaragod 671 124, Kerala, India

Behera, S.K.
ICAR-Indian Institute of Oil Palm Research, Pedavegi-534475, Andhra Pradesh, India

Chowdappa, P.
ICAR-Central Plantation Crops Research Institute, Kasaragod 671 124, Kerala, India

Chaturvedi, V.K.
ICAR-Central Plantation Crops Research Institute, Regional Station Kayamkulum 690533, Kerala, India

Dhanyalakshmi, K.H.
Department of Crop Physiology, University of Agricultural Sciences, GKVK, Bengaluru

Edwin Raj, E.
Plant Physiology Division, UPASI Tea Research Foundation, Tea Research Institute, Valparai, Nirar Dam PO, Coimbatore 642127, Tamil Nadu, India

Hebbar, K.B.
ICAR-Central Plantation Crops Research Institute, Kasaragod 671 124,Kerala, India

James Jacob
Rubber Research Institute of India, Kottayam-686009, Kerala, India

Jayashekar, S.
ICAR-Central Plantation Crops Research Institute, Kasaragod 671 124,Kerala, India

Jayasooryan, K.K.
Rubber Research Institute of India, Kottayam-686009, Kerala, India

John Sunoj, V.
Department of Agronomy, Kansas State University, 2004 Throckmorton Plant Science Center, Manhattan, Kansas State- 66506, USA

Kandiannan, K.
ICAR-Indian Institute of Spices Research, Kozhikode 673012, Kerala, India

Kasturi Bai, K.V.
Former Head, PB&PHT, ICAR-Central Plantation Crops Research Institute, Kasaragod-671 124, Kerala, India

Krishnakumar, R.
Rubber Research Institute of India, Kottayam-686009, Kerala, India

Krishnamurthy, K.S.
ICAR-Indian Institute of Spices Research, Kozhikode 673012, Kerala, India

Manorama, K.
ICAR-Indian Institute of Oil Palm Research, Pedavegi-534475, Andhra Pradesh, India

Muralikrishna, K.S.
ICAR- Central Plantation Crops Research Institute, Kasaragod, 671 124, Kerala, India

Mukesh Kumar Berwal
ICAR-Central Plantation Crops Research Institute, Kasaragod, 671 124, Kerala, India

Naresh Kumar, S.
Centre for Environment Science and Climate Resilient Agriculture, ICAR-Indian Agricultural Research Institute, New Delhi-110 012, India

Nataraja, K.N.
Department of Crop Physiology, University of Agricultural Sciences, GKVK, Bengaluru-560065, India

Prasada Rao, G.S.L.H.V.
Centre for Animal Adaptation to Environment and Climate Change Studies, Kerala Veterinary and Animal Sciences University, Mannuthy, Thrissur – 680 651, Kerala

Rajagopal, V.
Former Director, ICAR- Central Plantation Crops Research Institute, Kasaragod-671 124, Kerala, India

Ramesh, K.V.
CSIR-Centre for Mathematical Modelling and Computer Simulation (C-MMACS), NAL Belur Campus, Bangalore - 560 037, Karnataka, India.

Radhakrishnan, B.
The Director, UPASI Tea Research Foundation, Tea Research Institute, Valparai, Nirar Dam PO, Coimbatore 642127, Tamil Nadu, India

Raj Kumar, R.
The Director, UPASI Tea Research Foundation, Tea Research Institute, Valparai, Nirar Dam PO, Coimbatore 642127, Tamil Nadu, India

Rao, B.N.
ICAR-Indian Institute of Oil Palm Research, Pedavegi-534475, Andhra Pradesh, India

Ravi Bhat.
ICAR-Central Plantation Crops Research Institute, Kasaragod 671 124,Kerala, India

Rupa, T.R.
ICAR-Directorate of Cashew Research, Puttur - 574 202, Karnataka, India

Sajeevan, R.S.
Department of Crop Physiology, University of Agricultural Sciences, GKVK, Bengaluru-560065, India

Sujatha, S.
ICAR- Indian Institute of Horticultural Research, Hesaraghatta Lake Post, Bengaluru-560 089, India

Satheesh, P.R.
Rubber Research Institute of India, Kottayam-686009, Kerala, India

Suresh, K.
ICAR-Indian Institute of Oil Palm Research, Pedavegi-534475, Andhra Pradesh, India

Vara Prasad, P.V.
Department of Agronomy, Crop Physiology Lab, 2004 Throckmorton Plant Science Center, Kansas State University, Manhattan, KS 66506, USA

2017, Impact of Climate Change on Plantation Crops *Pages* **1–13**
Editors: **K.B. Hebbar, S. Naresh Kumar & P. Chowdappa**
Published by: **ASTRAL INTERNATIONAL PVT. LTD., NEW DELHI**

Chapter 1

Climate Change and Plantation Crops: An Overview

S. Naresh Kumar and P.K. Aggarwal

1. Introduction

Mean state of weather over a long period, usually 30 years is defined as the climate of a location. Ever since earth originated about 4.54 billion years ago, and over past 3 million years it's climate is dominated by glacial and interglacial cycles. During the last inter-glaciation period the global mean surface temperatures were 1-2°C warmer than present 15°C. Whereas during last glacial maximum the global mean surface temperatures were 4-7°C cooler than present. The gases (CO_2, H_2O, methane, oxides of nitrogen and sulphur, water vapour and aerosols) responsible for tapping the solar energy and warm the air are called greenhouse gases (GHGs). Without them, earth surface mean annual temperature would have been -17°C and greenhouse effect gives about 32°C (thus a global mean annual temperature of +14.56°C in 2013) warming allowing liquid H_2O and therefore living organisms to exist on the planet Earth.

The atmospheric concentration of GHGs was relatively low, but increased anthropogenic activities and industrialization led to fossil fuel combustion and the GHG emitting technologies on one hand, and deforestation and land use change on the other. The above activities caused the imbalance in carbon cycle with more emissions of CO_2 into atmosphere than its sequestration by the tropospheric activities. Apart from CO_2, other GHGs such as N_2O, CH_4, CFCs, NOx, SOx, water vapour and aerosols have differential potential to warm the earth surface, called global warming potential (GWP). The GWP is one for CO_2 while that of methane is 21 times and N_2O is 310 times of GWP of CO_2 in a 100 year period. The GWP of other GHGs is even very high and as the life-time of the GHGs is long, past and current emissions continue to warm the climate.

The CO_2 concentration has increased from a pre-industrial value of about 280 ppm to a global mean of 400.47 ppm in August, 2016. Similarly, the global atmospheric concentration of methane (<700 ppb to 1800 ppb), nitrous oxides (<290 ppb to current 330 ppb) and other important GHGs, has also increased considerably. In addition, use of urea is also contributing to the GHGs since urea production involves conversion of inert N_2 to reactive nitrogen form. Depending on the future developmental pathways, total GHGs emissions vary. Owing to uncertainty in future developmental pathways, several plausible developmental pathways are defined and broadly categorized into 4 groups. Initially, the future developmental pathways were defined as A1 (global, economic intensive developmental pathway with high energy requirement either from balanced sources (A1B), fossil fuels (A1F) or technology intensive (A1T); A2 (more regional and diversified but intensive economic developmental pathway); B1 (global economic pathway with due consideration for environmental protection) and B2 (regional and diversified economic pathway with due consideration for environmental protection) scenarios (IPCC 2001 and 2007).

In order to overcome the uncertainties and quantification issues associated with socio-economic related scenarios, the representative concentration pathways (RCPs) are being used (IPCC AR5, 2013 and 2014). These RCPs are based on radiative forcing of the GHGs and represent four GHG concentration trajectories for future. They are identified by their approximate total radiative forcing in year 2100 relative to 1750: 2.6 W m^{-2} for RCP2.6 (mitigation scenario), 4.5 W m^{-2} for RCP4.5 (stabilization scenario), 6.0 W m^{-2} for RCP6.0 (stabilization scenario), and 8.5 W m^{-2} for RCP8.5 (high GHG emission scenario). Current GHG concentration in the atmosphere has a radiation forcing of 2.3 W m^{-2}.

2. Defining Climate Change

United Nations Framework Convention on Climate Change (UNFCCC), in its Article 1, defines "climate change" as: "*A change of climate which is attributed directly or indirectly to human activity that alters the composition of the global atmosphere and which is in addition to natural climate variability observed over comparable time periods*". On the other hand, the Inter-Governmental Panel on Climate Change (IPCC) defines climate change as *a statistically significant variation in either the mean state of the climate or in its variability, persisting for an extended period (typically decades or longer). Climate change may be due to natural internal processes or external forcings, or due to persistent anthropogenic changes in the composition of the atmosphere or in land use.* The UNFCCC thus makes a distinction between "climate change" attributable to human activities altering the atmospheric composition, and "climate variability" attributable to natural causes.

2.1. Past Changes in Climate

The Inter-Governmental Panel on Climate Change Report (2013), for the first time delineated the rise in temperature due to anthropogenic activities and natural variability. The global mean temperatures during 1951-2010 period increased by 0.6-0.7°C out of which natural variability contributed to ± 0.1°C change in temperature. The analysis by Indian Institute of Tropical Meteorology, Pune indicated that in Indian region also annual mean maximum temperatures increased by about 0.71°C in past 100 years. The rate of increase in maximum temperature enhanced and in post 1970's and the warming has been at a rate of 0.17°C/10 years. Similarly, the minimum temperature has been rising at a rate of 0.29°C/10 years. During 1871-2016 period, India faced 28 deficit and 20 excess monsoon years. Out of these, as many as 16 deficit monsoon years and 6 excess monsoon years fell in post 1960 period.

2.2. Future Projections on Change in Climate at Global Level and for South Asia

Climate change has two inherent components i) increase in mean state of climatic parameters, ii) increased variability . For example increase in frequency of extreme events such as heat waves, cold waves, floods, droughts, frost events, extreme rainfall events, etc. The climate change projections involve several steps. Earlier approach was to use socio-economic models for deriving the GHG emissions in future developmental scenarios. As mentioned earlier, depending on the future plausible developmental paths adapted by the world, the scenarios are broadly categorized into A1, A2, B1 and B2 scenarios (IPCC, 2000). Depending on the plausible developmental pathway, the GHG emission scenarios also change. For instance, in a developmental pathway which intensively uses the fossil fuels (A1F scenario), the CO_2 concentrations were projected to increase in the range of 386 to 495 ppm in 2020 (which means a 30 year period from 2010 to 2039), 555 to 702 ppm in 2050 (*i.e.*, 2040-2069) and 786 to 958 ppm by 2080 (*i.e.*, 2070-99) as against 386 to 457 ppm by 2020, 482 to 518 ppm by 2050 and 530 to 540 ppm by 2080 in a developmental pathway (B1) which uses eco-friendly technologies. Similarly, the concentrations of other GHGs also change. These differences in GHG emissions depending on plausible developmental pathways, in turn, result in variations in the magnitude of rise in temperatures, changes in rainfall and other climatic parameters.

However, in recent years this approach is modified to overcome the uncertainties attached to future socio-economic scenarios. The Representative Concentration Pathways (RCPs) are now used as per IPCC AR5 (2013). Under this the CO_2, methane and N_2O concentrations are projected to vary for different RCPs. The CO_2 concentrations projected increase very less in RCP 2.6 (range from 412ppm in 2020 to 421 ppm in 2100) as compared to extreme RCP 8.5 (range from 416 ppm in 2020 to 936 ppm in 2100).

The future climate of the earth system is projected by the Global Climate Models (GCMs). About 54 GCMs are being run in different labs in the world. The global climate models project an increase of global mean surface temperatures in the range of 0.3°C to 1.7°C (RCP2.6), 1.1°C to 2.6°C (RCP4.5), 1.4°C to 3.1°C (RCP6.0), 2.6°C to 4.8°C (RCP8.5) for 2081–2100 relative to 1986–2005 period. The agreement among GCMs and confidence level as far as rise in temperature is quite high while the level of agreement among GCMs and confidence level on rainfall projections is low. The projected increase in temperature for south Asia is in the range of 0.5 to

1.2°C by 2020, 0.88 to 3.16°C by 2050 and 1.56 to 5.44°C by 2080, depending on the scenario of future development. In general, precipitation is projected to increase by 6- 14 per cent in 2080s for Indian region. Further, the IPCC AR5 report on climate change has projected an increase in frequency of droughts, floods, and extreme events of temperature and rainfall affecting terrestrial systems. The sea level rise and decrease in ocean water pH threaten the coastal ecosystems as well.

2.3. Climate Change Seasonal Projections for India

From agricultural point of view climate change annual projection will be of less use. Climate change scenarios from over 50 GCMs projections were analyzed at Environmental Modelling lab, Centre for Environment Science and Climate Resilient Agriculture, Indian Agricultural Research Institute, New Delhi to derive the seasonal climate change scenarios for agricultural analysis. The results include 1) rise in minimum temperatures is projected to be more than rise in maximum temperatures; 2) rise in temperatures to be more during rabi than during Kharif; 3) during Kharif, minimum temperatures to increase in the range of 0.946 - 4.067°C (2020 to 2080) in different RCPs over baseline temperatures (1976-2005 period), while the projected increase in rabi is in the range of 1.096 - 4.652°C (2020 to 2080); 4) similarly, maximum temperatures during kharif to increase in the range of 0.741 - 3.533°C (2020 to 2080) in different RCPs while the projected increase in rabi is 0.882- 4.01°C (2020 to 2080); 5) rise in temperatures are projected to be more in northern parts of India than in southern parts; 6) rainfall projections, though less robust, indicate an increase during Kharif and rabi seasons over Indian region; 7) Kharif rainfall is projected to increase in the range of 2.3-3.3 (2020), 4.9-10.1 per cent (2050), while rabi rainfall is projected to increase in the range of 12 per cent (2020), 12-17 per cent (2050); 8) rainfall increase (per cent) is projected to be more during rabi than during Kharif; 9) An indicator for variability, coefficient of variation (CV) for minimum and maximum temperatures during rabi is significantly more than

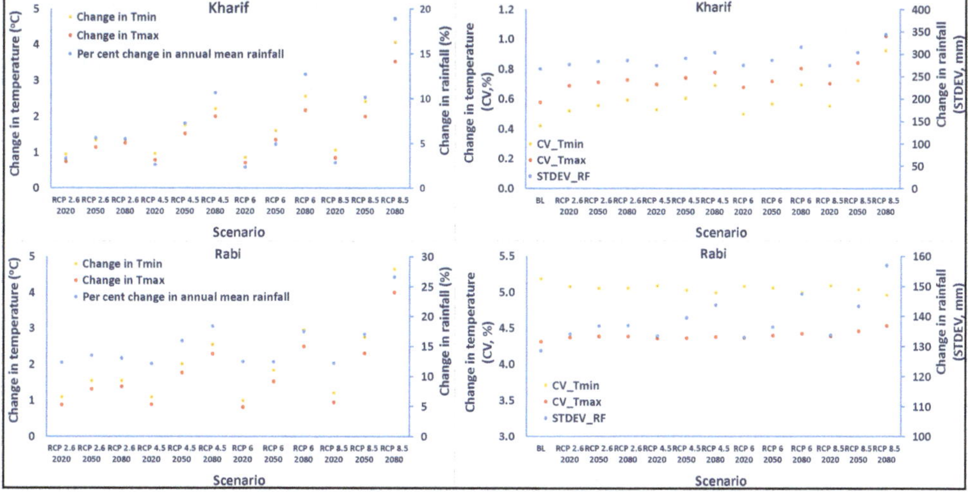

Figure 1.1: Projected Climatic Variability in *Kharif* and *Rabi* Seasons in India.

during *Kharif* (Figure 1.1) 10) the CV for maximum temperatures is projected to rise slightly during both seasons 11) on the other hand CV for minimum temperatures to rise during Kharif; while in rabi season it may remain high but similar in future climates; 12) the standard deviations for rainfall are more during kharif than during rabi season and these are projected to increase in 2020, 2050 and 2080 scenarios; 13) the CO_2 concentration are projected increase in the range of 419-432 µmol (2020), 441-572 µmol (2050) and 429 to 799 µmol (2080) in different RCPs. This analysis indicated a progressive climate change and variability in *Kharif* and *rabi* seasons with significant spatio-temporal variations in India towards the end of the century.

Coupled to these changes in mean values, the climatic variability is also projected to increase causing increased frequency of heat and cold waves, frost and heavy precipitation events, and flood and drought events. These stresses may occur even in single season at a location. It is also likely that future tropical cyclones will become more intense, with larger peak wind speeds and heavier precipitation events. Spatio-temporal variations also are projected for increase in temperature and change in rainfall. Seasonal minimum temperature is likely to increase more in Uttar Pradesh, Bihar, Rajasthan, Maharashtra and peninsular India, while the seasonal maximum temperature is likely to increase more in UP, Bihar and peninsular India. As far as kharif season rainfall is concerned, it is likely to increase more in eastern and central parts of the country. During rabi season, increase in seasonal mean minimum temperature is likely to be more in central and western parts of India, while the mean seasonal maximum temperature is likely to be more in central India. Rabi rainfall is likely to increase in north India, parts of Andhra Pradesh and Maharashtra. All these environmental changes are likely to increase the pressures on Indian agriculture including horticultural crops.

3. Challenges of Climate Change to Plantation Crops

Unlike the annual crops, on which the impact of climate change is well documented, research efforts and therefore information on plantation crops is scarce. Plantations being perennial in nature have to face the impact of climate change even during a single generation or in a standing plantation. Hence it is important that the impact of climate change on plantation crops is understood well to reduce the adverse impacts. Perennial crops such as coconut, rubber, tea, coffee, cashew, oil palm, arecanut and cocoa are grown in ecologically sensitive areas such as coastal belts, hilly areas and areas with high rainfall and high humidity. These plantation crops contribute substantially to the agricultural exports at global level and to the livelihoods of millions of farmers, farm workers and industrial workers at national level.

3.1. Current Sensitivity of Plantation Crops to Climate Change

Plantation crops are perennial in nature and are exposed to variable weather conditions during their life cycle. In India plantation crops *viz.*, coconut, cashew, rubber, tea, coffee, arecanut, cocoa, oil palm, etc. and spice crops like black pepper, cardamom, clove, etc. are grown in large areas of Kerala, Karnataka, Tamil Nadu, Andhra Pradesh, Maharastra, West Bengal and Assam (Table 1.1). These plantations provide sustenance to the millions of farmers across these states.

Table 1.1: Area Under Plantation Crops in India

Crop	Area	Production	Productivity	Major States
Coconut	19.75 lakh ha	20,439.6 million nuts	10345 nuts/ha/year	Karnataka, Kerala, TN and AP
Rubber	7.58 lakh ha	9.19 lakh tonnes	1351 kg/ha	Kerala, TN
Cashew	10.27 lakh ha	7.25 lakh tonnes	707 kg/ha	Maharastra, Karnataka, Kerala, Odisha, TN and AP
Tea	5.63 lakh ha	1233 million kg	2186 kg/ha	Assam, West Bengal, TN, Kerala
Coffee	3.4 lakh ha	3.48 lakh tonnes	~1014 kg/ha	Karnataka, Kerala, TN
Arecanut	4.45 lakh ha	7.29 lakh tonnes	1165 kg/ha	Karnataka and Kerala
Cocoa	78,000 ha	16,000 Tonnes of beans	4.75 q/ha	Karnataka and Kerala
Oil Palm	95,000 ha	~1 lakh tonnes	2-3 t oil/ha	AP and Karnataka
Black pepper	2.01 lakh ha	55,400 tonnes	275 kg/ha	Kerala and TN

Source: Department of Agricultural Statistics and Ministry of Commerce.

The plantations mentioned in Table 1.1 are mainly concentrated in the South India, in east coast and north-eastern India. A brief account of impacts, adaptation options and mitigation potential of plantation crops is given below.

3.1.1. Coconut

The coconut (*Cocos nucifera* L.) palm is one of the most economically important tree crops of the humid tropical regions in the world. Coconut is mainly grown in Indonesia, Philippines, India and Sri Lanka, which account for 8375 million hectares. As on any other crop, weather influences the growth and development of coconut (Prasada Rao, 1991; Peiris and Thattil, 1998; Rajagopal *et al.*, 1996; Rajagopal and Naresh Kumar, 2003, Naresh Kumar and Kasturi Bai, 2009; Naresh Kumar *et al.*, 2009 and 2013; Naresh Kumar *et al.*, 2016). When other external factors such as fertility, management, pest and diseases are non-limiting, the yield variation can be explained mainly based on influence of weather parameters such as rainfall, evapotranspiration, temperature, solar radiation, sunshine hours, relative humidity and wind velocity, major climatic variables that influence the coconut yield. The duration of dry spell during initiation of inflorescence primordium, ovary development and button-size nut stages, in that order have greater influence on nut yield than other stages (Rajagopal *et al.*, 1996). Droughts as well as dry spells affect nut yield in next four years to follow with maximum effect on fourth year (Rajagopal *et al.*, 1996; Naresh Kumar *et al.*, 2009). Consecutive droughts in Tamil Nadu (1998-2002), droughts in Karnataka and prolonged dry spells Kerala had affected the coconut plantations. Rising temperatures in east coast are affecting the nut yield. The average nut yield of districts of coastal Andhra Pradesh declined owing to rise in temperatures.

3.1.2. Black Pepper

Black pepper (*Piper nigrum* L.) is a plant of humid tropics requiring adequate rainfall and humidity. India is a major pepper producing country in the world. It successfully grows between 20° North and South latitude and from sea level up to 1500 m. (Radhakrishnan *et al.*, 2002). The rainfall requirement of the crop varies from 2000-3000 mm. Tropical temperature and high relative humidity with little variation in day length throughout the year is favorable to the crop. It does not tolerate excessive heat and dryness (Sivaraman *et al.*, 1999). Rainfall after stress induces profuse flowering (Pillay *et al.*, 1988). A relative humidity of 60-95 per cent is optimum at various stages of growth. The ideal temperature is 23°C-32°C with an average of 28°C. The late commencement of south-west monsoon during 1997-1998 had caused a delay in flower initiation of black pepper by about 3 weeks. Intensive shedding occurs during years in which heavy north-east monsoon showers are received after a spell of dry period followed by SW Monsoon (Sukumara Pillay *et al.*, 1977). Droughts in Waynad distrct of Kerala significantly affected pepper yield. Studies also quantified temperature effects on yield and quality of pepper (NPCC, 2007).

3.1.3. Cardamom

In Cardamom Hill Reserve (CHR) of Kerala, the cardamom plantations are exposed to extreme fluctuations in the local climate resulting in successive crop failures (Raj and Murugan, 2000).

3.1.4. Arecanut

Arecanut is mainly an irrigated plantation and is sensitive to water stress, and temperature and rainfall play a major role in its productivity. Droughts in plains of Karnataka affected the arecanut yield. A survey conducted to assess the impact of drought on arecanut cultivation in four major arecanut growing districts of Karnataka indicated that the estimated yield loss at district level varied from 9.3 per cent to 20.2 per cent causing an economic loss from Rs. 11970/ha to Rs. 26600/ha. Analysis indicated that increase in temperature even by 1°C will reduce productivity in Karnataka. The reduction will be more in plains of Karnataka as compared to the western ghat portions of Karnataka. Arecanut productivity in central Kerala is likely to be benefitted by about 100 kg chali/ha/year due to increase in temperature up to 1°C. On the other hand, a 10 per cent reduction in rainfall during July and August will improve the arecanut yields by about 70 kg chali/ha/year.

Analysis on change climatic suitability in A1B 2030 scenario for arecanut production indicates a reduction in suitable areas, particularly in Kerala, coastal Karnataka and North-Eastern region (Naresh Kumar., 2010). Most of the areas which are currently having unsuitable climate for arecanut production may remain unsuitable and hence the production in these areas will depend on the availability of irrigation-water source. Some districts of north West Bengal and Westren Assam are projected to become more suitable for arecanut production in 2030 scenario. Since the favourable climate for growth can be taken as a surrogate for productivity, current analysis indicates that future production of arecanut in India is projected reduce in 2030 scenario due to climate change.

3.1.5. Cocoa

Cocoa is grown as the intercrop under arecanut or under coconut. Cocoa, being a shade-loving crop, is being influenced only indirectly by the increase in atmospheric temperatures. The crop is maintained in irrigated conditions and presently the area under cocoa is confined to limited pockets of Karnataka and Kerala with an overall area of about 30 thousand hectares. Recently, cocoa area is expanding in places like coastal Andhra Pradesh. Analysis indicated that a rise in temperature by 1°C will be beneficial for the crop productivity. The improvement is likely to be about 100 kg of dry beans/ha. This benefit will be more in cocoa growing areas of Karnataka as most of the area is falling in the foot hills of Western Ghats area as compared to that in central Kerala. However, the current management and irrigation supply should be maintained or improved to exploit this benefit. Further, an increase in temperature beyond 3°C is likely to reduce the cocoa yields. Further analysis indicates that large areas in Kerala, coastal Karnataka, and some parts of Odisha, West Bengal and North-Eastern region are having suitable climate for cocoa production (Naresh Kumar., 2010). Since cocoa is grown as an intercrop under the shade of coconut or arecanut, the area of these plantations largely decide the area under cocoa. However, currently it is grown in only about 35,000 ha in India. Climatic suitability for cocoa is projected to remain same even in A1B 2030 climate change scenario (Naresh Kumar., 2010). Some of the areas in coastal AP and western Assam are projected to become more suitable. Hence, the future production of cocoa in India is likely to be determined by the area and management and less by the climatic factors.

3.1.6. Tea

In Central Africa reductions in tea production have been observed in the season following the ENSO event. In north east India, temperature difference and mean vapour pressure of the 2nd and 5th week and rainfall of the 4th week prior to plucking generally contributed the most to weekly crop yield (Biswas *et al.*, 1996). A high average air temperature in late January delayed sprouting in mid-February, whereas a high average air temperature after mid-February hastened it. Based on the growth responses to temperature, five temperature zones were distinguished: basal, ranging from 25 to 30°C; 2 hardening temperatures, 5 to 15°C and 35 to 40°C; and 2 lethal temperatures, -3 to -15°C and 45 to 55°C (Patarava, 1987). Shoot extension stopped below about 12.5 °C. Shoots of mature tea required 491 day degrees above the base temperature of 12.8 °C to grow to a harvestable size of 15 cm. (Tanton, 1982b). Increases in temperature, soil moisture deficit and saturation vapour pressure deficit in the low elevations will adversely affect growth and yield of tea. Exceptionally low rainfall in the period of March to November resulted in a reduction in yield during May to December. In tea, total root weight, weight of fibrous and conducting roots, total root volume and root respiration were markedly reduced by flooding. Low snow cover and the resulting frost damage especially in January and Febuary. However, due to an adequate snow cover the local tea plantations survive annually almost without damage. The ecological impact of night frost on the tea plant by freezing-stress to the young flush is serious as it normally leads to a crop loss for 6 months until full recovery (Domroes, 1997).

3.1.7. Coffee

In Mexico, yield from fruit to dry coffee was better in plantations established at altitudes greater than 650 m above sea level. The percentage of dry seed (19.27-9.20 per cent) was greater in plantations established at low altitudes. Yields from dry to green (oro) coffee (55.40-58.07 per cent) were higher in coffee plantations at higher altitudes than in plantations at low altitudes (Gonzalez Arcos *et al.*, 2001).

3.1.8. Cashew

Unseasonal rainfall during March, 2008 caused severe yield loss in cashew apart from affecting the quality.

3.1.9. Rubber

Rainfall and its distribution pattern, sunshine hours, mean temperature, wind velocity and evaporation are the crucial factors governing establishment, growth and yield of *Hevea*. Cumulative rainfall had significant positive influence on the yield and yield components (Sailajadevi *et al.*, 2000). Weekly weather conditions like maximum temperature of 30.4°C, minimum temperature of 22.8°C, 5.9 h sunshine and 72 mm of rain are associated with high yield. Any deviation from these meteorological conditions for up to a period of 6 months may significantly influence latex production. The North Konkan region experiences rainless period for about 5-6 months every year concomitant with high intensities of solar radiation and high temperatures, occasionally rising beyond 41°C during the day, while in the NE region, the winter season lasts for 2-3 months with the minimum temperature falling as low as 5°C at least for brief periods at night, but the days can be relatively warmer with abundant sunlight. In both these unfavourable regions, with suitable agro-management, polyclonal seedlings of *Hevea* establish reasonably well and with some added care, bud grafts of high yielding clones also could be grown. The long dry season affected the yield and the period of tapping in the first half-year in major rubber growing areas of Vietnam. The growth of *H. brasiliensis* in the Konkan region of India was affected as in summer months when ambient temperatures during the daytime exceeded 36°C, and soil moisture deficits were severe. Growth only occurred during the monsoon period. (Chandrashekar *et al.*, 1996).

4. Future Thrust

4.1. Adaptation Strategies in Plantations to Face Climate Change Impacts

The genotypic, agronomic and socio-economic adaptation measures become important not only to sustain but also to improve the yield of plantation crops.

4.1.1. Genetic Adaptation

Some of the strategies include:

- ✰ Identification of genes imparting stress tolerance
- ✰ Marker assisted selection

- ☆ Identification *in situ* tolerant plants and using them in population improvement programme
- ☆ Use of local or farmers varieties/plants as source of desirable genetic pool in breeding for stress tolerance
- ☆ Breeding for source-sink balance, root traits, quality traits, *etc.*

4.1.2. Agronomic Adaptation

- ☆ Improving input use efficiency (water, nutrients, labour, energy)
- ☆ Low-carbon technologies
- ☆ Resource conservation technologies

4.1.3. Socio-economic Adaptation

- ☆ Expanding crop insurance coverage
- ☆ Value addition and market interventions
- ☆ Convergence of local developmental departments to improve the support system.
- ☆ Income diversification and strengthening fallback mechanisms

4.1.4. Technology Interventions

- ☆ Development of decision support systems for technology dissemination and wide scale adaptation of improved technologies
- ☆ Convergence of technological interventions for anticipation and prevention of adverse conditions such as climatic risks and market risks.
- ☆ Weather forecast based agro-advisories for pest, nutrient and crop management

4.1.5. Policy Support

- ☆ Development of safety nets for volatile market (national and international) situations
- ☆ Incentivizing the plantation sector for its GHG mitigation role.

4.2. Exploitation of Mitigation Potential of Plantations Crops

Plantation crops are very good candidates for mitigation of GHGs. The carbon sequestration potential of coconut (Naresh Kumar, 2009), cocoa, arecanut (Balasimha and Naresh Kumar, 2013) and rubber have been quantified. Similarly, other plantation crops have immense potential which needs to be quantified. Further, soil carbon sequestration and economics of ecological services of plantation crops need attention of researchers. This information will help in value addition to plantation sector for increasing the income to the farmer.

References

Balasimha, D. and Naresh Kumar,S. (2013). Net primary productivity, carbon sequestration and carbon stocks in areca-cocoa mixed crop system. *J. Plantn Crops*, 41, 8-13.

Biswas, A.K., Karmokar, P.K. and Barbora, B.C. (1996). Crop-weather relationship for forecasting weekly crop of tea. *Journal of Plantation Crops*, 24, 107-114.

CPCRI Annual Report 2004. Central Plantation Crops Research Institute, Kasaragod, Kerala 671 124 pp.136, 2004.

Chandrashekar, T.R., Marattukalam, J.G. and Nazeer, M. A. (1996). Growth reaction of Hevea brasiliensis to heat and drought stress under dry subhumid climatic conditions. *Indian Journal of Natural Rubber Research*, 1996, 9(1), 1-5.

Domroes, M. T.I. (1997). Frost and its agroclimatic impact in Sri Lanka. *Tropical-Ecology*,1997, 38(2), 285-295.

Gonzalez Arcos, F., Almaguer Vargas, G. and Diaz Vicente, V.M. (2001). Determinationof yield of coffee, Coffea arabica L., and benefit of different altitudes of the Soconusco region, Chiapas, Mexico. *Proceedings of the 45th Annual Meeting*, Lima, Peru, 15-19 November 1999. Proceedings-of-the-Interamerican-Society-for-Tropical-Horticulture, 2001, 43, 135-138.

IPCC. (2000). Special Report on Emissions Scenarios (SRES). A report by the Intergovernmental Panel on Climate Change (IPCC).

IPCC. (2001). Assessment Report 3-WGI: Climate Change -The Scientific Basis and WG II: Impacts, Adaptation, and Vulnerability.

IPCC. (2007). Assessment Report 4-WGI: Climate Change - The Physical Science Basis and WG II: Impacts, Adaptation, and Vulnerability.

IPCC. (2013). Assessment Report 5-WGI: The Physical Science Basis.

IPCC. (2014). Assessment Report 5-WGII: Impacts, Adaptation, and Vulnerability.

Kasturi Bai, K.V., Naresh Kumar,S. and Rajagopal,V. (2009). Abiotic stress tolerance in coconut. CPCRI Pub., Kasaragod, India. p. 53.

NPCC, (2007). Final Report of the National Network Project on Climate Change. 'Climate change effects on growth and productivity of plantation crops with special reference to coconut and black pepper: Impact, adaptation and vulnerability and mitigation strategies' (Naresh Kumar *et al.*,). CPCRI report.

Naresh Kumar, S. (2009). Carbon sequestration in coconut plantations. In Global Climate Change and Indian Agriculture-case studies from ICAR Network Project (PK Aggarwal ed.), ICAR, New Delhi Pub., pp. 129-134.

Naresh Kumar, S.(2010). Assessment of Climate Change Impacts on Plantation Crops: Enabling Activity for Preparation of India's Second National Communication to UNFCCC: Submitted to Ministry of Environment and Forestry, Govt. of India, 2010.

Naresh Kumar, S. and Aggarwal,P.K. (2013). Climate change and coconut plantations in India: Impacts and potential adaptation gains. *Agril. Syst.* http://dx.doi.org/10.1016/j.agsy.2013.01.001.

Naresh Kumar, S. and Aggarwal, P. K. (2009). Impact of climate change on coconut plantations. In Global Climate Change and Indian Agriculture-case studies from ICAR Network Project (PK Aggarwal ed.), ICAR, New Delhi Pub., pp.24-27.

Naresh Kumar, S. and Kasturi Bai,K.V. (2009). Photo-oxidative stress in coconut seedlings: Early events to leaf scorching and seedling death. *Braz. J. Plant Physiol.* 21(3), 223-232.

Naresh Kumar, S., Kasturi Bai, K. V., Rajagopal, V. and Aggarwal, P.K. (2008). Simulating coconut growth, development and yield using InfoCrop-coconut model. *Tree Physiology,* 28, 1049–1058.

Naresh Kumar, S., Rajagopal V.,Cherian, V.K., Siju Thomas, T., Sreenivasulu, B., Nagvekar, D.D., Hanumanthappa, M., Bhaskaran, R., Vijaya Kumar, K., Ratheesh Narayanan, M.K. and Amarnath, C.H. (2009). Weather data based descriptive models for prediction of coconut yield in different agro-climatic zones of India. *Indian J. Hort,* 66 (1), 88-94.

Naresh Kumar, S., Rajagopal, V. and Kasturi Bai,K.V. (2016). Abiotic Stress Tolerance in Horticultural Crops: Coconut and Areca Nut **In:** *Abiotic Stress Physiology of Horticultural Crops* (eds N.K.S. Rao, *et al*) Springer Pub, pp. 269-306.

Peiris, T.S.G. and Thattil,R.O. (1998). The study of climate effects on the nut yield of coconut using parsimonious models. *Expl. Agric,*34, 189-206.

Prasada Rao, G.S.L.H.V. (1991). Agrometeorological aspects in relation to coconut production. *Journal of Plantation Crops,* 19 (2), 120-126.

Rajagopal, V., Shivashankar, S. and Jacob Mathew. (1996). Impact of dry spells on the ontogeny of coconut fruits and it's relation to yield. *Plantn. Res. Dev,*3(4),251-255.

Rajagopal, V. and Naresh Kumar, S. (2003b). 'Charecterization of drought in different coconut growing areas and drought management strategies" *Final Report-AP Cess fund project,* p. 105.

Raj, N.M. and Murugan, M. (2000). Ecological decline of cardamom hills -an analysis. Spices and aromatic plants: challenges and opportunities in the new century. Contributory papers. Centennial conference on spices and aromatic plants, Calicut, Kerala, India, 20-23 September, 2000. 172-176.

Radhakrishnan, V.V., Madhusoodanan, K.J., Kuruvilla, K.M. and Vadivel, V. (2002). Production technology for black pepper. *Indian Journal of Arecanut, Spices and Medicinal Plants,* 4(2), 76-80.

Sailajadevi, T., Nair, R.B., Dey, S.K., Devakumar, A.S., Licy, J., Kothandaraman, R. and Sethuraj, M.R. (2000). Impact of weather on yield and yield components in some elite Hevea clones. *Indian Journal of Natural Rubber Research,*13(1-2), 98-102.

Sivaraman, K., Kandiannnan, K., Peter, K.V. and Thankamani, C.K. (1999). Agronomy of black pepper (Piper nigrum). *Journal of spices and Aromatic Crops,* 8(1), 1-18.

Sukumara Pillay, V., Sasikumaran, S. and Venugopalan Nambiar, P.K. (1977). A note on preliminary observation of spike shedding in pepper. *Arecanut and Spices Bulletin,* 8(4), 93-94.

2017, Impact of Climate Change on Plantation Crops *Pages 15–44*
Editors: **K.B. Hebbar, S. Naresh Kumar & P. Chowdappa**
Published by: **ASTRAL INTERNATIONAL PVT. LTD., NEW DELHI**

Chapter 2

Coconut

S. Naresh Kumar, V. John Sunoj, K.S. Muralikrishna, K.B. Hebbar, V. Rajagopal, K.V. Kasturi Bai and P. Chowdappa

1. Causes of Climate change

Over exploitation of fossil fuels, deforestation, land use change and energy use inefficient technologies have led to rapid accumulation of greenhouse gases (GHGs) in the atmosphere. The CO_2 concentration has increased from a pre-industrial value of about 280 ppm to 401 ppm in 2016. Similarly, the global atmospheric concentration of methane and nitrous oxides and other GHGs has also increased considerably. In addition, globally about 190 Mt of urea is used in agriculture. As a consequence emission of N_2O from agricultural fields led to rise in the atmospheric concentration of N_2O. Since the life-time of the CO_2, methane and N_2O are about 120, 10 and 150 years, respectively, past and current emissions continue to warm the climate and cause climate change even if GHG emissions are reduced. Climate change is defined as "A change of climate which is attributed directly or indirectly to human activity that alters the composition of the global atmosphere and which is in addition to natural climate variability observed over comparable time periods" (United Nations Framework Convention on Climate Change (UNFCCC). The UNFCCC thus makes a distinction between "climate change" attributable to human activities altering the atmospheric composition, and "climate variability" attributable to natural causes.

2. Past Trends in Global and Indian Climate

The global mean temperatures during 1951-2010 period increased by 0.6-0.7°C out of which natural variability contributed to ±0.1°C change in temperature (IPCC, 2013, 2014). This indicated that the anthropogenic rise in temperature is about 0.6°C during 1951-2010 period. In Indian region, annual mean maximum temperatures increased by about 0.71°C in past 100 years. The rate of increase in temperature enhanced and in post 1970's the warming has been at a rate of 0.17°C/10 years. Similarly, the minimum temperature has been rising at a rate of 0.29°C/10 years (IITM, 2011). Rise in temperatures alter the hydrological cycle and thus the precipitation. Since 1871, India faced 28 deficit and 20 excess monsoon years and out of these, as many as 16 deficit monsoon years and 6 excess monsoon years fell in post 1960 period.

3. Projected Climate Change Scenarios

Increase in atmospheric concentration of GHGs vary depending upon the current and future developmental pathways. For instance, rapid economic growth without due consideration for environmental protection will lead to unabated accumulation of GHGs in atmosphere. These GHGs can cause the warming of 8.5 $W.m^{-2}$ (this scenario is called representation concentration pathway 8.5 -RCP 8.5). This will lead to a very high increase in global temperatures and disturbances in hydrological cycle and thus precipitation. On the other hand, eco-friendly technologies and environment-friendly developmental strategies will have least accumulation of GHGs in atmosphere leading to not so much warming (about 2.6 $W.m^{-2}$ –RCP 2.6).

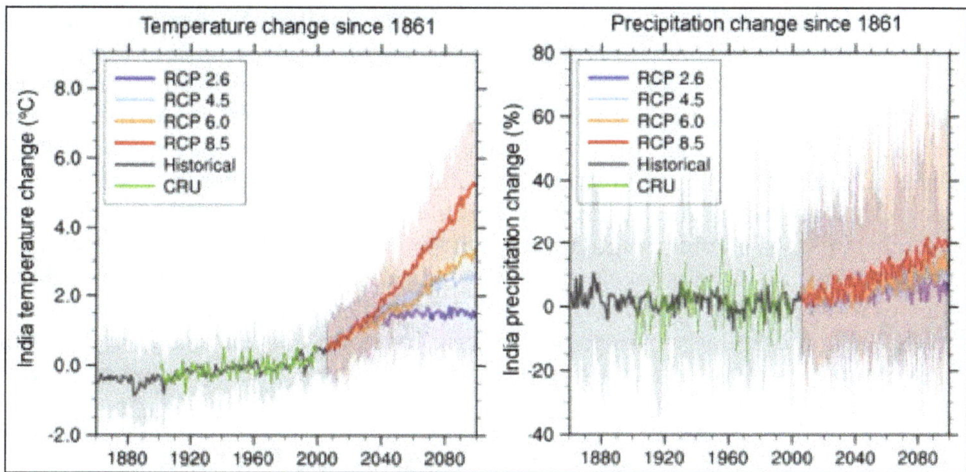

Figure 2.1: Projected Change in Mean Annual Temperature and Precipitation Over India in different Climate Scenarios (*Source*: Chaturvedi *et al.*, 2012).

Future climatic scenarios are simulated using the global climate models (GCMs) and regional climate models (RCMs). More than 56 GCMs project an increase of global mean surface temperatures in the range of 0.3°C to 1.7°C (RCP 2.6), 1.1°C to 2.6°C (RCP 4.5), 1.4°C to 3.1°C (RCP 6.0), 2.6°C to 4.8°C (RCP 8.5) for 2081–2100 relative to 1986–2005 period (Figure 2.1). The projected increase in temperature for south Asia is in the range of 0.5 to 1.2°C by 2020, 0.88 to 3.16°C by 2050 and 1.56 to 5.44°C by 2080, depending on the scenario of future development. Further, the IPCC AR5 (2014) on climate change has projected an increase in the frequency of droughts, floods, and extreme events of temperature and rainfall. The report also projected an increase in the area encompassed by the monsoon phenomena. These climatic changes further increase the pressures on Indian agriculture including plantation crops.

4. Climatic Change and Agriculture

Plant growth, development and yield are highly influenced by the climatic factors apart from other environmental factors. Among many climatic factors that influence crop performance temperature, rainfall, CO_2 and light are the most important. Temperatures influence the growth rates and high temperatures affect photosynthesis (Blum *et al*, 1994; Hatfield and Prueger, 2015) growth and development, shorten growth period (Porter and Gawith, 1999), affect sink parameters (Hatfield and Prueger, JH (2015) and thus yield (Assange *et al.*, 2010, Naresh Kumar *et al.*, 2013, 2014 a and b). Similarly very low temperatures cause tissue damage, slow metabolic rates and affect the growth, development and yield of the plants.

Projected increase in frequent droughts and heavy precipitation events in future climates means plants will be exposed to multiple stresses even in a single season (Naresh Kumar *et al.*, 2012). Impacts of water stress range from cell water potential loss to leaf and plant wilting causing severe loss to yield as a consequence of reduced photosynthesis, growth and canopy area. Similarly the excess light causes photo-oxidative stress damaging cell membrane resulting in leaf scorching and plant death (Naresh Kumar and Kasturi Bai, 2009). However, increase in atmospheric CO_2 can be beneficial to plant photosynthesis, particularly in C3 plants (Ainsworth and Long, 2005; Kimball *et al.*, 1995, 2002). Though, elevated CO_2 may not benefit C4 plants in normal management conditions, but under water stressed conditions, it indirectly increases photosynthesis and yield by reducing water use and delaying drought stress via stomatal regulation (Ghannoum *et al.*, 2000). Thus climate change impacts crops at various levels and influences the yield and quality of produce. For instance, elevated CO_2 may reduce the protein concentration (Ainsworth and Long, 2005), particularly in nitrogen limiting environments. Similarly, contents of anthocyanin, carotene, starch, oil, flavonoids, lycopein, *etc.* are affected in various crops (Ainsworth and Long, 2005).

5. Observed Impacts on Agriculture: Some Examples

Indian agriculture is often called as a 'gamble of monsoon' and is prone to climatic risks. The climate related aberrations have been significantly affecting the crop productivity in India. Several examples on climatic-stress related yield loss exist for annual crops and are relatively well documented. Unfavorable monsoon has been affecting the productivity of Kharif season crops. The terminal heat stress and early heat stress is lowering wheat yields while extreme weather events are affecting almost all crops. For instance, heavy rainfall in Madhya Pradesh during pod maturation affected the soybean yields during 2013 monsoon season. In 2014 and 2015, hailstorms in Maharashtra affected many horticultural crops. However, only a few examples are documented for perennial crops. For instance, unseasonal rainfall in March 2008 affected the quality and quantity of cashew yield, which can be taken as an example of extreme weather event. On the other hand, consecutive droughts during 1998-2002 in Tamil Nadu and 2006-2007 in Karnataka have affected the coconut and arecanut yields, respectively, which can be an example of inter-annual extreme weather conditions (Naresh Kumar, 2011). Shift in apple cultivation from 1250 mamsl to 2500 mamsl to in Himachal Pradesh is partly attributed to the non- fulfillment of chilling requirement for apple due to rise in temperature (Bhagat *et al.*, 2009). This is one of the classical examples of mean change in climatic condition of a region.

6. Projected Impacts on Agriculture and Adaptation Gains

Globally, climate change is projected to affect the productivity of wheat, maize and soybean (Esterling *et al.*, 2007; Asseng *et al.*, 2015, IPCC 2014). Analysis at Environmental Modelling lab, Indian Agricultural Research Institute indicated that, on all India basis, the impacts of climate change on yields in 2030s range from -2.5 to -12 per cent for crops such as rice, wheat, maize, sorghum and mustard (Naresh Kumar *et al.*, 2011, 2012, 2013, 2014a and b). On the other hand, yields of some crops such as soybean, potato (in north- west India) and coconut (in west coast and in north east India) are projected to gain due to climate change (Naresh Kumar *et al.*, 2012; Naresh Kumar and Aggarwal, 2013). However, these impacts have spatial variation. For instance, positive impacts are projected for potato yields in north-west India while potato in central India may be affected. Adaptation can enhance the rice (+20 per cent), wheat (+11 per cent), maize (+21 per cent), mustard (+25 per cent), sorghum (+8 per cent), soybean (+12 per cent), potato (+8 per cent, all India) and coconut (+33 per cent all India) yield in future as indicated by the simulation analysis (Naresh Kumar *et al.*, 2011, 2012, 2013, 2014a and b; 2015; Naresh Kumar and Aggarwal, 2013).

7. Effects of Climate Change on Coconut

In India coconut is mainly grown in Kerala, Karnataka, Tamil Nadu, Andhra Pradesh, Maharashtra, West Bengal and Assam (Table 2.1). These plantations provide sustenance to the millions of the farmers across these states. Kerala and North-eastern states are generally characterized by coconut based home-stead gardens, forming the backbone for the livelihood security of millions. Thus it is important to quantify the impact of climate change on coconut plantations and formulate the suitable adaptation strategies. Even though coconut is grown in over 200 districts in India, the major producing area is confined to just about 20 districts (Figure 9.2), which contribute almost 70 per cent to national production. Eleven out of 14 districts of Kerala together contribute ~ 22 per cent to the national production. Contribution from Tamil Nadu (~31 per cent), Karnataka (~23 per cent) and Andhra Pradesh (~8 per cent) mainly comes from 10 districts. Thus, majority of coconut production is confined to a very small area making it's production more dependent on climatic factors.

It is important to understand the impact of climate change on plantation crops which, being perennial in nature, face climate change and variability even during a single generation or in a standing plantation. For example a seedling of coconut, with all likelihood, will face the increased CO_2 concentrations, temperatures, changed

Table 2.1: Relative Contribution of Coconut Area, Production, Productivity by different States in India

States/Union Territories	Coconut (2013-14)				
	Area ('000 ha)	Contribution to Total (per cent) Area	Production (Million nuts)	Contribution to Total (per cent) Production	Productivity (Nuts/ha/ year)
Andaman and Nicobar	21.9	1.02	129.97	0.60	5935
Andhra Pradesh	121.92	5.70	1828.46	8.44	14997
Assam	20.23	0.95	136.61	0.63	6753
Bihar	15.25	0.71	141.42	0.65	9273
Chhattisgarh	1.52	0.07	22.1	0.10	14539
Goa	25.75	1.20	128.13	0.59	4976
Gujarat	31.63	1.48	295.03	1.36	9328
Karnataka	517.29	24.17	5041.15	23.27	9745
Kerala	797.21	37.24	5968.01	27.55	7486
Lakshadweep	2.57	0.12	70.91	0.33	27591
Maharashtra	28.08	1.31	187.47	0.87	6676
Mizoram	0.03	0.00	0.12	0.00	4000
Nagaland	1.45	0.07	16.32	0.08	11255
Odisha	50.78	2.37	324.93	1.50	6399
Puducherry	1.96	0.09	34.09	0.16	17393
Tamil Nadu	465.11	21.73	6917.25	31.93	14872
Telengana	1.61	0.08	24.09	0.11	14963
Tripura	6.91	0.32	28.3	0.13	4096
West Bengal	29.3	1.37	370.83	1.71	12656
All India	**2140.5**		**21665.19**		**10122**

Source: Coconut Development Board, Kochi.

rainfall amount and pattern, etc. in next 60 years of its economic yield producing life span. Keeping in view the fact that coconut plantations are grown in ecologically sensitive regions such as coastal regions, the research on understanding the impacts of climatic stresses and climatic change on coconut was initiated at Central Plantation Crops Research Institute, Kasaragod, India during 2002. World over, this has been the first comprehensive climate change study on coconut.

7.1. Approaches for Climate Change Studies

Quantification of impact of major climate change parameters such as increased temperature and CO_2, and change in rainfall on crops is done using three approaches, *viz.*, field or controlled environment experiments, meta-analysis and simulation models. In field experimentations, quantifications are done by growing crops in Open Top Chamber (OTC) or in Free Atmospheric Carbon dioxide Enrichment

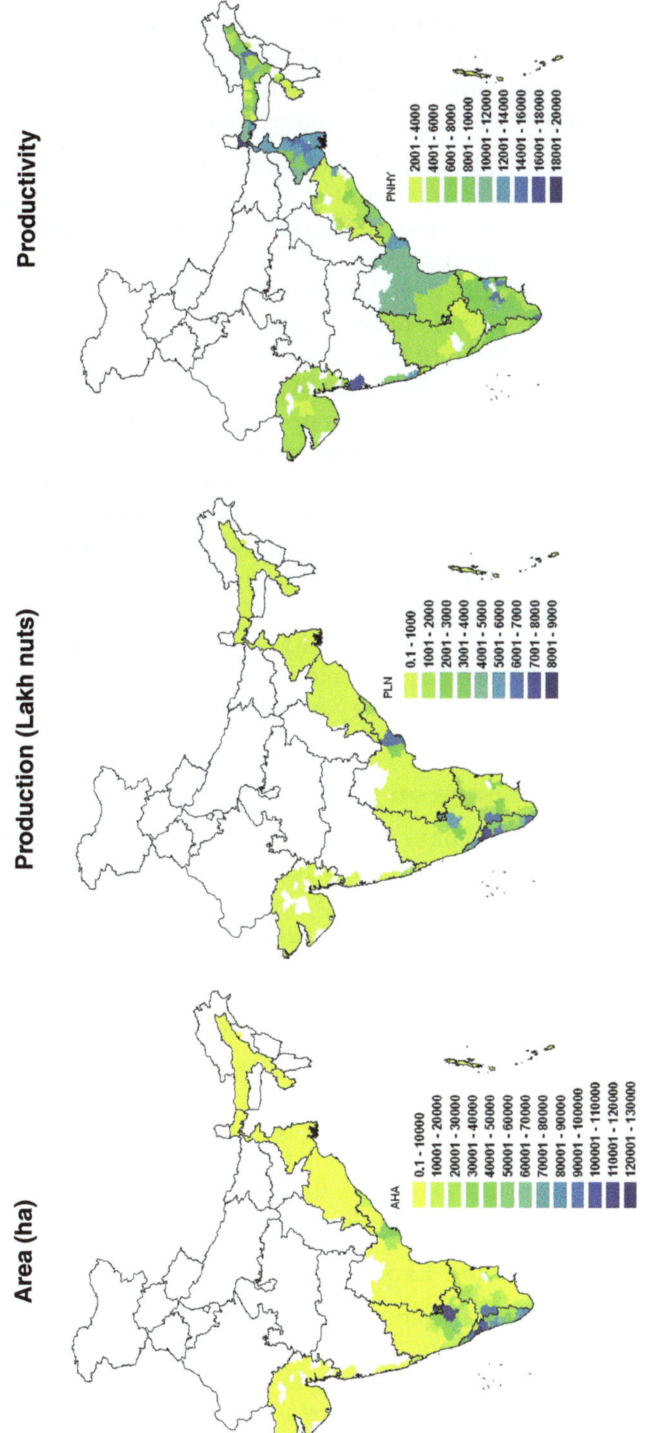

Figure 2.2: Spatial variation in area production and productivity of coconut in India

(FACE) rings, in Free Atmospheric Temperature Elevation (FATE) rings and in T-FACE facilities (where both temperature and CO_2 are modified). These studies provide vital information on crop response to elevated CO_2 or temperature or to both, in spite of being costly and time taking procedure. However, the information generated cannot be generalized for regional/state or national level.

Meta-analysis and analysis of past data indicates the climate related impacts but it is difficult to effectively desegregate confounding effects of technological and other interventions even by different de-trending methods. Another major limitation is that CO_2 effects cannot be delineated and thus derived future projections are not reliable. The other approach is to use well validated simulation models. Simulation models are strong tools which provide opportunity to use various climate change scenarios in combination with different management parameters for analyzing the regional impacts, adaptation and vulnerability. Apart from these, well developed crop simulation models provide opportunity to assess the impacts of climatic variability on crop growth. Such studies provide the guidelines for 1) relative advantage of adaptation strategies to be followed for minimizing the climate change impacts 2) best possible adaptation strategy 3) technology development needs 4) research gaps 5) regional spatial and temporal variation for yield gap 6) regional policy direction and so on. Further, crop simulation models coupled with remote sensing and geographical information systems (RS-GIS) prove to be strong tools for not only impact, adaptation and vulnerability assessments, but also for land use change and land use plan as well as for yield forecasting, crop monitoring and management. Use of more than one strategy and coupling of strategies provide more robust information and assessments.

7.2. Impact of Climate Change on Coconut

In assessing the impact of climate change on coconut and for deriving the adaptation strategies, a five-pronged strategy *viz.*, 1) quantification of response of coconut seedlings to elevated CO_2 (550 and 700 ppm) and temperature (+2°C over control OTC) in Open Top Chamber system, 2) field experiments, multi-location experiments, 3) lab experiments, 4) surveys and 5) simulation analysis using InfoCrop-COCONUT model was used (Naresh Kumar et al, 2007a). With the development of a simulation model for coconut (Naresh Kumar *et al.*, 2008), it became possible to simulate effects of various climate change scenarios in combination with different management parameters for analyzing the regional impacts. Coconut model has immense potential for plantation management decisions in India and in other parts of the world. At CPCRI, Kasaragod, an Open-Top Chamber facility is used for quantifying the impacts of elevated CO_2 and temperature on coconut and areca nut seedlings and cocoa grafts (Naresh Kumar et al, 2007a).

7.2.1. Results from the Open Top Experiments

The experiments conducted at Central Plantation Crops Research Institute, India using the Open Top Chambers (Figure 2.3) from 2006 provided the first estimates of the response of coconut seedlings to elevated CO_2 and temperature (Naresh Kumar, 2007-2010; 2011; Naresh Kumar *et al.*, 2011, Muralikrishna *et al.*, 2009; 2013, Muralikrishna 2012, Sunoj, 2012; Sunoj *et al.*, 2009, 2013, 2014, 2015; Hebbar et

Figure 2.3: Open Top Chamber Facility for Studying the Effect of Elevated CO_2 and Temperature at CPCRI, Kasaragod, India.

al. 2013). Seedlings of coconut (2 hybrids and 3 cultivars), arecanut (5 cultivars) and cocoa (3 hybrids and 4 cultivars) were grown in Open Top Chamber facility with SCADA controlled pumping and monitoring of CO_2 to set level of 550 and 700 ppm and hot air for elevated temperature (+2°C).

After exposure of seedlings for three years, results indicated that in general, elevated [CO_2] at 550 and 700 µmol mol^{-1} benefited the growth and development of coconut seedlings (Figure 2.4) as they assimilated more carbon dioxide, due to higher photosynthetic rates and larger leaf area, resulting in significant increase in shoot and root biomass (Table 2.2). The leaf water balance was maintained by thicker cuticle, by reduced stomatal density and by regulation of stomatal conductance in both elevated [CO_2] and temperature. The growth of seedlings, grown in open top chamber with elevated temperature of 3°C above 31°C (the annual mean maximum

Figure 2.4: Response of Coconut Seedling to Elevated CO_2; (a) Control (b) 550 ppm CO_2 and (c) 700 ppm CO_2.

temperature of experimental site, was significantly reduced (Figure 2.5) consequent to the reductions of leaf area (>50 per cent), chlorophyll content (>40 per cent) net photosynthesis as compared to control. However, there was higher deposition of epicuticular wax on the leaves.

Figure 2.5: Response of Coconut Seedling to High Temperature; (a) Control (b) Ambient +3°C (c) 550 ppm CO_2 + Ambient +3°C.

In general, plants are more sensitive to high temperature during reproductive phase than vegetative stage. Pollen grains are highly sensitive to high temperature. Coconut cultivars CGD, MGD, MYD, MOD and WCT indicated a wide variability in optimum temperature requirement for pollen germination and tube length growth. At 20°C WCT had a germination of 90 per cent and reduced to 42 per cent at 40°C (Figure 2.6). Across all the temperatures, pollen germination was highest in WCT followed by MGD and was the least in MYD (Figure 2.7) (Hebbar and Chaturvedi, 2015). Cultivar differences for heat tolerance exist in some crops such as rice, cowpea, and peanut, but knowledge about the effects of extreme high temperatures is very limited in coconut because diverse germplasm has not been extensively screened.

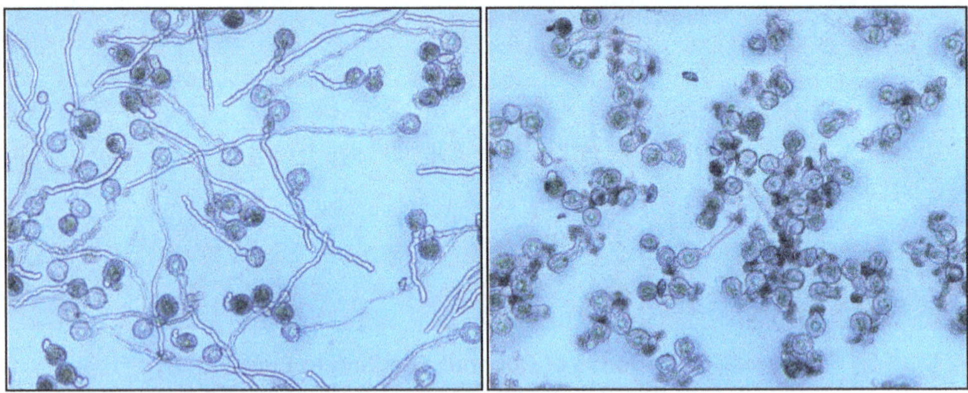

Figure 2.6: Pollen Germination of WCT at 25 (Left) and 40°C (Right).

Table 2.2: Influence of Elevated Temperature and CO_2 on Anatomical, Physiological and Biochemical Parameters of Coconut Arecnaut and Cocoa Seedlings (Source: Naresh Kumar 2007a)

Parameter	*% change from chamber control*		
	Temp. (2+°C)	*CO_2 (550 ppm)*	*CO_2 (700 ppm)*
Coconut			
Specific leaf weight	3.6	28.5	2.9
Net photosynthetic rate	-15.8	21.0	85.6
Stomatal conductance	-20.7	-53.3	-25.9
WUE	1.6	124.7	81.2
Leaf water potential	-54.4	11.6	2.3
Chlorophyll a/b ratio	1.9	7.7	1.4
Stomata density	-5.9	-10.7	-14.0
Lower cuticular thickness	16.7	23.5	26.0
Phenols concentration in leaf	-28.5	-2.1	-2.1
Proline in leaf tissue	5.4	121.7	97.1
Total soluble sugars in leaf tissue	-29.7	204.	41.4
Reducing sugars in leaf tissue	-44.7	29.9	58.0
Amino acids in leaf tissue	5.3	26.4	109.6
Starch in leaf tissue	-10.4	428.2	366.1
SOD activity	67.4	17.4	11.3
PPO activity	-67.7	-80.6	-77.4
Nitrogen in shoot tissue	-15.0	6.0	24.6
Nitrogen in root tissue	-7.2	31.9	13.0
CN ratio shoot	17.4	-6.5	-20.1
CN ratio root	8.2	-24.5	-11.0
Arecanut			
Total dry matter	-8.6	8.4	22.1
SLW	-33.6	-44.0	-38.1
Net photosynthetic rate	13.4	15.0	109.1
Leaf water potential	-10.6	24.6	10.0
Chl a/b ratio	27.8	17.6	10.2
Leaf thickness	-0.5	10.9	12.9
Amino acids in leaf tissue	-24.8	95.6	12.9
Total soluble sugars in leaf tissue	49.3	17.6	7.7
Starch in leaf tissue	54.8	154.5	86.3
Phenols in leaf tissue	25.4	20.2	0.4
Nitrogen in shoot tissue	4.4	6.8	7.3
Cocoa			
Root/shoot ratio	2.5	-12.0	-11.5
Total dry matter	-31.5	10.4	33.5
Number of roots	-48.3	-21.0	-15.9
Net photosynthetic rate	15.7	73.9	97.6
WUE	-3.8	20.7	60.4
Leaf water potential	-6.2	12.5	26.6
Chl a/b ratio	3.9	2.8	10.6
Total chlorophyll	25.1	16.1	35.7
SLW	-2.2	7.5	4.3
Leaf let thickness	-3.5	10.7	12.9
Total soluble sugars in leaf tissue	-3.7	-13.8	-13.6
Starch in leaf tissue	37.3	154.5	78.4
Phenols in leaf tissue	-4.0	-11.2	-8.7

The biochemical analysis indicated that elevated [CO_2] increased the concentrations of leaf biochemicals such as soluble sugars, reducing sugars, free

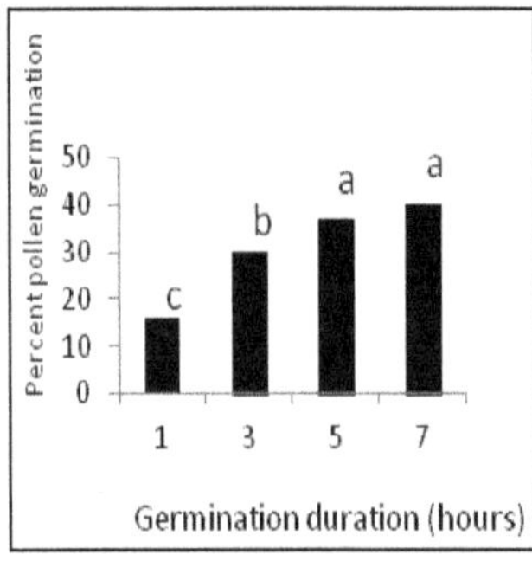

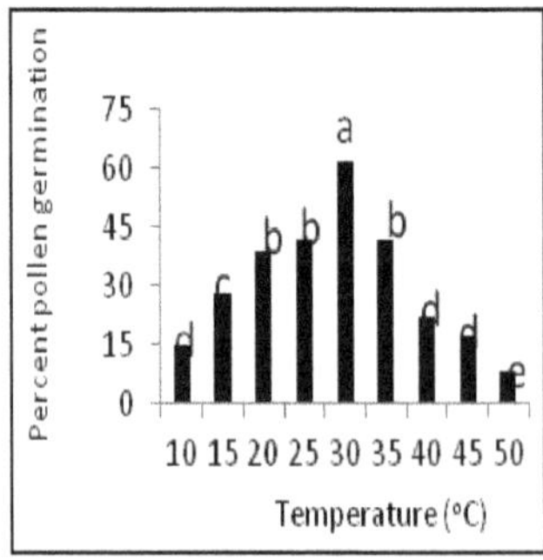

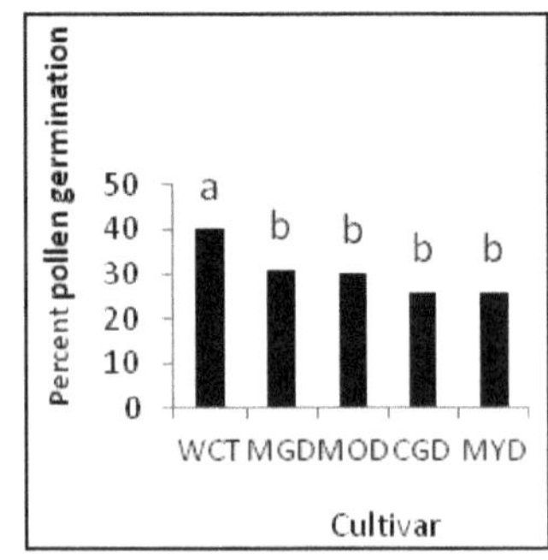

Figure 2.7: Effect of Temperature on Pollen Germination, (a) Percentage of pollen germination in WCT at 20°C, (b) Percentage of pollen germination at different temperatures in WCT for the period of 5 Hr, (c) Percentage of pollen germination of varieties (CGD, MGD, MYD, MOD and WCT) at different temperatures.

amino acids, starch and proline. The specific activities of superoxide peroxidase and catalase also increased. On the other hand, elevated [CO_2] reduced the concentration of total phenols, polyphenol oxidase activity and shoot and root CN ratio. Elevated [CO_2] also increased the activities of rhizosphere soil amylase, cellulase due to reduction in soil pH. Elevated temperature significantly increased concentration of leaf total free amino acids, proline and heat stable protein fraction. Apart from these the CN ratio in shoots and roots increased as also the activities of superoxide dismutase, catalase and soil invertase and sol cellulase. On the other hand, elevated temperature significantly reduced concentrations of leaf total soluble sugars, reducing sugars, total phenols, starch and reduced the activities of poly phenol oxidase and peroxidase. Coconut plantations may become more prone to diseases and pests due to reduction on the concentration of phenols in future climates.

Availability of soil nutrients in elevated [CO_2] and temperature condition, in general shown a reduction, except that of Na, K and Mg in elevated CO_2. This is an indicat or of a possibile higher requirement of fertilizer application in future scenarios. Results further indicated that in spite of variations in leaf and soil biochemical concentrations, coconut seedlings have capacity to adapt to future climate through changes in biochemical mechanisms that make some cultivars more suitable than others.

7.2.2. Impact of Elevated Temperatures on Quality of Coconut Copra and Oil

The copra yield/nut, oil percentage and fatty acid profiles of various cultivars grown at different agro-climatic zones was analyzed for their relationship with the weather during nut growth periods for the nuts harvested during January, April, July and October months (Naresh Kumar, 2005). The analysis indicated impact of Tmax, Tmin, Rainfall, RH on the copra yield/nut, oil per cent and fatty acid composition in coconut oil. The saturated to unsaturated fatty acid ratio increased rise in minimum temperature, while it slightly decreased with rise in maximum temperature (Figure 2.8). Copra weight/nut decreased while oil percentage slightly increased with rise in maximum temperatures.

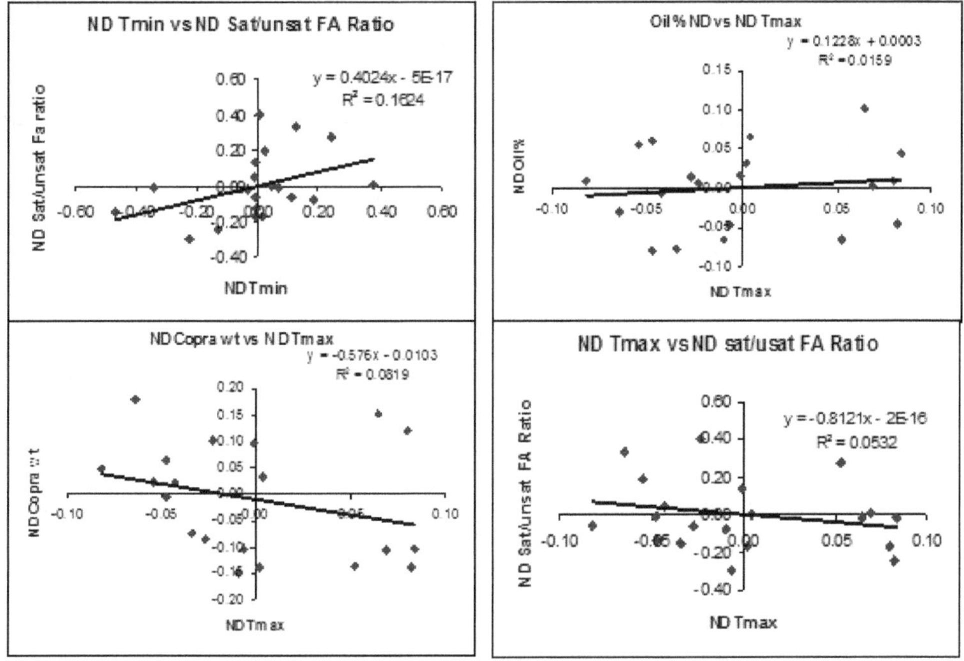

Figure 2.8: Normalized Deviation Analysis of Coconut Quality Parameters in Relation to Temperature Deviations.

7.2.3. Observed Impacts of Weather Extremes on Coconut Plantations

Examples

A study on the impact of climatic risk events like cyclones and droughts on coconut indicated that consecutive drought years in Coimbatore district of Tamil Nadu and Tumkur district of Karnataka in India caused not only severe reduction in yield but also left about 2 lakhs palms dead in both the districts (Naresh Kumar, 2011). More than 6 lakh palms were severely affected. Consecutive droughts in Coimbatore district reduced the coconut production by about three lakh nuts/ year for four years. Productivity loss was to the tune of about 3500 nuts/ha/year (Naresh Kumar *et al.*, 2010, Naresh Kumar 2011). Length and frequency of dry spell has negative impact of coconut yields in different agroclimatic zones (Naresh Kumar *et al.*, 2007, Naresh Kumar 2011).

In Godavari districts of Andhra Pradesh, cyclone in 1995 badly affected the coconut yields. The yields drastically reduced in 1995-96 and then it took four to five years for the recovery of affected gardens. Reduction in coconut yields was to the tune of 3350 lakh nuts/year for 6 years. The loss in East Godavari district was to the tune of about 2200 lakh nuts/year for six years. The productivity was reduced by 6200 nuts/ha/year in East Godavari district and by ~4100 nuts/ha/ year in Andhra Pradesh. During 2003-04 only the production levels could reach pre-cyclone period thus severely affecting the production for six years (Naresh

Kumar *et al.*, 2008, Naresh Kumar 2011). The impact of super cyclone in Odisha during 1999 caused severe loss to coconut plantations resulting in perennial loss of income to the farmer.

7.2.4. Climate Change Projections for Coconut Growing Regions in the World and in India

Climate change becomes particularly important for coconut plantations on three counts, in addition to those mentioned earlier. 1) Of the 12.9 Mha area under coconut plantations in the world, majority of area (~70 per cent) falls in the coastal zone, 2) Among the 90 countries that grow coconut, majority are islands. 3) Majority of the plantations are rainfed. As all regions of the world, coconut growing regions also are projected to face climate change. The climate models project (IPCC, 2014) the increase in temperature of 1.8-5.1°C in across these regions. The precipitation is projected to change -15 to 25 per cent in these regions. In addition, increase in the heavy rainfall events, a decline in winter rainfall, sea level rise, frequent intense tropical cyclones and change in *El Nino* are projected. All these potentially damage the coastal ecosystems and affect coastal agriculture including coconut plantations.

In India, during 1960-1990 period, the annual mean maximum temperatures ranged from 25.3 to 36°C and annual mean minimum temperatures ranged from 15.6 to 26°C in the coconut growing areas. Annual rainfall ranged from 600 to 4500 mm in these areas. Currently, east-coast is warmer and receives less rainfall as compared to the west-coast areas. Coconut growing areas in eastern and north-eastern region are characterized by hot summers and cold winters. North-east region also receives high rainfall. Climate change is projected to increase the temperatures by 1.26-2.4°C and change in annual rain fall in the range of -13 to 20 per cent by 2030 in the coconut growing areas. By 2080 temperatures are projected to rise by 2.95 to 4.9°C and annual rain fall is projected to change in the range of -15 to 30 per cent in these areas. However, these changes will have spatial and temporal variations.

7.2.5. Projected Impacts of Climate Change on Coconut Plantations

A review of literature indicates that mainly two methods are applied for analyzing the climate change impacts on coconut productivity. In the first approach, using the statistical models, it was reported that changes in monsoon rainfall pattern and increase in maximum air temperature are two key factors on the variability of coconut production in the principal coconut growing regions of Sri Lanka (Peiris *et al.*, 2006). They indicated that the projected coconut production after 2040 in all climate scenarios, when other external factors are non-limiting, will not be sufficient to cater the local consumption for the increased population.

The second approach is to use simulation models. Development of InfoCrop-COCONUT model (Naresh Kumar *et al.*, 2008), which can simulate the growth, development and yield of coconut taking into account of soil characteristics, weather, management and varietal characteristics, enabled researchers to use the later approach for coconut (Naresh Kumar and Aggarwal, 2009, 2013) InfoCrop-COCONUT is being released as CocoSim.

The simulation analysis on all India basis projected an increase in coconut productivity up to 4 per cent during 2020, up to 10 per cent in 2050 and up to 20

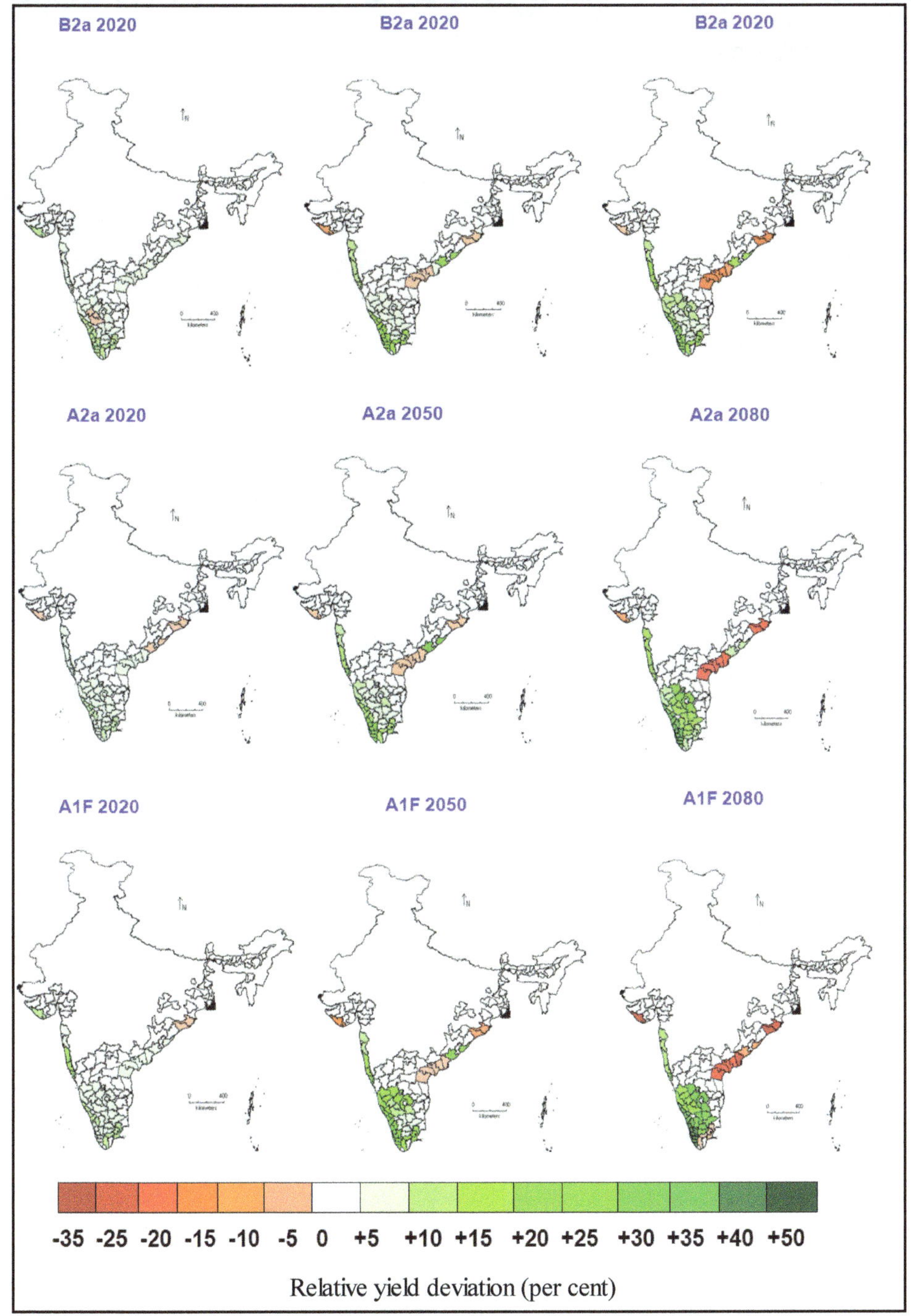

Figure 2.9: Impact of Climate Change on Coconut Productivity in 2020, 2050 and 2080 Climate Scenarios Based on HadCM3 Scenarios. The yield change is relative to mean productivity of coconut for five years (2000-2005) period (*Source*: Naresh Kumar, 2010).

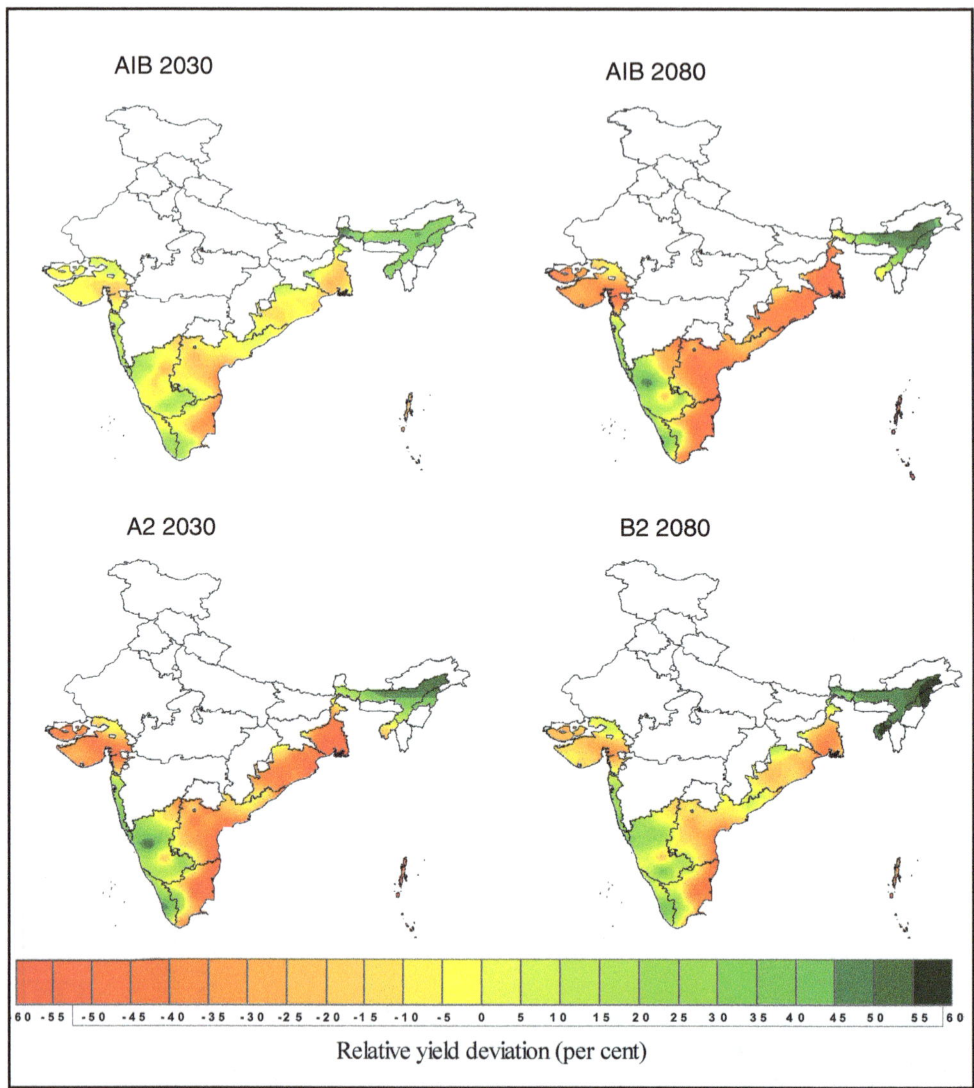

Figure 2.10: Spatial Variation in Impact of Climate Change on Coconut Productivity in India Based on High Resolution PRECIS A1B 2030 and 2080; A2-2080 and B2-2080 scenarios. The yield change is relative to mean productivity of coconut for five years (2000-2005) period (*Source*: Naresh Kumar and Aggarwal, 2013).

per cent in 2080 over current yields in HadCM3 A2a, B2a and A1F and PRECIS A1b scenarios (Figures 2.9 and 2.10). In west coast, yields are projected to increase by up to 10 per cent in 2020, up to 16 per cent in 2050 and up to 39 per cent by 2080 whereas in east coast yields are projected to decline by up to 2 per cent in 2020, 8 per cent in 2050 and 31 per cent in 2080 scenario over current yields (Naresh Kumar and Aggarwal, 2009).

Table 2.3: Impact of Climate Change in Major Coconut Growing States different Scenarios

State	*A1b2030*	*B1b2080*	*A2 2080*	*B2 2080*
	(per cent change from current production)			
Andaman and Nicobar	7.4	–10.3	4.9	15.0
Andhra Pradesh	–1.4	–31.4	–21.2	–3.4
Assam	59.1	>60.0	>60.0	>60.0
Goa	35.6	42.5	52.6	46.7
Gujarat	–9.0	–38.6	–38.0	–19.2
Karnataka	5.3	–8.8	–9.5	–11.2
Kerala	6.5	18.5	21.2	14.0
Lakshadweep	6.5	33.5	34.5	16.5
Maharashtra	15.7	25.6	30.5	26.3
Nagaland	>60.0	>60.0	>60.0	>60.0
Odisha	–6.5	–48.2	–40.7	–16.9
Puducherry	–25.7	–45.2	–41.0	–30.8
Tamil Nadu	–1.6	–10.3	–5.3	–2.8
Tripura	54.6	>60.0	>60.0	>60.0
West Bengal	2.9	–39.6	–24.2	–3.9
All India	4.3	1.9	6.8	5.7

Source: Naresh Kumar, 2010; Naresh Kumar and Aggarwal, 2013.

Subsequent fine resolution analysis projected increase in coconut productivity in western coastal region, Kerala, parts of Tamil Nadu, Karnataka and Maharashtra (provided current level of water and management is made available in future climates as well) and also in North-Eastern states, islands of Andaman and Nicobar and Lakshadweep and yield decrease for Andhra Pradesh, Odisha, West Bengal, Gujarat and parts of Karnataka and Tamil Nadu due to climate change. On all India basis, coconut productivity is projected to increase by 4.3 per cent in A1B 2030, 1.9 per cent in A1B 2080, 6.8 per cent in A2 2080 and 5.7 per cent in B2 2080 scenarios of PRECIS over mean productivity of 2000–2005 period.

7.3. Adaptation Strategies to Climate Challenges on Coconut Plantations

In plantation crops, management of crop becomes very important during adverse conditions in order to sustain the productivity. Since a standing plantation crop will face the climate change and variability effects during their life period due to perennial nature, the multi-pronged strategy may be adapted to reduce the adverse impacts of climate change and also to maximize the positive effects of climate change (Figure 2.11).

Simulation analysis indicated that agronomic adaptations like soil moisture conservation, summer irrigation, drip irrigation, and fertilizer application cannot

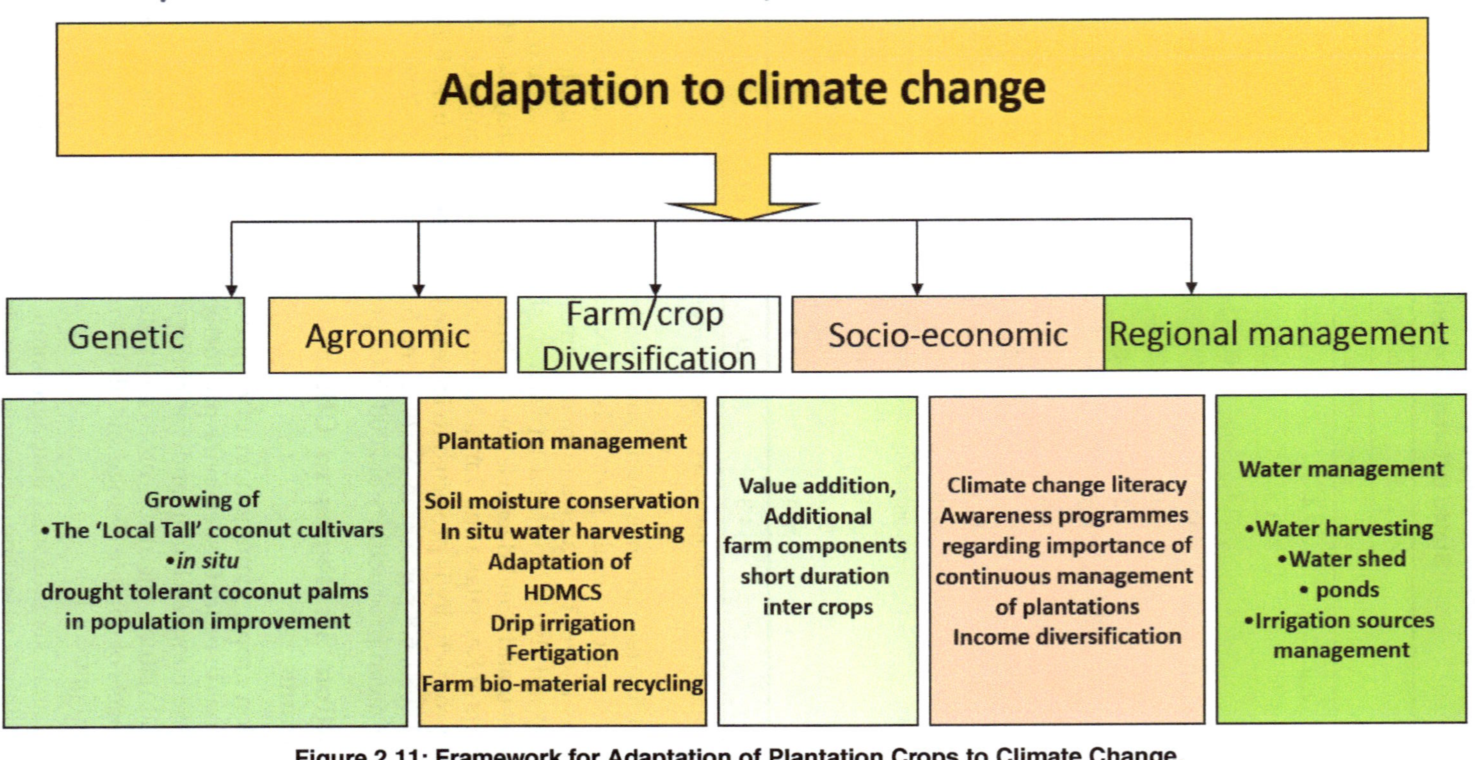

Figure 2.11: Framework for Adaptation of Plantation Crops to Climate Change.

only can minimize losses in majority of coconut growing regions, but also improve productivity substantially. Further, genetic adaptation measures like growing improved local Tall cultivars and hybrids under improved crop management is needed for long-term adaptation of plantation to climate change, particularly in regions that are projected to be negatively impacted by climate change (Figures 2.12 and 2.13). Such strategy can increase the productivity by about 33 per cent in 2030, and by 25–32 per cent in 2080 climate scenarios in India (Naresh Kumar and Aggarwal, 2013). In fact, productivity can be improved by 20 per cent to almost double if all plantations in India are provided with above mentioned management even in current climates (Naresh Kumar and Aggarwal, 2013). In places where positive impacts are projected, current poor management may become a limiting factor in reaping the benefits of CO_2 fertilization, while in negatively affected regions adaptation strategies can reduce the impacts. Thus, intensive genetic and agronomic adaptation to climate change can substantially benefit the coconut production in India (Naresh Kumar and Aggarwal, 2013). This analysis indicates that in changing climates, coconut productivity i) increase in some regions ii) some regions need adaptation to offset the yield loss iii) other regions remain vulnerable despite adaptation (Figure 2.10; Table 2.3). This information becomes extremely useful to initiate focused research activities for developing the integrated and novel adaptation strategies so that the i) beneficial effects of climate change can be harnessed and ii) projected vulnerable regions sustain yields in future climates as well. Similar analysis is needed for all coconut growing areas of the world as well as for other plantation crops so that effective adaptation strategies can be developed and implemented. Moreover, such analysis also provides a direction to researchers for developing the crop improvement and management strategies.

Data indicate substantial gains obtainable by adapting to climate change. In Kerala, providing more fertilizers along with summer time irrigation coupled with soil moisture conservation can further improve the positive gains due to climate

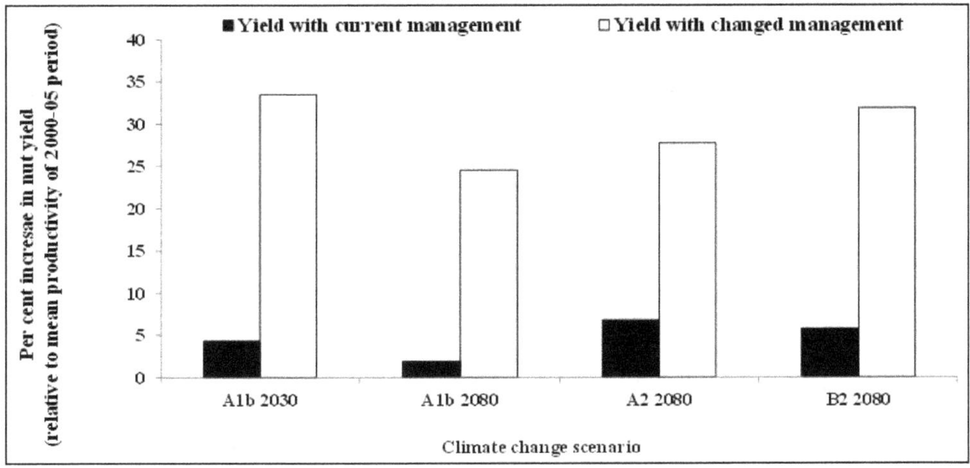

Figure 2.12: Change in all India Average Coconut Yield Due to Climate Change with and Without Adaptation (*Source*: Naresh Kumar and Aggarwal, 2013).

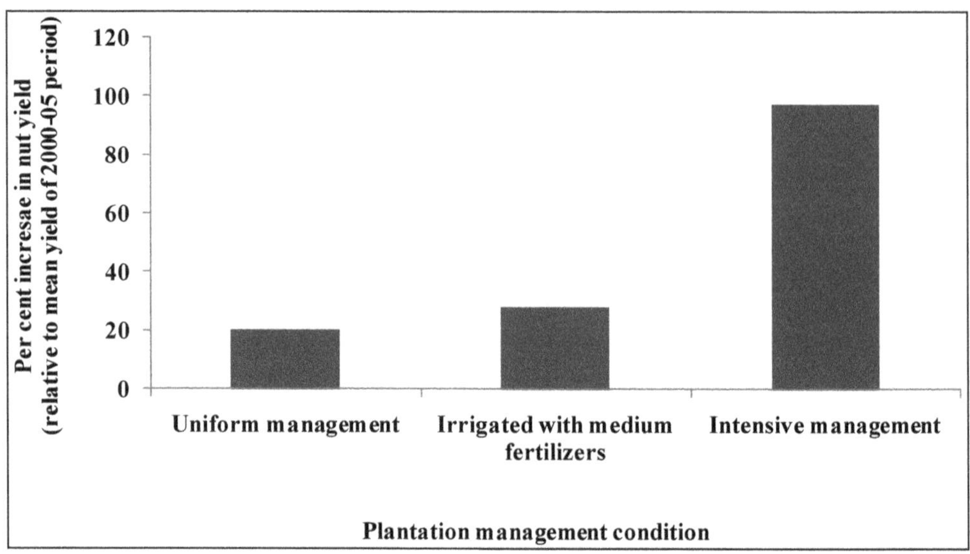

Figure 2.13: Gains in Nut Yield due to Adoption of Improved Management in Current Climates in India. The yield change is relative to mean productivity of coconut for five years (2000-2005) period. In irrigated with medium fertilizers, all farmers in India are assumed to provide assured irrigation and supply at least 2/3rd of the location specific recommended dose of fertilizers. (*Source*: Naresh Kumar and Aggarwal, 2013).

change by about 40 per cent in different scenarios. In Karnataka, West Bengal, Gujarat, Maharashtra and Odisha assured irrigation and providing more fertilizers not only can off-set the negative impacts but can result in higher yields. Adaptation gains in these states ranged from 2-47 per cent in different scenarios. In order to offset the negative impacts of climate change in Andhra Pradesh and Tamil Nadu, intensive management of plantations is needed. This includes planting of improved and tolerant varieties as well. In North-Eastern states, providing summer irrigation and even low dose of fertilizers can further improve the positive impacts of climate change. Similarly, coconut plantations in islands, if managed scientifically –by proper spacing or by canopy management- and by providing summer irrigation and even low dose of fertilizers, yield can be improved by 70 per cent in future climates (Naresh Kumar and Aggarwal, 2013).

Drought in current year not only affects the current year yield but also that of subsequent years. In coconut, the drought impact is worst in 3 and 4th year to follow (Rajagopal *et al.*, 1996; Naresh Kumar *et al.*, 2007b). Apart from this recovery will take 3-4 years thus causing a perennial loss in farm income. Thus, crop management becomes very important aspect. Multi-location trails on soil moisture conservation practices indicated very significant improvement in yield due to prolonged retention of soil moisture (Naresh Kumar *et al.*, 2006). The low-cost measures include basin mulching with husk/leaves/other farm biomass, burial of composted coir pith, hush burial, which acts as the moisture storage and retention mechanism in palm/ plant basin (Naresh Kumar, 2004). This helps to reduce the amount and frequency of irrigation. Drip irrigation has proved to be very effective method of saving coconut plantations in drought condition. Drip irrigation could save coconut plantations from consecutive drought impacts in Tamil Nadu during 1998-2002 period (Figure 2.14).

Figure 2.14: Effect of Drip Irrigation on Coconut Plantations during Consecutive Drought Years (1998-2002) in Pollachi Area of Tamil Nadu. Plantation in the left side is with drip irrigation.

Fertigation (Subramanian *et al.,* 2012) along with soil moisture conservation practices is one of the important water management strategies in climate change scenarios. By adapting this strategy and by providing higher nutrients, one can not only can reduce the adverse impacts of climate change but also can maximize the positive impacts of climate change.

7.3.1. Genetic Improvement for Changing Climates

Genetic improvement is the most important aspect of sustaining coconut productivity, particularly in areas that are identified as vulnerable to climate change. For instance, in India east coast is projected to be vulnerable in spite of some adaptation measures. Cultivars having heat tolerance, pollen viability at high temperatures, pistillate flower formation and nut retention at high temperatures form some of the important parameters for selection criteria.

Growing of the 'Local Tall' coconut cultivars (*e.g.* WCT and LCT (for west coast), Tiptur Tall (for Karnataka), SakhiGopal Tall (for Odisha), ECT (for AP), in areas prone to water scarcity is one option. Utilization of identified *in situ* drought tolerant coconut palms (Figure 2.15) in population improvement programme is very important for making the crop more resilient to climate change conditions (Naresh Kumar 2004; Naresh Kumar *et al.,* 2006). High biomass production capability (Kasturi

Figure 2.15: *In situ* Drought Tolerant Palms Identified in Farmers' Fields in different Agro-climatic Zones and being Used in Population Improvement Studies (*Source*: Naresh Kumar, 2004).

Bai *et al.*, 1996) with high harvest index for oil, copra and nuts also should form the criteria for selection of the varieties in changing climates. Pest and disease defense should continue to be included in crop improvement efforts, as coconut palms may become vulnerable to pests and diseases in future.

7.3.2. Other Important Adaptation Options

- ✰ Crop diversification and high density multi-cropping system
- ✰ Water harvesting and recycling, water-cooperatives, drip irrigation and fertigation
- ✰ Increasing farmer awareness on farm natural resource management and input use efficiency
- ✰ Identification and adoption of low carbon technologies
- ✰ Area expansion in positive impact zones
- ✰ Value addition and income diversification
- ✰ Weather forewarning and weather-based agro-advisories
- ✰ Market-linkage and cooperative initiatives
- ✰ Income insurance safety nets such as crop insurance

Mitigation Potential of Coconut Plantations

The interaction between coconut and climate are two ways. While climate change influences coconut plantations, the climate itself is modified by the plantations. In climate change terminology, mitigation refers to i) reduction in GHG emission ii) capture of GHGs for long (typically over 30 years) period. Coconut plantations are very good candidates for mitigating the GHG emission 1) by acting as a source of biofuel and energy source, 2) carbon sequestration 3) micro-climate modification, and 4) air purification by O_2 release. Apart from these, coconut plantations and other plantation crops have multi-dimensional functions in the ecosystem and social system (Figure 2.16).

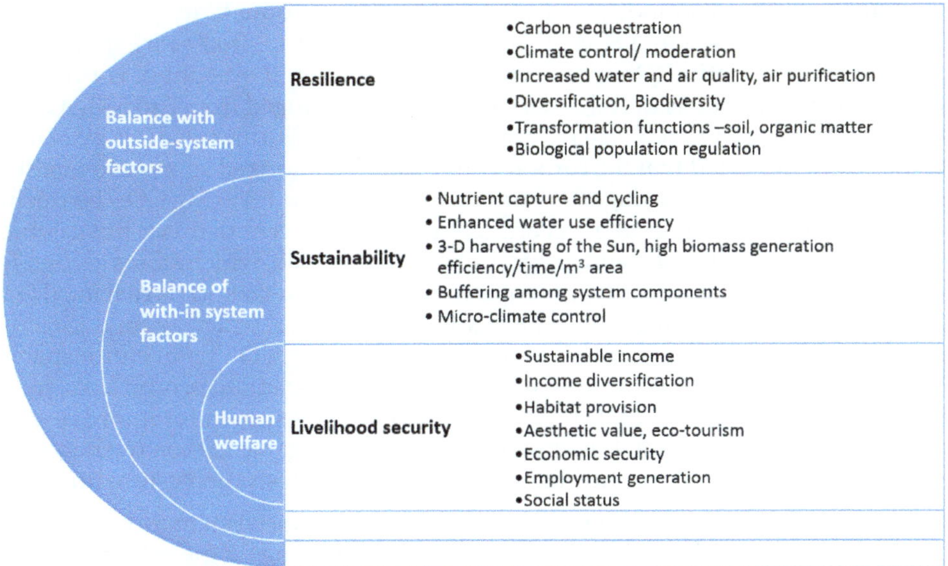

Figure 2.16: The Multi-dimensional Functions of Plantation Crops Systems (Naresh Kumar, 2015).

Coconut oil as biodiesel reported to reduce GHG emissions by about 60 per cent compared to the diesel (Pascual and Tan 2004). The net CO_2 emission from the carbon balance is equivalent to 77-104 g/MJ of diesel displaced by biodiesel (Tan *et al.*, 2004). Studies conducted in Vanuatu islands indicated possibilities of at least 9 per cent more profit, if covered under clean development mechanism over copra sales for coconut plantations (Roupsard *et al.*, 2004). The biodiesel conversion factor for coconut is taken at 130 L/Mg (FAO 2008, Johnston 2009). Utilization of coconut residues such as shells and husks for heat and power generation in the diesel production increases carbon savings in the range of 0.47-0.72 MgC/ha/yr (Vega, 2011).

Coconut plantations have immense potential to mitigate the GHG emissions by sequestering the carbon. The net primary production estimations of coconut monocrop in different agro-climatic zones of India indicated that carbon sequestration potential of coconut above ground biomass ranged from 8 Mg CO_2/ha/year to 32 Mg CO_2/ha/year depending on cultivar, agro-climatic zone, soil type and management (Naresh Kumar, 2009). Annually sequestered carbon stocked in stem is in the range of 0.3 to 2.3 Mg CO_2/ha (Figure 2.17a). Annual carbon sequestration by coconut plantation is higher in red sandy loam soils and lowest in littoral sandy soils. The standing carbon stocks ranged from 15 Mg CO_2/ha to 60 Mg CO_2/ha in a 20 year old plantation (Figure 2.17b). Further, using coconut simulation model the carbon stocks in stem of coconut in its economic life span of 60 years is estimated to be around 120 Mg CO_2/ha (Naresh Kumar, 2009). It is estimated that coconut plantations in India have carbon sequestration potential of about 7.54 Mt C. y^{-1}, which is equivalent to about 27.6 Mt $CO_2.y^{-1}$ and approximately 0.013 per cent of the total GHG emissions from India in 2010 (Naresh Kumar, 2015). Total GHG emissions from India in 2010 were 2,136.8 million tonnes of CO_2eq (BUR, 2015).

Other studies also indicate that net ecosystem productivity of 20-year-old coconut plantation is about 8 Mg C/ha/year (~29 Mg CO_2/ha/year) at optimum fertilizers and irrigation (Roupsard *et al.*, 2006). Further, if amount of copra taken out of plantations is considered, a net sequestration is ranged from 3 to 5 Mg C/ha/year (~11-18.4 Mg CO_2/ha/year) (Roupsard *et al.*, 2008). Further analysis by Naresh Kumar, 2009 indicated that the carbon in nut yield, categorized as short to medium duration (~2 to 5 years) storage, is estimated to be around 1,350 Mg CO_2/ha during the entire economic life span of coconut. Around 50 per cent of this is husk, which can be recycled into the soil and around 22 per cent is shell, which stores carbon for longer duration. Remaining 28 per cent is removed as copra, which is rich in oil (~65 per cent). Besides, carbon storage in leaf and inflorescence, which again is for short to medium duration, is estimated to be around 1000 Mg CO_2/ha in entire span of coconut economic life of 60 years. Currently, many value added products are made out of husk, shell and leaves. If litter is recycled into the system for improving the soil organic carbon content, the improvement in soil organic carbon over a period of time will also qualify for the inclusion in carbon sequestration estimations.

Coconut plantations influence the microclimate of the system thus regulating the weather and climate of the place and region depending on the extent spread of the plantation. The estimated microclimate regulation by the coconut plantations are about 1°C at a regional scale (Naresh Kumar, 2015). Release of O_2 by coconut plantation is estimated to be around 36 Mg O_2/ha/year (Naresh Kumar, 2015). For full commercial exploitation, potential clean development mechanism projects in coconut should include biomass carbon sequestration, renewable energy generation, substitution and greenhouse gas mitigation. Though, currently coconut is not included as one of the species for carbon sequestration in carbon markets, the immense potential of coconut plantations in managing the climate change should be exploited for the benefit of all.

In the socio-economic front, the livelihood security of coconut farmers is greatly dependent on the remuneration from the coconut based farming and products. Geo-textiles also provide a major scope for increased income in view of the increasing green-technology scenarios. Coconut plantations also serve as the green-buffer zones for adverse weather events. For instance during super-cyclone, which stuck the cost of Odisha in India, many people survived for days by climbing on coconut palms and consuming the nuts. Coconut grooves are attractive tourist places. Therefore focused and concerted research efforts are required to quantify the ecosystem services and exploit immense potential of coconut plantations in the tropics.

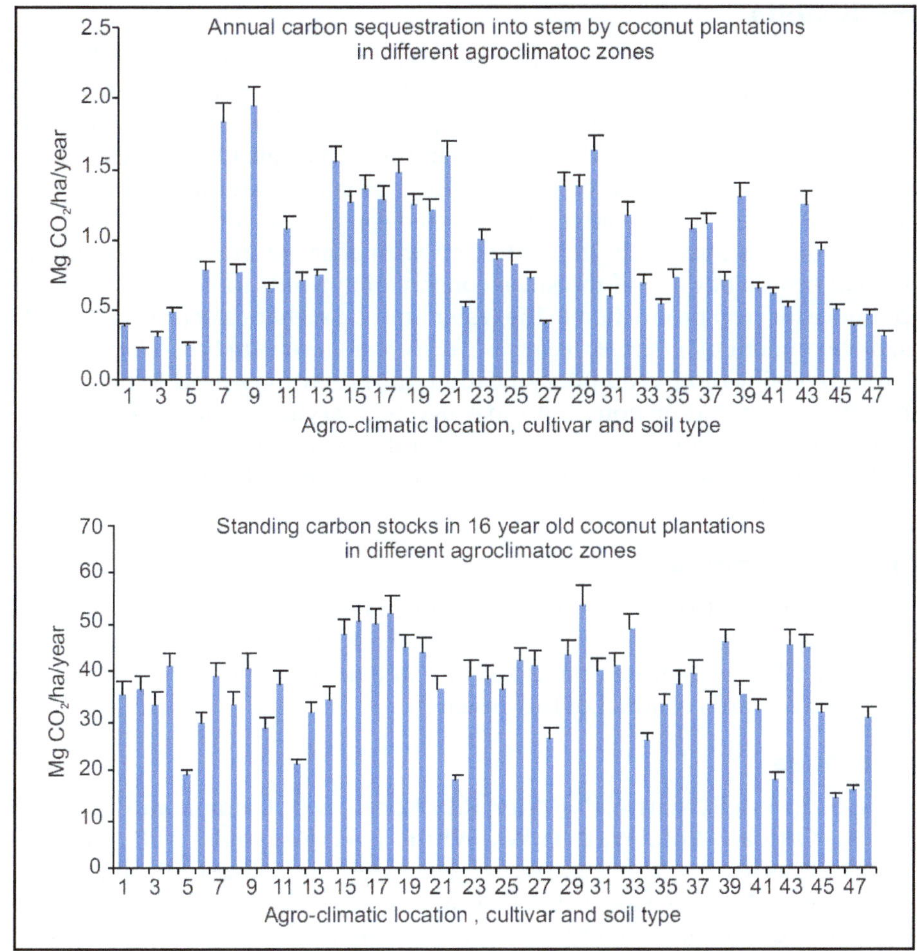

Figure 2.17: Annual Carbon Sequestration in Coconut Stem (a) and Carbon Stocks (b) in Coconut Plantations in different Agro-climatic Zones and Soil Types.

a: y-axis-I-irrigated; DI-drip irrigation; BCL- black clay loam soil; RSL-red sandy loam soil; RCL –red clayey loam soil; RL- red laterite soil; S-sandy soil; LS- coastal sandy soil; b, c and d -y axis – 1-5 Ambajipeta, clay loam, Irrigated (1-WCT, ECT, LCT, ECT× GBGD, GBGD); 6-14 Ratnagiri, drip irrigation, sandy loam (6-WCT, 7- ECT, 8-LCT, 9-ECT × GBGD, 10-GBGD, 11-Pratap, 12-COD, 13-WCT × COD, 14-COD × WCT); 15-28 Aliyarnagar, red sandy loam, irrigated (15-ECT × GBGD, 16-LCT × GBGD, 17-ECT ×MYD, 18-Pratap, 19-GBGD × ECT, 20 – ECT, 21-LCT, 22-COD, 23- WCT, 24- COD × WCT, 25 -WCT × COD, 26- Tiptur Tall, 27- GBGD, 28-Arasampatty Tall); 29-38 Ariskeri, red clay and black clay, drip irrigated (29-LCT, 30- Arsikere Tall, 31-WCT (black clay), 32-WCT (red clay), 33-Tiptur Tall, 34-GBGD, 35-WCT × COD (black clay), 36-WCT × COD (red clay), 37-COD × WCT, 38- WCT × GBGD); 39-42 Kidu, red laterite, drip irrigated (39-ECT, 40-WCT, 41-LCT, 42-COD); 43-46 Veppankulum, sandly loam, irrigated (43- ECT, 44-WCT, 45-LCT, 46-COD); 47-48 Kasaragod, coastal sandy (47-WCT rainfed, 48- WCT-drip irrigated). Nems of coconut cultivars WCT- West coast Tall; ECT- East Coast Tall; LCT- Laccadive Ordinary Tall; GBGD-Gangabondam Dwarf; COD- Chowghat Orange Dwarf; MYD-Malayan Yellow Dwarf). (*Source*: Naresh Kumar, 2009).

Conclusions

Past adverse climate events such as droughts and cyclones severely affected the coconut yields. Coconut simulation model based projections indicate beneficial influence of climate change on coconut in western coast of India provided the current water sources are made available in future as well. On the other hand the plantations in irrigated areas of Tamil Nadu, Karnataka, Andhra Pradesh, Odisha and Gujarat are likely to be adversely affected causing loss of production. Quality of the produce is also likely to be affected with changes in nut composition and fatty acid profile of oil. Soil moisture conservation needs to be given paramount importance for crop production in the water limiting scenarios of future. Apart from this, development of *'green technologies'*, *'climate resilient varieties'*, *'climate proofing'* by crop diversification, crop intensification, mixed farming need to be promoted. The carbon sequestration potential and ecosystem services of plantation crops need to be exploited for acquiring *'green credits'* for the overall benefit of the farmer. Use of technology and policy support are the need of the hour to improve profitability to the coconut farmer.

Acknowledgements

Sincere thanks to the Director, Central Plantation Crops Research Institute and DDG (Hort) ICAR. This work is funded by the ICAR-National Network project on Climate Change. Sincere thanks also are due to all the field and lab staff those contributed towards conduct of this research.

References

Asseng, S., Ewert, F., Martre, P., Rötter, R.P., Lobell, D.B., Cammarano, D., Kimball, B.A., Ottman, M.J., Wall, G.W., White, J.W., Reynolds, M.P., Alderman, P.D., Prasad, P.V.V., Aggarwal, P.K., Anothai, J., Basso, B., Biernath, C., Challinor, A.J., De Sanctis, G., Doltra, J., Fereres, E., Garcia-Vila, M., Gayler, S., Hoogenboom, G., Hunt, L.A., Izaurralde, R.C., Jabloun, M., Jones, C.D., Kersebaum, K.C., Koehler, A.K., Müller, C., Naresh Kumar, S., Nendel, C., Leary, G.O., Olesen, J.E., Palosuo, T., Priesack, E., EyshiRezaei, E., Ruane, A.C., Semenov, M.A., Shcherbak, I., Stöckle, C., Stratonovitch, P., Streck, T., Supit, I., Tao, F., Thorburn, P., Waha, K., Wang, E., Wallach, D., Wolf, J., Zhao, Z. and Zhu. Y. (2015). Rising temperatures reduce global wheat production. Nature Climate Change, http://dx.doi.org/10.1038/nclimate2470.

Ainsworth, E.A. andLong, S.P. (2005). What have we learned from 15 years of free air CO_2 enrichment (FACE)? A meta-analytic review of the responses of photosynthesis, canopy properties and plant production to rising CO_2. *New Phytologist*, 165, 351-372.

Bhagat, R.M. Rana. And Kalia,R.S.V.(2009). Weather changes related shift in apple belt in Himachal Pradesh; in Global Climate Change and Indian Agriculture-case studies from ICAR Network Project (ed.), Aggarwal P K (ICAR, New Delhi Pub), pp. 48-53.

Blum, A., Sinmena, B., Mayer, J., Golan, G. and Shpiler, L. (1994). Stem reserve mobilization supports wheat-grain filling under heat stress. Aust. *J. Plant Physiol*, 21, 771-781.

BUR (2015). India: First Biennial Update Report to the United Nations Framework convention on Climate Change. Ministry of Environment, Forest and Climate Change, Government of India. p. 180.

Easterling, W., Aggarwal, P.K., Batima, P., Brander, K., Erda, L., Howden, M., Kirilenko, A., Morton, J., Soussana, J.F., Schmidhuber, S. and Tubiello, F. (2007). Food fibre and forest products In: Parry M L Canziani O F Palutikof J P van der Linden P J Hanson C E (Eds) Climate Change 2007: Impacts Adaptation and Vulnerability Contribution of Working Group II to the Fourth Assessment Report of the Intergovernmental Panel on Climate Change Cambridge University Press, pp. 273-313.

Ghannoum, O., Von Caemmerer, S., Ziska, L.H. and Conroy, J.P.(2000). The growth response of C4 plants to rising atmospheric CO_2 partial pressure: A reassessment. *Plant Cell Environ*, 23, 931-942.

Hatfield, J.L. and Prueger, J.H. (2015). Temperature extremes: Effect on plant growth and development. *Weather and Climate Extremes*, 10, 4–10.

Hebbar, K B., Sheena, T L., Shwetha Kumari, K., Padmanabhan, S., Balasimha, D., Mukesh Kumar. and George V. Thomas. 2013. Response of coconut seedlings to elevated CO2 and high temperature in drought and high nutrient conditions. *Journal of Plantation Crops* 41: 118.

Hebbar, K.B. and Chaturvedi, V.K. (2015). Impact and adaptationstrategiesof coconut to climate change. In Proc. of Kerala Environment Congress-2015, being held at Centre for Environment and Development, Thiruvananthapuram during May 06-08, 2015, pp. 73-78.

IITM, (2011). Indian Institute of Tropical Meteorology. In: INCCA Report, Ministry of Environment and Forests. GOI.

IPCC, (2013). Climate Change 2007: Climate Change Physical basis. Summary for Policymakers (Inter-Governmental Panel on Climate Change).

IPCC, (2014). Climate Change 2014: Climate Change Impacts, Adaptation and Vulnerability. Summary for Policymakers (Inter-Governmental Panel on Climate Change).

Kasturi Bai, K.V., Rajagopal, V., Prabha. C. D., Ratnambal. M. J. and George. M.V. (1996 a). Evaluation of coconut cultivars and hybrids for dry matter production. *J. Plantn. Crops*, 24, 23-28.

Kimball, B.A., Pinter, P.J. Jr., Garcia, R.L., LaMorte, R.L., Wall, G.W., Hunsaker, D.J., Wechsung, G., Wechsung, F. and Kartschall, Th. (1995). Productivity and water use of wheat under free-air CO_2 enrichment. *Global Change Biology*, 1(6), 429-442.

Kimball, B.A., Kobayashi, K. and Bindi, M. (2002). Responses of agricultural crops to free-air CO_2 enrichment. *Advances in Agronomy*, 77, 293–368.

Muralikrishna, K.S., Naresh Kumar, S. and John Sunoj V.S. (2010). Elevated CO_2 and temperature affect leaf anatomical characteristics in coconut (Cocos nucifera L.). International workshop on climate change and Island Vulnerability" admat Island, U.T of Lakshadweep, 28-31 October 2010.

Muralikrishna, K.S. (2012). Physiological response of coconut (*Cocos nucifera* L.) seedlings to elevated [CO_2] and temperature. By Ph.D Thesis submitted to Under Department of Bioscience, Mangalore University, 2007-2012.

Muralikrishna, K.S, Naresh Kumar S. and John Sunoj. V. S. (2013). Elevated CO_2 and Temperature Affect Leaf Anatomical Characteristics in Coconut (*Cocos nucifera* L.). In: Climate Change and Island and Coastal Vulnerability (Eds- J. Sundaresan, S. Sreekesh, AL. Ramanathan, L. Sonnenschein and R. Boojh). ISBN 978-94-007-6015-8, Springer Pub. pp 141-153.

AP Cess Fund Project: Variability in content and composition of fatty acids in coconut oil due to genetic and environmental factors. Submitted to ICAR, New Delhi-2005

Naresh Kumar, S. (2005). Variability in content and composition of fatty acids in coconut oil due to genetic and environmental factors. AP Cess Fund Project Final Report Submitted to ICAR, New Delhi.

Naresh Kumar, S., Govindakrishnan, P.M., Swarooparani, D.N., Nitin, C.H., Surabhi, J. and Aggarwal P.K. (2015). Assessment of impact of climate change on potato and potential adaptation gains in the Indo-Gangetic Plains of India. *International Journal of Plant Production,* 9(1), 151-170.

Naresh Kumar, S.,Pramod Kumar, A., Kumar, U., Jain, S., Swaroopa Rani,D. N.,Nitin Chauhan. and Rani Saxena. (2014a). Vulnerability of Indian mustard (Brassica juncea (L.) Czernj. Cosson) to climate variability and future adaptation strategies. Mitigation and Adaptation Strategies to Global Change, 10.1007/ s11027-014-9606-Z.

Naresh Kumar S, Aggarwal, P. K., Swarooparani, D. N., Rani Saxena., Nitin Chauhan. and Surabhi Jain. (2014b). Vulnerability of wheat production to climate change in India. Climate Research, doi: 10.3354/cr01212.

Naresh Kumar, S., Aggarwal, P. K., Rani Saxena., Swaroopa Rani, D.N., Surabhi Jain. and Nitin Chauhan. (2013). An assessment of regional vulnerability of rice to climate change in India. Clim. Change, DOI 10.1007/s10584-013-0698-3.118 issue 3-4 June 2013. p. 683 – 699.

Naresh Kumar, S. and Aggarwal,P.K. (2013). Climate change and coconut plantations in India: Impacts and potential adaptation gains. Agril. Syst, http://dx.doi.org/10.1016/j.agsy.2013.01.001.

Naresh Kumar, S., Aggarwal, P. K., Swaroopa Rani, D.N., Surabhi Jain., Rani Saxena. and Nitin Chauhan. (2011). Impact of climate change on crop productivity in Western Ghats, coastal and northeastern regions of India. *Current Sci,* 101 (3),33-42.

Naresh Kumar, S., Kasturi Bai, K. V., Rajagopal V. and Aggarwal, P.K. (2008). Simulating coconut growth, development and yield using Info Crop-coconut model. Tree Physiology (Canada), 28:1049–1058.

Naresh Kumar, S. and Aggarwal, P. K. (2009). Impact of climate change on coconut plantations. In Global Climate Change and Indian Agriculture-case studies from ICAR Network Project (PK Aggarwal ed.), ICAR, New Delhi Pub. pp.24-27.

Naresh Kumar, S., Rajagopal, V., Siju Thomas,T. and Cherian, K.V.(2006). Effect of conserved soil moisture on the source-sink relationship in coconut (*Cocos nucifera* L.) under different agro-climatic conditions in India. The Ind. J. Agril. Sci, 76(5), 277-281.

Naresh Kumar, S. (2015). Carbon sequestration and Beyond by Plantations: Exploring scope for Added Income. In: Agroforestry: Present Status and Way Forward (Dhayani, SK., Newaj, R., Alam B., Dev, I eds). Biotech Books Pub., pp.193-205.

Naresh Kumar, S.,Sairam, C. V., Jose, C.T.,Kasturi Bai, K.V., Palnaiswamy, C., Rajagopal, V., Krishnamurthy, K.S., Kandiannan K. and Champakam. B. (2007a.)Climate change effects on growth and productivity of plantation crops with special reference to coconut and black pepper: Impact, adaptation and vulnerability and mitigation strategies- ICAR Network Project on Impact of Climate Change on Indian Agriculture entitled Final Report, pp.21.

Naresh Kumar, S. (2010). Assessment of Climate Change Impacts on Plantation Crops: Enabling Activity for Preparation of India's Second National Communication to UNFCCC. Final Report submitted to Ministry of Environment and Forests, Governament of India. pp.48.

Naresh Kumar, S., Singh, A.K., Aggarwal., P.K., Rao, V.U.M. and Venkateswarlu, B. (2012). Climate change and Indian Agriculture: Salient achievements from ICAR network project. IARI Pub., 32p.

Naresh Kumar, S. (2008). Modeling impacts and adaptations in coconut to climate change in different agro-climatic zones and field response of coconut to increased temperatures and CO_2. Project Annual Report.

Naresh Kumar, S. (2011). Climate change and Indian agriculture: Current understanding on impacts, adaptation, vulnerability and mitigation. *J. Plant Biol*, 37 (2), 1-16.

Naresh Kumar, S. (2013). Network Project on Climate Change: Final Report, pp. 390.

Naresh Kumar, S. and Kasturi Bai, K.V. (2009). Photo-oxidative stress in coconut seedlings: Early events to leaf scorching and seedling death. Braz. *J. Plant Physiol*, 21(3), 223-232.

Naresh Kumar, S. (2004). Drought management in coconut gardens. (English)– CPCRI publication, pp.16.

Naresh Kumar, S., Rajagopal, V., Siju Thomas, T., Vinu K, Cherian., Ratheesh Narayanan, M. K., Ananda, K. S., Nagawekar, D. D., Hanumanthappa, M., Vincent, S. and Srinivasulu, B. (2007b). Variations in nut yield of coconut

(Cocos nucifera L.) and dry spell in different agroclimatic zones of India. *Ind. J. Hort*, 64 (3), 309-313.

Pascual, L.M. and Tan,R.R. (2004). Comparative life cycle assessment of coconut biodiesel and conventional diesel for Philippine automotive transportation and industrial boiler application. LCA/LCM 2004, 11-24 July, 2004, LCAcentre.org.

Peiris, T. S. G. (2006). Impact of climate change on coconut industry in Sri Lanka. Paper presented at the Third International Conference on Climate Impact and Assessment (TICCIA), 24-27 July, Cairns, Australia.

Porter, J.R. and Gawith, M. (1999). Temperatures and the growth and development of wheat: a review. *Eur. J Agron*, 10, 23–36.

Rajagopal, V., Shivashankar, S., Jacob Mathew. (1996). Impact of dry spells on ontogeny of coconut fruits and it's relation to yield. *Plantn. Res. Dev*, 3, 251-255.

Subramanian, P., Dhanpal, R., Mathew, A.C., Palaniswami, C., Upadhyay, A.K., Naresh Kumar, S. and Reddy, D.V.S. (2012). Effect of fertilizer application through micro-irrigation technique on nutrient availability and coconut production. *J. Plantation Crops*, 40, 168-173.

Sunoj, J.V.S. 2012. Biochemical response of coconut (*Cocos nucifera* L.) seedlings to elevated [CO_2]and temperature. By John Ph.D Thesis submitted to Under Department of Bioscience, Mangalore University, 2007-2012.

Sunoj, J.V.S., Naresh Kumar,S. and Muralikrishna, K.S. 2010. Variation in total phenolic concentration in coconut (*Cocos nucifera* L.) seedlings under elevated CO_2 and temperature in different seasons International conference on climate change and environment' at CUSAT, Kochi during 24-26 October 2010.

Sunoj, J.V.S., Naresh Kumar,S. and Muralikrishna, K.S. (2013). Variation in total phenols concentration in coconut (*Cocos nucifera* L.) seedlings under elevated CO_2 and temperature in different seasons In: Climate Change and Environment (eds: Sundaresan, J., Sreekesh, S., Ramanathan, A.L., Sonnenschen, Lenand Boojh, Ram,) Scientific Pub, 286 p.

Sunoj, J.V.S., Naresh Kumar, S., Muralikrishna,K.S. and Padmanabhan, S.(2015). Enzyme Activities and Nutrient Status in Coconut (*Cocos nucifera* L.) Seedling Rhizosphere Soil after Exposure to Elevated CO_2 and Temperature. *Journal of the Indian Society of Soil Science*, Vol. 63, No. 2, pp 191-199.

Sunoj, J.V.S., Naresh Kumar, S. and Muralikrishna, K.S. (2014). Effect of elevated CO_2 and temperature on oxidative stress and antioxidant enzymes activity in coconut (*Cocos nucifera* L.) seedlings. *Indian J Plant Physiol*, DOI 10.1007/ s40502-014-0123-6.

Tan, R.R., Culaba, A.B. and Purvis, R.I. (2004). Carbon balance implications of coconut biodiesel utilization in the Philippine automotive transport sector. *Biomass and Energy*, 26, 579-585.

Roupsard, O., Bonnefond, J.M., Irvin, M., Berbigier, P., Nouvellon, Y., Dauzat, J., Taga, S., Hamel, O., Jourdan, C., Saint-Andre, I. Miallet-Serra, J.P., Labouisse, D., Epron, R., Joffre, S., Braconnier, R., Navarro, M. A. and Bouillet,J.P. (2006).

Partitioning energy and evapo-transpiration above and below a tropical palm canopy. *Agril. Forest Meterol,* 139(3-4), 252-268.

Roupsard, O., Lamanda, N., Jourdan, C., Navarro, M., Miallet-Serra J. I., Dauzat, and Tiata Sileye. (2008). Coconut carbon sequestration. Part I: Highlights on carbon cycle in coconut plantations. *Coconut Res. Dev,* 24, 1-14.

Vega, E. V. (2011). Cleaner Production Opportunities for Improvement of Carbon Saving in the Production of Coconut Biodiesel, *International Journal of Chemical and Environmental Engineering,* 2 (5), 356-361.

2017, Impact of Climate Change on Plantation Crops Pages 45–60
Editors: **K.B. Hebbar, S. Naresh Kumar & P. Chowdappa**
Published by: **ASTRAL INTERNATIONAL PVT. LTD., NEW DELHI**

Chapter 3

Physiological and Biochemical Response of Coconut to Climate Change Variables

K.B. Hebbar, Mukesh Kumar Berwal, M. Arivalagan and V.K. Chaturvedi

1. Introduction

The threat of climate change is projected to be more in coastal tract and hilly areas of India where plantation crop like coconut is the predominant crop which provides sustenance to more than 10 million people and contributes Rs. 83000 million annually to the Gross Domestic Product (GDP) of the country. Coconut is grown between 20° N and 20° S latitude. It can be grown even at 26° N latitude but the temperature is the main limitation. The optimum weather conditions for good growth and nut yield in coconut are well distributed annual rainfall between 130 and 230 cm, mean annual temperature of 27 °C, abundant sunlight ranging from 250 to 350 Wm^{-2} with at least 120 hours per month of sun shine period. Since, it is humid tropical crop it grows well above 60 per cent humidity (Child 1974, Murray 1977). The climate change will affect coconut plantation through higher temperatures, elevated CO_2 concentration, precipitation changes, increased weeds, pests, and disease pressure. In this chapter the effect of climate change variables and their interaction effect on coconut growth and development and the strategies need to be adopted to combat these variables are discussed.

2. Responses to Moisture Deficit

Coconut has been considered as extravagant in water consumption. Daily water requirement is estimated between 30 and 120 L (Jayasekara and Jayasekara

1993), 55 and 115 L (Yusuf and Varadan 1993) by an adult coconut depending on soil moisture content and evaporative demand of the atmosphere. Mean *E* varied from 0.09 to 1.52 L day^{-1} m^{-2} leaf area in 3.5-year-old dwarf coconut palms, as estimated by measurements of xylem sap flux density (Araujo 2003). Taking the mean value of Araujo (2003) (0.8 L day^{-1} m^{-2}) and a leaf area of 140 m^2 in dwarf varieties (Ramadasan and Kasturi Bai 1999), the calculated transpirational water loss (114 L day^{-1}) agrees with the values reported by Jayasekara and Jayasekara (1993) and is only slightly higher than that reported by Yusuf and Varandan (1993) for tall varieties. Compared to the tall varieties, some evidence suggests that dwarf varieties use water more extravagantly (IRHO-CIRAD, 1992).

Drought stress affects coconut production in almost all coconut growing countries, since it is mainly a rainfed plantation (Coomans 1975; Mathes 1988; Bhaskara Rao *et al.*, 1991). Hence, the productivity is low in these areas by ~50 per cent of irrigated gardens. Coconut faces summer dry spells each year apart from the frequent occurrence of drought. Coconut plantation during the last 3 to 4 years is facing the severe threat of climate change in Karnataka, Tamil Nadu, Kerala and Andhra Pradesh which are the major coconut growing states. Lakhs of coconut trees were withered during the summer months of 2013 and 2014 in south interior Karnataka due to scanty rainfall. Almost similar fatality happened in some districts of Tamil Nadu. During the summer of 2016 vast tracts of coconut withered in Northern Kerala due to extended drought. Though some trees recover with the arrival of monsoon but the production will be affected at least for 3 years. On the other hand in the east coast of Andhra Pradesh and Odisha large number of trees was uprooted due to the cyclones. In 2015 all along the west coast, plants had scorching of leaves due to salt spray effect. This is projected to increase further as the long term climatological data for 140 years in the humid tropics of India indicate cyclic pattern in rainfall with a declining trend in annual and southwest monsoon rainfall during the past 60 years (Krishna Kumar *et al.*, 2008). Being perennial in nature, coconut palm had a long duration from the initiation of inflorescence primordia to nut maturity (~44 months) with longer pre-fertilization period (~32 months) than post-fertilization (12 months) period. Hence, the impact of drought occurring at any of the critical stages of the development of inflorescence affects nut yield (Rajagopal *et al.*, 1996; Rajagopal *et al.*, 2000) not only in current year but also in next three years to follow, thus makes the problem more severe (Naresh Kumar, 2002).

The effects of water deficit on the physiology, growth and productivity of coconut have been widely documented (Repellin *et al.*, 1994; 1997; Rajagopal and Kasturi Bai 1999; Prado *et al.*, 2001; Azevedo *et al.*, 2006; Gomes *et al.*, 2007). Compared to the tall varieties, evidence suggests that dwarf varieties use water more extravagantly due to its elevated transpiration rate (IRHO-CIRAD 1992), greater number of stomata per unit leaf surface (stomatal frequency) and lower wax content on the leaf surface (Rajagopal *et al.*, 1990), as well as a poorer stomatal control of water loss (Passos and Silva 1990). In contrast, tall varieties show a more conservative water use (Voleti *et al.*, 1993). Kasturi Bai *et al.* (1997) observed that West African Tall (WAT) behaves relatively better than the hybrids under drought conditions, due to lower g_s (0.10 mol m^{-2} s^{-1}) and, as a consequence, improved tissue water conservation.

WUE has been shown to vary among varieties of talls and dwarfs and also among ecotypes of the same variety (Prado *et al.*, 2001; Gomes *et al.*, 2002; Hebbar *et al.*, unpublished data). In a WUE study at CPCRI Kasaragod, Hebbar and chaturvedi (2015) observed that tall genotypes Kalpadhenu and FMST had high WUE under 100 per cent FC (Figure 3.1) due to their higher root biomass. On the other hand under water deficit stress dwarf maintained higher WUE due to higher stomatal conductance. Further, it was observed that talls had higher stomatal sensitivity compared to dwarfs (Figure 3.2). It indicated that both root growth and stomatal sensitivity are the important traits governing drought tolerance in coconut. Passos *et al.* (1999), comparing three dwarf genotypes, observed that Malayan Yellow Dwarf (MYD) showed better WUE than the other two genotypes (Malayan Red Dwarf, MRD and Brazilian Green Dwarf, BGD). The authors attributed the superiority to (1) higher stomatal sensitivity to changes in leaf water potential, (2) higher g_s during the rainy season, which resulted in higher P_N and better leaf cooling and nutrient uptake capacity due to transpiration, (3) a more developed root system, which leads to higher water-uptake efficiency.

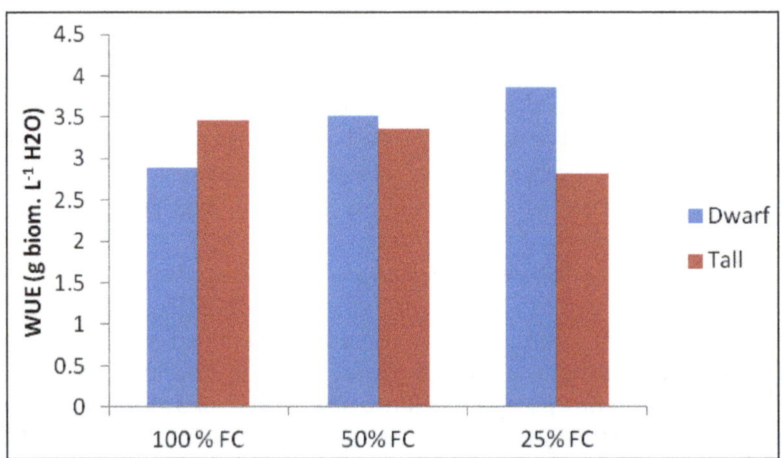

Figure 3.1: WUE of Tall and Dwarf Genotypes at different Moisture Regimes (CD at 5 per cent : 0.8).

3. Responses to Temperature Variations

Temperature influences the growth and development of all crops, shaping potential yield throughout the growing season. Current temperatures in the coastal tracts of India are optimum for production, while in east coast temperatures already exceed the optimum. The ideal mean annual temperature for coconut growing is usually considered to be in the region of 29°C (27 - 32°C), with abundant sunshine and a well-distributed annual rainfall. Temperature events higher than normal are expected to reduce coconut yield (Rajagopal and Kasturi Bai 1990). Temperature

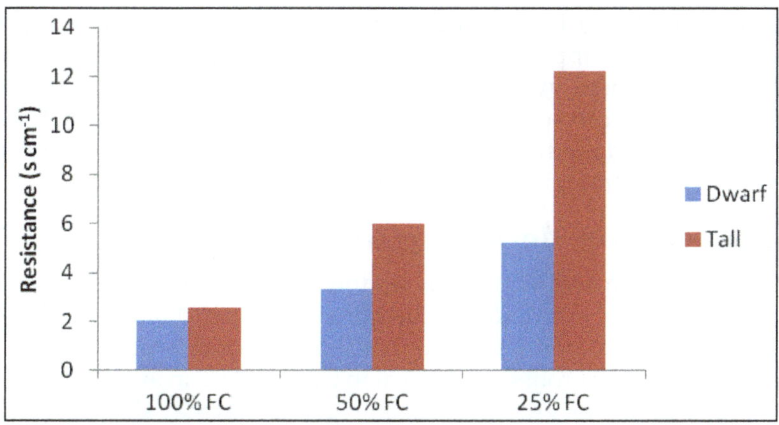

Figure 3.2: Stomatal Resistance of Tall and Dwarf Genotypes at different Moisture Regimes (CD at 5 per cent : 2.1).

can affect photosynthesis through modulating the rates of activity of photosynthetic enzymes and the electron transport chain (Sage and Kubien 2007) and, in a more indirect manner, through leaf temperatures defining the magnitude of the leaf-to-air vapor pressure difference, a key factor influencing stomatal conductance.

Crops are sensitive to HT, particularly during flowering time. High temperature can have both negative and positive impacts on growth and production in coconut. The negative impacts such as added heat stress, especially in areas at low to mid-latitudes already at risk today such as south interior Karnataka and Tamil Nadu, but they also may lead to positive impacts in currently cold-limited high-latitude regions of Assam and West Bengal. High temperature increases both photorespiration and the dark respiration and thus the total biomass production go down. Regression analysis indicated increase in T_{min} increased the leaf emergence rate: increase in T max increased inflorescence emergence rate; pistillate flower production has curvilinear relationship with rainfall/month (150mm/month-opt), nut retention has curvilinear relationship with T_{max} (32°C-opt) and T_{min} (20°C-opt). Frequent but short periods of temperature below 15°C result in fruit abnormalities such as bicarpelate nuts and lack of pollination under North Indian conditions (Naresh Kumar *et al.*, 2008).

4. Reponses to CO_2 Concentrations

In almost all plants with C3 photosynthetic pathway, such as coconut the rate of photosynthesis is limited by the atmospheric CO_2 level. For these crops, higher CO_2 will allow greater photosynthetic production. Results of thousands of studies of the effects of the increased CO_2 level on photosynthesis and crop growth are summarized in the CO_2 Science website. For most of the crops the benefit varied from 32 to 49 per cent (Idso and Idso 2000). A further benefit may result from the fact that stomatal aperture, and hence transpiration is reduced under high CO_2 (Ainsworth and Rogers, 2007). This should lead to improved water use efficiency *i.e.*, the amount of biomass produced per unit water transpired will increase and could be important for future climate when water supply is projected to be scarce.

Thus, climate change has far reaching implications for plant growth, production and food security, and approaches are required for adapting to new climates. Two primary approaches broadly exist for adapting plantation crops to these conditions: 1) improving existing crop cultivars and developing new varieties, 2) devising new cropping systems and methods for managing crops in the field. These approaches include the specific strategies discussed below.

5. Anatomical and Morphological Traits

The coconut stem acts as a water conductor and capacitor which enables it to withstand water stress was demonstrated in a study by Villalobos *et al.* (1992). The fibrous root system (homorhizic) of an adult coconut can protrude as far as 3.0 m from the trunk, but most roots reach 1.5 m in length (Avilan and Rivas 1984; Cintra *et al.*, 1992; 1993). The root growth of coconut genotypes may shift to deeper sites in response to dehydration of superficial soil layers (Cintra *et al.*, 1993). In a cropping system where vegetables, fruit trees, medicinal plants were grown coconut produced better root laterals with root hairs which are very important for the absorption of water and nutrients (Subramanian *et al.*, 2010).

Several water stress adaptive features were found in coconut leaflets. Waxy cuticle on the upper epidermis, thicker cuticle at the edge, water tissue with thin-walled cells at the upper and lower angles of the straightened leaflet margin, xylem tracheids with thick lignifications, fibrous sheet encircling seven to eight large vascular bundles in a strong midrib, and tracheids with scalariform thickening in diminutive vascular bundles (Naresh Kumar *et al.*, 2000). Drought tolerant cultivars had more scalariform thickening on tracheids and large sub-stomatal cavities.

6. Leaf Photosynthesis

Under non-limiting conditions coconut develops a large and highly productive canopy, being capable of an estimated 51 ton ha^{-1} year^{-1} of total dry matter production (Foale 1993). Short-term responses of coconut to water stress such as low g_s and water potential which often impair P_N and E have been extensively documented (Repellin *et al.*, 1994; 1997; Rajagopal and Kasturi Bai 2002). Carbon assimilation rate is impaired in both tall (Reppellin *et al.*, 1997; Prado *et al.*, 2001) and dwarf genotypes (Gomes *et al.*, 2007) in response to atmospheric and soil water deficit. Reductions of P_N from 7 to 47 per cent and from 12 to 67 per cent have been reported for dwarf and tall genotypes, respectively. Drought-induced photosynthetic reductions are initially attributable to limited CO_2 diffusion from the atmosphere to the intercellular spaces as a result of stomatal closure (Reppellin *et al.*, 1994; 1997). Non-stomatal factors have been demonstrated to contribute to the reduction in P_N both during a period of severe water deficit and during the recovery phase after resuming irrigation (Gomes and Prado 2007; Gomes *et al.*, 2007). In addition, fluorescence measurements recorded by Kasturi Bai *et al.* (2006) indicated reduction in FV/Fm (photochemical efficiency) with decreasing water potential suggesting damage to photosynthetic apparatus under stress.

7. Biochemical Responses and Osmotic Adjustment

Information concerning the protoplasmic tolerance to drought stress has led to the conclusion that coconut leaves have highly efficient systems that protect cell membranes and their intracellular components. Lipid composition, lipid peroxidation level, and the activities of enzymes related to oxidative stress are good indicators of dehydration tolerance in leaves of coconut. Water deficit induced a reduction in total leaf lipid content, mainly that of the chloroplast membranes, an effect particularly expressive in the less drought-tolerant genotypes (Repellin *et al.*, 1994). In addition, an increase in the degree of lipid unsaturation in response to severe drought was also observed, which seems to be related to the maintenance of membrane fluidity, mainly in the chloroplasts (Repellin *et al.*, 1997). Coconut cultivars considered drought tolerant showed a lower level of lipid peroxidation and higher activity of catalase, superoxide dismutase, and peroxidase than cultivars empirically classified as drought susceptible. Indeed, peroxidation level was negatively correlated ($R^2 > 0.73$) with activity of antioxidant enzymes (Shivashankar *et al.*, 1991; Chempakam *et al.*, 1993). Epicuticular wax content was less in both control and drought induced plants, but increased with drought in dwarf while there was no change in talls. On the other hand super oxide dismutase (SOD) specific activity increased till 50 per cent FC and decreased with further increase in stress *i.e.* at 25 per cent FC as compared to control plants (Figure 3.3) (Hebbar *et al.*, unpublished data). It was high in FMST a drought tolerant variety.

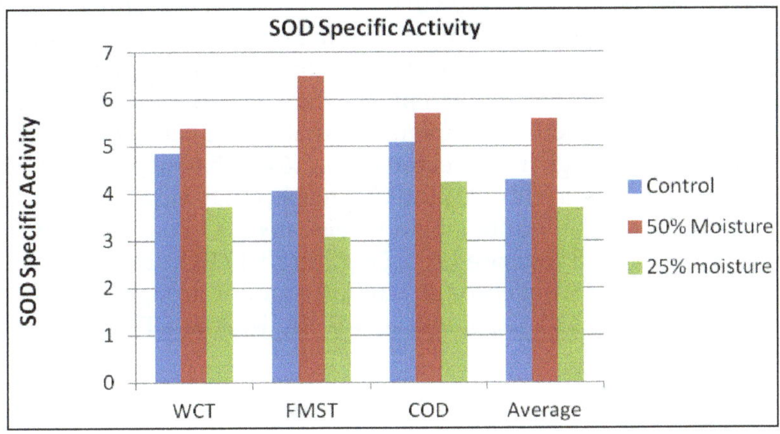

Figure 3.3: SOD Specific Activity under Induced Drought Stress.

8. Biomass Production and Water Use Efficiency

In almost all crops with C3 photosynthetic pathway, such as coconut the rate of photosynthesis is limited by the atmospheric CO_2 level. For these crops, higher CO_2 will allow greater photosynthetic production. In an Open top chamber (OTC) study at 550 and 700 ppm CO_2 biomass increased by 8 and 25 per cent respectively against ambient CO_2 concentration of 380 ppm. (Hebbar *et al.*, unpublished data). The higher growth rate of plants under elevated CO_2 was closely associated with the photosynthesis (PN). Photosynthesis was highest 14.4 at 700 ppm CO_2 as against 10.14 under ambient condition. Plants grown under [ECO_2] had higher PN to Tr ratio and thus had high WUE. However, chlorophyll fluorescence data measured in the same leaf where the PN was measured indicated that Fv/Fm (dark adapted values) which reflects the potential quantum efficiency of PSII, was on par in ambient and elevated CO_2 plants. For a given amount of water, plants under elevated CO_2 produced higher biomass and thus had higher whole plant WUE (water use efficiency). WUE was 2.53 g/litre under ambient condition and it had increased to 3.14 at 700 ppm CO_2.

9. Interaction Effect of CO_2, High Temperature and Moisture Deficits

The interaction effect of climate change variables CO_2 and elevated temperature (ET) with drought on growth and development of coconut seedlings was studied in an Open Top Chamber (OTC) at CPCRI, Kasaragod (Hebbar *et al.*, 2013b). High temperature (3°C above ambient) decreased the biomass by 10 per cent. High temperature in addition to drought had a compounded effect and reduced the biomass by 16 per cent. To certain extent the elevated CO_2 could offset the negative effect of temperature in coconut. The stimulatory effect of CO_2 under drought and high temperature was less and it could increase the biomass by only 8 per cent with 700 ppm CO_2.

Crown initiation and crown growth was slow at ET. It was only 1.3 cm/day with ET and 1.7 cm/day with elevated CO_2+ET as against 1.8cm/day of plants grown in ambient condition. Similarly, crown growth rate significantly reduced in drought plants. Leaf splitting was faster when plants were grown under elevated CO_2 and was slow with drought and ET treatments. Stomatal conductance and transpiration on the other hand were high in ET treatments 0.216 and 5.63 as against 0.125 moles/m^2/s and 2.58 moles/m^2/s with 700 ppm CO_2 respectively indicating better intrinsic tolerance of plants to water limitation under elevated CO_2 concentration. The WUE was low at ET (2.28) and increased to 2.56 g/litre ET+CO_2 indicating higher CO_2 could offset the effect of ET in coconut. Similarly, under drought too the WUE was the highest at 700 ppm CO_2 (2.70) and it was the least at ET (2.144) Hebbar and Chaturvedi 2015).

10. Strategies Adopted for Improving Existing Cultivars and Developing New Varieties

10.1. Integration of Beneficial Traits into Existing Crops through Use of Germplasm Accessions

At ICAR-CPCRI Kasaragod germplasm lines are collected from different agro-climatic zones and maintained and evaluated at International gene bank at Kidu Karnataka and used to develop cultivars adapted to climate change. There are all together 434 germplasm accessions and they have been evaluated for different biotic and abiotic stress characters. The tolerant traits are incorporated to get climate resilient varieties. Some of the drought tolerant varieties developed at CPCRI are Chandra Kalpa, Kalpatharu, Kera Keralam, Kalpa Mitra, Kalpa Dhenu, Kera Sankara, and Chandra Laksha.

10.2. Identified Germplasm/Variety and Traits that Tolerate Drought and Heat

Coconut yield drops when it experiences drought, excessive heat, deviating from the optimum for growth during key stages, including pollination, flowering, and nut development periods. Drought and heat are the important abiotic stresses affecting coconut yield.

11. Field-level Evaluations of Crop Germplasm/Varieties

To differentiate drought resistance genotypes, several selection indices have been suggested on the basis of a mathematical relationship between favorable and stress conditions (Clarke *et al.*, 1984; Huang 2000). Tolerance (TOL) (Clarke and McCaig 1982; Clarke *et al.*, 1992), mean productivity (MP) (Clarke and McCaig 1982), stress susceptibility index (SSI) (Fischer and Maurer 1978), geometric mean productivity (GMP) and stress tolerance index (STI) (Fernandez, 1992) have all been employed under various conditions. These indices have been tested at Arsikere, Karnataka which was reeling under severe drought for the years 2011, 2012 and 2013. Yield data from rainfed and irrigated coconut orchards of experimental farm at Arasikere for the genotypes as shown in the Table 3.1 (Hebbar *et al.*, unpublished data) was collected for the years 2011, 2012 and 2013 and the different indices were calculated.

Table 3.1: The Relative Yield Performance of Coconut Genotypes in Rainfed and Irrigated Conditions of Arasikere, Karnataka and the Calculation of Tolerant Indices.

Cultivar	*Coconut/Plant*		*Per cent Reduction*	*DSI*	*Geometric Mean*
	Irrigated	*Rainfed*			
WCT	52	4	92	1.14	14
LCT	63	19	70	0.86	35
ADOT	54	5	91	1.12	16
Sanramon	74	2	97	1.20	12
WCTXGBGD	180	25	86	1.06	67
BS1	44	21	52	0.65	30
PHOT	73	0	100	1.23	0
WCTXCOD	85	5	94	1.16	21
CODXWCT	72	3	96	1.18	15
TPT	62	25	60	0.74	39
Zanzibar	130	36	72	0.89	68
Java	47	20	57	0.71	31
Mean	**78**	**14**	**81**	**1.00**	**29**

The drought intensity index (DII) as calculated using the formula given by Ramírez-Vallejo and Kelly (1998) was 0.8. Values exceeding 0.7 would indicate severe drought.Schneider *et al.* (1997) showed the Geometric mean (GM) which is square root of the product from rainfed and irrigated for an individual genotype. From the above table GM was high for Zangiber and WCT x GBGD which also had high yield under stress and non-stress conditions.Ramírez-Vallejo and Kelly (1998) also concluded that the most effective approach to breed for resistance to drought would be based first on selection for high geometric mean seed yields followed by selection for low Fischer Maurer drought susceptibility index values. The Fischer and Maurer drought susceptibility index (DSI) is calculated as follows: DSI= (1-Ys/ Yi)/DII(Fischer and Maurer, 1978).

Though, drought susceptibility index (DSI) was low for BSI and Java, but they were low nut production. Caution in using this index is advised as certain genotypes with the lowest DSI rankings had the lowest overall yield potential (White and Singh 1991). Small yield differences between the stress and non-stress treatments produce low DSI values even though the potential yield of the line is low. Therefore in coconut GM is the best indicator of drought tolerance under field condition and can be used in breeding programs across different environments. WCT x GBGD selected for drought tolerance had higher photosynthesis better retention of nuts under stress.

12. Devised Cropping/Farming Systems to Alleviate the Effect of Climate Change

Coconut is mostly grown in coastal and hilly areas where the rainfall is very high and the soil is poor in nutrients. The soil is sandy or laterite which has very low water holding capacity. Studies conducted at CPCRI and elsewhere indicated that coconut based farming system approach is the best adaptation strategy to overcome the effect of climate change. Appropriate, site specific cropping system management practices have been developed which help alleviate the effects of abiotic and biotic stresses on crop productivity and yield. Coconut is a tree which has no branches and grows straight vertically upwards providing more space under its canopy. Its leaves are such that it allows sun light to the crops grown under it. Between two coconut trees, fruit trees such as lime, lemon, guava, pomegranate, custard apple, cocoa, nutmeg, clove crops are planted at 15 -20 ft distance. These are medium sized crops both in height as well as canopy and can easily fit in between two adjacent coconut trees. They can be planted simultaneously or after the coconuts are established. It takes 8 to 10 years for coconut trees to start yielding properly. Whereas a number of the above mentioned crops start yielding well within 3 -5 years and last only 15 -20 years. By that time the coconut will be in its peak yield stage and will be about 20 ft high. The intercrops may be replaced by any other crop including vegetables and grasses and another cycle of medium sized intercrops can be established. Coconut farming systems have dramatic powers to stabilize eroding farmland, especially sloping lands. Practices like using nitrogen fixing perennials, ploughing, and intensive livestock rotation have fantastic soil building abilities. Plantings of useful trees can protect coastlines from damage caused by increased storm activity.

12.1. Cultural Practices, Soil Conservation and Water Management Techniques are Evolved to Manage the Drought

12.1.1. Optimize Land Use

Intensifying yields sustainably on existing arable land uses land more efficiently with better soil management. Soil management techniques like mulching of basin with coir dust at 50kg/palm, burial of husks in 3 or 4 layers, application of green manures or organic manures (FYM) at 50 to 100 kg/palm, spreading dried coconut leaves and other organic residues (mulching effect), addition of tank silt at 100 to 200 kg/palm and organic agriculture to increase soil's water retention capacity are some of the ways to improve the productivity from unit land and reduce the climate change effect. Similarly, soil conservation measures *viz.* terracing the palm basins in sloppy lands to interrupt run off of water and to enhance soil moisture, rain water harvesting: in-*situ* (land configuration, mulching etc.) and *ex-situ* (Ponds, micro water harvesting structure -jalkund *etc.*), bunding the field to prevent runoff of water. These measures would help in rainfed orchards.

12.1.2. Optimize Water-use Efficiency

With climate change, water supplies are expected to become threatened in certain regions of coconut cultivation, but water management strategies, such as drip irrigation, can conserve water and protect from water shortages. To achieve "more crop per drop", water management techniques like pitcher irrigation (bury two or three earthen pots/hollow bamboos and fill them with water to moisten subsoil), drip irrigation (two or three drippers per palm to wet subsoil layer) or if adequate water is available irrigate with 200 liters water/palm once in four days and mulching the basin with dry leaves facilitate the retention of soil moisture and achieve the "more crop per drop".

12.1.3. Use Crop Models in Decision-Making

Crop models can be used to compare crop management strategies, assist producers weigh both economic and environmental considerations as they make decisions about crop varieties, cropping dates, and management practices (Jones *et al.*, 2003 ; Hebbar et al., 2013c). Infocrop model of coconut (Naresh Kumar *et al.*, 2008) indicated that negative impacts of climate change can be overcome by adaptation strategies such as assured irrigation through drip system, soil moisture conservation, and by providing fertilizers/nutrients through organic and inorganic source in doses higher than those currently applied by the farmers (Naresh Kumar and Aggarwal, 2013). This practice in Kerala could improve the positive gains due to climate change by 7 to 21 per cent in different scenarios. Similarly, in Karnataka, West Bengal, Gujarat, Maharashtra and Odisha these practices not only off-set the negative impacts but also could result in higher yields. In North-Eastern States, providing summer irrigation and even low dose of fertilizers could further improve (in the range of 10-33 per cent) the positive impacts of climate change. Coconut plantations in islands, if managed scientifically by proper spacing, canopy management, summer irrigation and even with low dose of fertilizers the productivity could be enhanced to an extent of 2-25 per cent (Naresh Kumar and Aggarwal, 2013).

13. Coconut is an Excellent Tree Crop for Climate Change Mitigation

13.1. Carbon Sequestration and Carbon Stocks in Coconut

Plantation crops has significant potential for offsetting and reducing the projected increases in green house gas (GHG) emissions and regarded as an important option for greenhouse gases mitigation. Above ground biomass in coconut varied from 15 CERs to 35 CERs depending on cultivar, agroclimatic zone, soil type and management. Annually sequestered carbon stocked in the stem is in the range of 0.3 to 2.3 CERs. Standing C stocks in 16 year old coconut cultivars in different agro-climatic zones varied from 15 CERs to 60 CERs (Naresh Kumar, 2009). C sequestration by coconut plantation is higher in red sandy loam soils and lowest in littoral sandy soils.

13.2. Coconut can Check Erosion and Wind Speed

Probably coconut is the only crop next to mangroves grows well in coastal areas. It is the best suited crop for climate change situations as it can withstand temporary water logging conditions like floods and tides with special adaptability against strong winds, storms and cyclones. It has a fibrous root system spread over few meters which not only takes up water and nutrients and anchors the plant but also helps in checking the erosion in high rainfall areas. Coconut orchards also act as strong wind breaks and reduce storms and cyclones.

14. Strategies for the Future

The existing scientific knowledge can address to adapt cropping systems to climate change in the short-term. However, uncertainties and limited predictability in the long-term require an infrastructure that drives innovation and implements crop adaptation strategies in a sustainable manner. In particular, research investments and efforts are needed to further:

- Understand the physiological, genetic, and molecular basis of adaptation to drought, heat and biotic stresses likely resulting from climate change;
- Develop region specific farming models that integrate genetic and management technology.
- Give more thrust on value addition to avoid volatile price and stability in income.
- Translate new knowledge into new agricultural systems that integrate genetic and management technologies (*i.e.*, both breeding and agronomy will contribute to adaptation); and
- Transfer knowledge effectively and make technologies and innovations widely available to increase food production and stability.
- Ensure effective collaboration and communication between both public and private sector research and development to create knowledge, and develop and transfer new technologies. Although the contributions of government, universities, and industry may vary with crop, region, and

time, the roles of each can be tailored to develop crop varieties, cropping systems, and agricultural management strategies appropriately.

References

Ainsworth, E.A. and Rogers, A. (2007). The response of photosynthesis and stomatal conductance to rising CO2: mechanisms and environmental interactions. *Plant Cell and Environment*, 30, 258-270.

Araujo, M.C. (2003). Demanda hidricae distribuicao de raizes do coqueiro anao verde (*Cocos nucifera* L.) na regiao norte fluminense. Campos dos Goytacazes, Universidade Estadual do Norte Fluminense. M.Sc. Thesis.

Avilan, L.A. and Rivas, N. (1984). Study of the root system of coconut (*Cocos nucifera* L.). Oleagineux, 39, 13-23.

Azevedo, P.V., Sousa, I.F., Silva, B.B. and Silva, V.P.R. (2006). Water-use efficiency of dwarf-green coconut (*Cocos nucifera* L.) orchards in northeast *Brazil. Agricultural Water Managemant*, 84, 259-264

Bhaskara Rao, E.V.V.B., Pillai, P.V. and Mathew, J. (1991). Relative drought tolerance and productivity of released coconut hybrids. In: Silas EJ, Aravindhakshan M, Jose AI (ed.)Coconut Breeding and Management, KAU, Vellanikkara Thrissur, India.

Chempakam, B., Kasturi Bai, K.V. and Rajagopal, V. (1993). Lipid peroxidation and associated enzyme activities in relation to screening for drought tolerance in coconut (*Cocos nucifera* L.). *Plant Physiology and Biochemistry*, 20, 5-10.

Child, R. (1974). Coconut 2nd ed. Longman, London.

Cintra, F.L.D., Leal, L.S. and Passos, E.E.M. (1992). Evaluation of root system distribution in dwarf coconut cultivars. *Oleagineux*, 47, 225-234.

Cintra, F.L.D., Passos, E.E.M. and Leal, L.S. (1993). Evaluation of root system distribution in Tall coconut cultivars. *Oleagineux*, 48:453-461.

Clarke, J.M. and McCaig, T.N. (1982). Evaluation techniques for screening for drought resistance in wheat. *Crop Science*, 22, 503–506.

Clarke, J.M., DePauw, R.M. and Townlet-Smith, T.F. (1992). Evaluation of methods for quantification of drought tolerance in wheat. *Crop Science*, 32, 723–728.

Clarke, J.M., Townley-Smith, T.M., McCaig, T.N. and Green, D.G. (1984). Growth analysis of spring wheat cultivars of varying drought resistance. *Crop Science Journal*, 24, 537-541.

Coomans, P. (1975). Influence des facteurs climatiques sur les fluctuations saisonnieres et annuelles de la production du cocotier. *Oleagineux*, 30, 153-159

Fernandez, G.C.J. (1992). Effective selection criteria for assessing stress tolerance. In: Kuo C.G. (Ed.), Proceedings of the International Symposium on Adaptation of Vegetables and Other Food Crops in Temperature and Water Stress, Publication, Taiwan, pp. 257-270.

Fischer, R.A. and Maurer, R. (1978). Drought resistance in spring wheat cultivars: I. Grain yield responses. *Australian Journal of Agricultural Research*, 29, 897-912.

Foale, M.A. (1993). Physiological basis for yield in coconut. In: Nair MK, Khan HH, Gopalasundaran P, Bhaskara Rao EVV (eds), Advances in Coconut Research and Development, Oxford and IBH Publishing Co. PVT., New Delhi.

Gomes, F.P. and Prado, C.H.B.A.(2007). Ecophysiology of coconut palm under water stress. *Brazilian Journal of Plant Physiology*, 19(4), 377-391.

Gomes, F.P., Mielke, M.S., Almeida, A.A.F. and Muniz, W.S. (2002). Leaf gas exchange in two dwarf coconut genotypes in the southeast of Bahia State, Brazil. *Coconut Research and Development*, 18, 37-55.

Gomes, F.P., Oliva, M.A., Mielke, M.S., de Almeida, A.F., Leite, H.G. and Aqino, L.A. 2008. Photosynthetic limitations in the leaves of young Brazilian green dwarf coconut (*Cocos nucifera* L. 'nana') palm under well-watered conditions or recovering from drought stress. *Environmental and Experimental Botany*, 62, 195-204.

Hebbar, K.B. and Chaturvedi, V.K. (2015). Impact and adaptationstrategiesof coconut to climate change. In Proc. of Kerala Environment Congress-2015, being held at Centre for Environment and Development, Thiruvananthapuram during May 06-08, 2015, pp. 73-78.

Hebbar, K.B., Balasimha, D. and G.V.Thomas. (2013a). Plantation crops response to climate change: Coconut Perspective. In Climate-resilient Horticulture: adaptation and mitigation strategies by H.P. Singh, Srinivas Rao and Shivashankara K.S (eds.). Springer Publications. DOI 10.1007/978-81-322-0974-4_16, © Springer India 2013

Hebbar,K.B., Sheena,T.L., Shwetha, K., Padmanabhan, S.,Balasimha,D., Mukesh Kumar and Thomas, G.V. (2013b). Response of coconut seedlings to elevated CO2 and high temperature in drought and high nutrient conditions. *Journal of Plantation Crops*, 2013, 41(2): 118-122

Hebbar, K.B., Venugopalan, M.V., Prakash A.H. and P.K.Aggarwal. (2013c). Simulating the impacts of climate change on cotton production in India. *Climatic Change*, 118, 701-713 (DOI: 10.1007/s10584-012-0673-4)

Huang, B. (2000). Role of rot morphological and physiological characteristics in drought resistnce in plants. In: Wilkinson R.E. (ed.). *Plant-Environment Interaction*. Marcell-Dekker, New York. pp. 39-63.

Idso, C.D., Idso, K.E. (2000). Forecasting world food supplies: The impact of the rising atmospheric CO_2 concentration. *Technology*, 75, 33-55.

IRHO-CIRAD. (1992). Coconut-Study of yield factors. *Oleagineux*, 47, 324-337.

Jayasekara, K.S. and Jayasekara, C. (1993). Efficiency or water use in coconut under different soil/plant management systems. In: Nair MK, Khan HH, Gopalasundaran P, Bhaskara Rao EVV (eds) Advances in Coconut Research and Development, Oxford and IBH Publishing Co Pvt. Ltd, New Delhi.

Jones, J.W., Hoogenboom, G.C., Porter, H., Boote, K.J., Batchelor, W.D., Hunt, L.A., Wilkens, P.W., Singh, U., Gijsman, A.J. and Ritchie, J.T. (2003). The DSSAT cropping system model. *European Journal of Agronomy*, 18, 235-265.

Kasturi Bai, K.V. (2010). Impact of climate change and adaptation strategies in coconut. International conference on coconut biodiversity for prosperity, 25-28 October, 2010, CPCRI Kasaragod India.

Kasturi Bai, K.V., Rajagopal, V., Balasimha, D. and Gopalasudaram, P. (1997) Water relations, gas exchange and dry matter production of coconut (*Cocos nucifera* L.) under irrigated and non-irrigated conditions. *Coconut Research and Development*, 13, 45-58.

Katuri Bai, K.V., Rajagopal, V. and Naresh Kumar, S. (2006). Chlorophyll fluorescence transients with response to leaf water status in coconut. *Indian Journal of Plant Physiology*, 11, 410-414.

Krishnakumar, K.N., Rao, G.S.L.H.V.P. and Gopakumar, C.S. (2008). Climate change at selected locations in the humid tropics. *Journal of Agrometeorology*,10, 59-64.

Mathes, D.T. (1988). Influence of weather and climate on coconut yield. *Coconut Bulletin* 5, 8-10.

Murray, D.V. (1977). Coconut palm. In. Alvim TA, Kozlowski TT (eds.) Ecophysiology of Tropical Crops. Academic Press, New York.

Naresh Kumar, S. (2009). Carbon sequestration in coconut plantations. In Global Climate Change and Indian Agriculture-case studies from ICAR Network Project (PK Aggarwal ed.), ICAR, New Delhi Pub., pp.129-134.

Naresh Kumar, S. (2015). Carbon sequestration and Beyond by Plantations: Exploring scope for Added Income. In: Agroforestry: Present Status and Way Forward (Dhayani, SK., Newaj, R., Alam B., Dev, I eds). Biotech books Pub., pp193-205

Naresh Kumar, S. and Aggarwal, P.K. (2013). Climate change and coconut plantations in India: Impacts and potential adaptation gains. *Agricultural Systems*, http://dx.doi.org/10.1016/j.agsy.2013.01.001.

Naresh Kumar, S., Kasturi Bai, K.V., Rajagopal, V. and Aggarwal, P.K. (2008). Simulating coconut growth, development and yield with the Info Crop-coconut model. *Tree Physiology*, 28, 1049-58.

Naresh Kumar, S., Rajagopal, V. and Karun, A. (2000). Leaflet anatomical adaptations in coconut cultivars for drought tolerance, pp.225-229. Recent Advances in Plantation Crops Research, CPCRI contribution.

Naresh Kumar, S., Rajagopal, V., Siju Thomas, S., Vinu Cherian, K. M., Hanumanthappa, M., Anil Kumar, B., Srinivasulu, B.and Nagvekar, D. D. (2002). Identification and characterization of *in situ* drought tolerant coconut palms in farmers' fields in different agro-climatic zones. In: Sreedharan K, Vinod Kumar PK, Jayaram Basavaraj MC (eds.) *Proceedings of PLACROSYM XV*, Kerala.

Passos, E.E.M. and Silva, J.V. (1990). Fonctionnement des stomates de cocotier (*Cocos nucifera*) au champ. *Canadian Journal of Botany*, 68, 458-460.

Passos, E.E.M., Prado, C.H.B.A. and Leal, M.L.S. (1999). Condutancia estomatica, potencial hidrico foliar e emissao de folhas e inflorescencias em tres genotipos de coqueiro anao. *Agrotropica*, 11, 147-152.

Prado, C.H.B.A., Passos, E.E.M. and Moraes, J.A.P.V. (2001). Photosynthesis and water relations of six tall genotypes of *Cocos nucifera* in wet and dry seasons. *South African Journal of Botany*, 67, 169-176.

Rajagopal, V. and Kasturi Bai, K.V. (1999). Water relations and screening for drought tolerance. In: Rajagopal V, Ramadasan A (eds), Advances in Plant Physiology and Biochemistry of Coconut Palm, Asian and Pacific Coconut Community. Jakarta.

Rajagopal, V. and Kasturi Bai, K.V. (2002). Drought tolerance mechanism in coconut. *Burot Bulletine*, 17, 21-22.

Rajagopal, V., Kasturi Bai, K.V. and Voleti, S.R. (1990). Screening of coconut genotypes for drought tolerance. *Oleagineux* 45, 215-223.

Rajagopal, V., Kasturi Bai, K.V. and Naresh Kumar, S. (2000). Adaptive mechanism of coconut palms in the changing environment conditions for higher production. In: Extended summaries Vol 2. Natural Resources- Agrobiodiversity, International conference on managing natural resources for sustainable agricultural production in the 21st century, New Delhi.

Rajagopal, V., Shivashankar, S. and Mathew, J. (1996) Impact of dry spells on the ontogeny of coconut fruits and its relation to yield. Plant Rech Dévelopment, 3:251-255.

Ramadasan, A. and Kasturi Bai, K.V. (1999). Leaf area, dry matter production and yield. In: Rajagopal V, Ramadasan A (ed) Advances in Plant Physiology and Biochemistry of Coconut Palm, Asian and Pacific Coconut Community, Jakarta

Ramirez-Vallejo, P. and Kelly, I.D. (1998). Traits related to drought resistance in common bean. *Euphytica*, 99, 127-136.

Repellin, A., Daniel, C. and Zuily-Fodil, Y. (1994). Merits of physiological tests for characterizing the performance of different coconut varieties subjected to drought. *Oleagineux*, 49, 155-168.

Repellin,A., Pham Thi, A.T., Tashakorie, A., Sahsah, Y., Daniel, C. and Zuily-Fodil,Y. (1997). Leaf membrane lipids and drought tolerance in young coconut palms (*Cocosnucifera* L.). *European Journal of Agronomy*, 6:25-33.

Sage, R.F. and Kubien, D.S. (2007). The temperature response of C-3 and C-photosynthesis. *Plant, Cell and Environment*, 30, 1086-1106.

Schneider, K. A., Brothers, M. E. and Kelly, J. D. (1997). Marker assisted selection to improve drought resistance in common bean. *Crop Science*, 37, 51-60.

Shivashankar, S., Kasturi Bai, K.V. and Rajagopal, V. (1991). Leaf water potential, stomatal resistance and activity of enzymes during the development of moisture stress in coconut palm. *Tropical Agriculture*, 68, 106-110.

Subramanian, P., Dhanapal, R., Palaniswami, C. (2010). Cropping system for coastal sandy soil management *In* Coconut based cropping/Farming Systems eds. (George V. Thomas, V. Krishnakumar, H.P. Maheswarappa and C. Palaniswamy) CPCRI, Kasaragod 231 p.

Villalobos, E., Umana, C.H. and Chinchilla, C. (1992). Estado de hidratacion de la palma aceitera, en respuesta a la seguia en Costa Rica. *Oleagineux*, 47, 1-7.

Voleti, S.R., Kasturi Bai, K.V. and Rajagopal, V. (1993). Water potential in the leaves of coconut (*Cocos nucifera* L.) under rainfed and irrigated conditions. In: Nair MK, Khan HH, Gopalasundaran P, Bhaskara Rao EVV (eds) Advances in Coconut Research and Development, Oxford and IBH Publishing, New Delhi.

White, J.W. and Singh, S.P. (1991). Breeding for adaptation to drought. InA. Van Schoonhoven and O. Voysest (Eds.) Common Bean: research for Crop Improvement, C.A.B. Int. Wallingford, UK and CIAT, Cali, Colombia. pp. 501-560.

Yusuf, M. and Varadan, K.M (1993). Water management studies on coconut in India. In: Nair MK, Khan HH, Gopalasundaran P, Bhaskara Rao EVV (eds) Advances in Coconut Research and Development, Oxford and IBH Publishing, New Delhi.

2017, Impact of Climate Change on Plantation Crops *Pages* **61–74**
Editors: **K.B. Hebbar, S. Naresh Kumar & P. Chowdappa**
Published by: **ASTRAL INTERNATIONAL PVT. LTD., NEW DELHI**

Chapter 4

Arecanut and Cocoa

S. Sujatha, Ravi Bhat and P. Chowdappa

1. Introduction

Agriculture is affected by climate change and weather variability. Climate change has long term negative impact on agricultural productivity all over the world (Nellemann *et al.*, 2009). A small climatic instability can cause some devastatingsocio-economic consequences in many developing countries due to the dominant role of agriculture and its primary dependence on rainfall. Thus, accurate aassessment of yield response to future climate is needed to prioritize adaptation strategies. Proliferation of pests and diseases, reduced recovery and low resource use efficiency are the imminent consequences of climate change scenario. Nelson *et al.* (2009) reported that an increase in temperature is mainly due to global warming, which reduces crop yields and encourages pest proliferation. Changes in temperature affect the crop yield mainly through phenological development process. Phenology is a good indicator of global warming (Chmielewski and Rötzer 2000).Therefore, it necessitates a cases tudy to identify the relation between climate and crop growth and yield, focusing on a regionor climatic geography. Weather variability influences the yield and sustainability of perennial plantations considerably as economic yielding life spreads over several decades. In perennial crops, the productivity is influenced not only by rainfall and temperature but also by other weather parameters like relative humidity, evaporation and sunshine hours. In regions where perennial crops are economically and culturally important, improved assessments of yield responses to future climate are needed to prioritize adaptation strategies.

Arecanut (*Areca catechu* L.) and cocoa (*Theobroma cacao* L.) are the two major cash crops in humid tropics of India. These perennial plantations are sensitive to various biotic and abiotic stresses. Arecanut, which belongs to family *palmae*, grows to height of 10-15 m with a crown of 8-9 leaves. In arecanut, flowering initiates in 4th year and yield stabilizes by 8th year. Cocoa belongs to family *Malvaceae* (formerly *Sterculiaceae*) and grows to a height of 2.0-2.5 m with 15-20 m^2canopy area. In cocoa, flowering initiates in 3rd year and is segmental in repeated phases. The flowering to harvesting period is one year in arecanut and 150-180 days in cocoa. The average yield levels are 1.5-3.0 kg $palm^{-1}$ in arecanut and 1.0 - 2.0 kg $tree^{-1}$ in cocoa in West coast region of India. The arecanut-cocoa system is efficient and economically feasible. In the tropical belt where arecanut and cocoa are grown (28° N and S of equator), precipitation is confined to six months from June to November with average rainfall of 3700mm. Insufficient water has been a major limiting factor in post monsoon season (December-May) due to high evaporative demand of arecanut. In India, it is cultivated in 0.45 m hectares with a production of 0.73 m tonnes and productivity of 1400 kg ha^{-1} (GOI, 2015). The tropical plant, cocoa (*Theobroma cacao* L.), which is the source of chocolate is endemic to Amazon basin. Its cultivation has subsequently extended to tropical and subtropical regions of South and Central America, West Africa and Asia-Pacific. Cocoa is cultivated as a component crop in arecanut, coconut and oil palm in 78, 000 ha with a production of 16, 050 tonnes and productivity of 475 kg ha^{-1} (GOI, 2015). National statistics clearly indicated that area of arecanut exhibited an upward trend over time, but the productivity showed stagnant trend during last decade. The productivity of arecanut (kg ha^{-1}) fluctuated from 857 in 1970 to 1379 in 2002 and 1195 in 2012. Though area and production of cocoa showed upward trend, productivity fluctuated registering stagnant trend. These plantation crops have high economic value and provide sustenance to the millions of people in India. The low productivity of arecanut and cocoa is due to climatic, crop and soil constraints and strategies are developed to improve productivity. As perennial plantations remain productive for several decades, weather variability influences the yield. But the studies on the impact of climate change on arecanut and cocoa are scarce in India. With this background, an attempt was made in this chapter to assess overall changes in weather pattern in humid tropics and its impending influence on productivity of arecanut and cocoa based on data acquisition from published reports. Further, other reports are also discussed.

2. Climatic Conditions in Plantation Belt

The parameters like heavy rainfall, high relative humidity and low temperatures are the major climatic constraints in arecanut and cocoa growing regions of humid tropics. In other regions, low rainfall and high temperatures are the major problems. Heavy rainfall leads to leaching of potassium and calcium and high relative humidity is congenial for proliferation of pests and diseases. Low temperatures at high altitude areas lead to softness of kernel and low nut recovery. Untimely rains and heavy rainfall events are commonly noticed in all arecanut growing regions. The resource use efficiency reduces considerably due to water stagnation, run off, soil erosion and leaching of nutrients due to heavy rainfall events in laterite soil belt of humid regions especially in areas undulating topography. Drought results in yield loss of 15-75 per cent (Bhat and Sujatha, 2004). Heavy rainfall events inflict a yield loss up to 90 per cent due to fruit rot (Jose *et al.*, 2008; CPCRI, 2015). This emphasizes the need for thorough understanding of impact of climate change in order to streamline the adaptation strategies for sustainability of perennial plantations.

2.1. Suitable Climate for Arecanut

Arecanut is grown in varied climatic conditions. However, it is sensitive to extreme climatic conditions (Bhat and Abdul Khader, 1982). In general, climate influences the yield to the extent of 50 per cent. The crop is grown within 28° North and South of equator. The rainfall, relative humidity, altitude, evaporation and temperature affect the yield of arecanut (Vijaya Kumar *et al.*, 1991; Sunil *et al.*, 2011). The lower temperature at higher elevation is unsuitable for arecanut. In north east region of India the crop is grown in plains. Even though the crop can be grown at altitudes of 1000 masl, it is seen that the quality of the nut deteriorates as altitude increases (Nambiar, 1949).

The temperature range of 14°C-36°C is optimum for better growth of arecanut. In India, the crop is being grown in temperatures ranging from 5°C (as in places like Mohitnagar, West Bengal) and at 40°C (Vittal in Karnataka and Kannara in Kerala). The temperature below -2.8°C caused damage to foliage and even death of palms (Smith, 1958). Severe foliar damage is noticed at low temperature with low humidity (Bhat and Sujatha, 2004). The extremes of temperature and wide diurnal variations are detrimental to growth of the crop. Increase in minimum temperature during flowering stage (January to March) has positive influence on arecanut yield (Sunil *et al.*, 2011). Arecanut is grown in high rainfall areas such as Malnad of Karnataka (≥ 4500 mm) and in low rainfall areas like plains of Karnataka or parts of Coimbatore district in Tamil Nadu (750 mm). Higher rainfall during nut development stage (June to July) reduces the yield. Annual rainfall above 2000 mm has detrimental effect on arecanut yield, while this crop needs higher relative humidity during the morning hours throughout the year (Sunil *et al.*, 2011). High humid conditions provide congenial conditions for the rapid spread of diseases like fruit rot, bud rot etc. It has considerable influence on evapotranspiration hence on the water requirement of arecanut. Thus, climate change scenario would influence growth and yield of arecanut.

2.2. Suitable Climate for Cocoa

Cocoa is a major cash crop in many tropical countries, where the climate shows relatively little variation throughout the year, especially in terms of temperature, solar radiation and day length. Seasonal variations in both rainfall and temperature influence the pod setting. The microclimate existing in arecanut and coconut plantations is congenial for cocoa cultivation in India. In rainfed coconut gardens, the drought intensity is more pronounced in northern regions of Kerala and coastal Karnataka extending up to 5-6 months subjecting cocoa to severe stress. The situation is better in arecanut gardens, which is an irrigated crop.

For cocoa, the distribution of rainfall is more important than total rainfall. Intensity and distribution of rainfall decide the pattern of cropping in mature cocoa. Two dissimilar crop patterns are observed under rainfed and irrigated conditions in India. The annual rainfall in most of the cocoa growing areas lies between 1,250 and 3,600 mm. If rainfall is less than 1,250 mm, the crop needs to be irrigated during the rainless period. High rainfall in excess of 2,500 mm may lead to problems such as black pod disease due to high humidity. Cocoa growing areas have uniformly high humidity, often 100 per cent during night falling to 70-80 per cent by day.

Cocoa can be successfully grown up to 300 masl. However, it can be grown up to 1100-1200 m altitude. Cocoa can tolerate a mean monthly maximum temperature up to 33°C and the optimum range is 30-32°C. The temperature in most of the cocoa growing areas lies between a maximum of 30-32°C and a minimum of 18-21°C. Low temperatures have an inhibiting effect on cambium growth, which is linked to flowering. High temperature reduces the pod growing period and in turn the yield and bean size. Thus, the impact of climate change will be severe on cocoa. In West Africa, Brazil, other Latin American countries, Malaysia and Sri Lanka high correlations between rainfall and yield have been reported. Other climatic factors such as temperature, light intensity and day length normally are not limiting factors for cocoa yields except in Brazil which experiences nearly four months of low temperature during winter.

2.3. Data Acquisition

The weather data of 43 years recorded at ICAR-Central Plantation Crops Research Institute, Regional Station, Vittal, Karnataka, India (12° 15′N latitude and 75° 25′E longitude, 91 m above MSL) is utilized for assessing the impact of climate change on arecanut and cocoa. The climate of the location is humid tropical with average annual rainfall of 3686 mm. The yield data of arecanut and cocoa were collected from records of different experiments in arecanut and cocoa at the Institute. For computing average yield of arecanut in different years, data acquisition was attempted from the published reports of ICAR-CPCRI (CPCRI, 1996; Bhat and Mohapatra, 1989; Balasimha, 2007; Balasimha, 2009; Bhat *et al.*, 1999; Sujatha *et al.*, 1999; Bhat *et al.*, 2007a; Sujatha *et al.*, 2010; Sujatha *et al.*, 2011a; Sujatha and Bhat, 2013a and b). The average yield for a particular year was computed from all the records to nullify treatment or technology effect as technological and resource constraints are likely to limit productivity (Mendelsohn and Dinar, 2003).

2.3.1. Time Trend of Weather Variables in Coastal Humid Tropics

The descriptive statistics of weather variables at Vittal is given in Table 4.1. The normal weather variable was taken as the average of last 43 years. The changes in weather variables during 2000-2012 compared to previous years are given in Table 4.2. Precipitation trends indicated very low variability for total rainfall and rainy days among different years. Inter-annual variability of rainfall is generally large in the tropics.For the period 1970–2012, the mean annual precipitation is 3686 mm and ranges from 2114 mm in 1987 to 5610 mm in 1994. The total rainfall (RF) decreased by 531 mm *i.e.*, 14 per cent during 2000-2012 compared to 1970-1999. With respect to air temperature (T) changes, years explained higher variability for T_{max} (33 per cent) than T_{min} (9 per cent). The trends of temperature increase are +0.4°C for mean maximum ($P < 0.001$) and + 0.4°C for mean minimum during the last decade ($P < 0.002$). Thus, for the 43-year period, the observed difference between maximum and minimum is +0.8°C. Years explained 36 and 6 per cent variability in relative humidity (RH) at 7.30 and 14.20 hrs, respectively. The RH at morning time increased, while RH at afternoon reduced during the observed period. Years showed maximum variability for number of sunshine hours per day (66 per cent) and pan evaporation (43 per cent) among all weather parameters. Both sunshine (SS) hrs and pan evaporation reduced during 2000-2012 compared to preceding years.

Table 4.1: Descriptive Statistics of Weather Variables Averaged for 1970-2012 in Humid Tropics at Vittal

Variable	*Mean*	*Minimum*	*Maximum*	*Std. Deviation*
Annual rainfall (mm)	3686± 95	2113	5610	624
Maximum temperature (°C)	32.4 ±0.06	31.2	33.2	0.42
Minimum temperature (°C)	22.0±0.06	21.2	22.8	0.37
Rainy days	139 ± 2.04	111	165	13.4
RH at 7.30 hrs	93.8 ±0.17	91.3	96.1	1.10
RH at 14.20 hrs	60.8 ±0.28	57.1	64.7	1.87
Sunshine hours	6.6 ±0.10	5.1	7.6	0.67
Evaporation (mm)	4.0 ±0.08	3.1	5.0	0.54

Table 4.2: Weather Variability during 1970-2012

Variable	*1970-99*	*2000-2012*	*Change in 2000-2012 Over 1970-99*
Minimum temperature	22.2	21.8	+ 0.4
Maximum temperatute	32.3	32.7	+ 0.4
RH at 7.30 hrs	93.5	94.5	+ 1.0
RH at 14.20 hrs	60.8	60.8	–
Sun shine hours	6.9	6.1	– 0.8
Evaporation	4.3	3.7	– 0.6
Total Rainfall	3846	3315	– 531
Total rainy days	141	134	– 7

2.3.2. Relation between Yield of Arecanut/Cocoa and Weather Variables

Positive and significant correlations are observed between arecanut yield and weather variables such as T_{max} (r=0.48), T_{min} (r=0.16) and RH (r=0.32 to 0.49) (Table 4.3). Negative correlations are noticed between arecanut yield and rainfall/sunshine hours (r = -0.20 to -0.21), while no relation is observed for evaporation and rainy days. Simple correlations showed that kernel yield is more closely related to RH and T_{max} than T_{min}, SS hrs and RF (Table 3).Significant positive impact of SS hrs, PE, T_{min} and RF on dry bean yield of cocoa is noticed (Table 3), while correlations are negative for T_{max} and $RH_{forenoon}$.

2.3.3. Weather Variability and Impact on Arecanut and Cocoa

Perennial systems are slow to adapt and more vulnerable to climate change (Rosenzweig and Hillel, 1998; Burton and Lim, 2005).Botharecanut and cocoa are perennial in nature with high commercial value. Thus, climate change will have

Table 4.3: Correlation between Yield and Weather Variables

Variable	*Arecanut*	*Cocoa*
Minimum temperature	0.16*	0.30*
Maximum temperatute	0.48**	–0.14*
RH at 7.30 hrs	0.32**	–0.30*
RH at 14.20 hrs	0.49**	–0.02
Sun shine hours	–0.21*	0.42**
Evaporation	–0.05	0.37**
Total Rainfall	–0.20*	0.13*
Total rainy days	0.00	0.08

agronomic impacts on yields and also generate economic effects on prices, demand and trade. Despite development of efficient technologies, the productivity levels remained more or less stagnant during 1990-2010. This explains the influence of climate on yield. The significant relation of weather parameters with yield further substantiates the impact of weather (Table 4.3). The yield reduction might be due to reduced recovery and changes in phenology. The changes in weather variables might influence net photosynthesis, evapotranspiration, flowering, pollination and yield. Several reports indicated similar impact of weather changes on several crops (White *et al.*, 1999; Kramer *et al.*, 2000; Chuine *et al.*, 1999) and in cocoa (Joly and Hahn 1989; Balasimha *et al.*, 1991). In many cases, high precipitation is associated with a reduction in yields due to reduced pollination and increased incidence of diseases in wetter years. High rainfall (5610 mm in 1994) with high intensity rains results in spread of fruit rot (*Phytophthora palmivora*) and water stagnation leading to yield reduction in arecanut (CPCRI, 1996; Sujatha *et al.*, 1999). In 2007, the yield loss of 40 per cent is reported in arecanut due to continuous rainfall of >2500 mm in July-September, high RH and less sunshine hrs (Jose *et al.*, 2009). The results give indication that continuous and heavy rainfall as in July (>1000 mm) is not ideal for arecanut as it creates waterlogging, higher incidence of *Phytophthora* diseases and hampers pollination/nut development. The emergence of two new pests *viz.*, palm aphid and whitefly in arecanut is a consequence of either climate change or pest resurgence in perennial ecosystem (Joseph Rajkumar, 2013). The variations in rainfall from May to November and relative humidity at afternoon hours significantly affect the arecanut yield in Western Ghat region of Karnataka (Tejaswani *et al.*, 2014). Results of survey in South Konkan region of Maharashtra revealed that arecanut yield is directly proportional to rainfall above 4300 mm with maximum humidity (Salvi *et al.*, 2015) and arecanut prefers high relative humidity particularly during morning hours throughout its growth period.

The impact of rainfall on incidence of black pod disease is not visible in cocoa as pod development stage escapes high monsoon rainfall in humid tropics of India. However, cocoa production is reduced by 15-20 per cent in Nigeria in 2011 due to high rainfall and inability of the farmers to buy chemicals to combat black pod disease (Oredein, 2011). Lawal and Emaku (2007) analyzed the impact of climate change on cocoa production at the Cocoa Research Institute of Nigeria (CRIN) between 1985 to 2004 and stated that the standard deviation of cocoa output is 4.69. Further, it is stated that increase in temperature enhances cocoa production and increase in relative humidity decreases it. Oyekele (2012) stated that the sensitivity of cocoa production to hours of sunshine, rainfall, soil conditions and temperature makes it vulnerable to climatic change. Changing climate can also alter the development of pests and diseases and modify the host's resistance. Further, it is reported that the black pod disease is a major threat to cocoa production when the relative humidity is very high.

2.3.4. Differential Response of Arecanut and Cocoa

Attainable yield is mainly limited by water or nutrient supply (van Ittersum and Rabbinge, 1997; Stewart *et al.*, 2005). The results indicate that the cocoa is more affected by climate variability than arecanut. This might be due to conspicuous changes in phenology and increased incidence of pests and diseases like tea mosquito bug,mealy bug andblack pod (Personal communication). Both arecanut and cocoa are highly cross pollinated and thus weather variability might influence phenology. On an average, infestation of tea mosquito bug is about 25 per cent in cocoa during the last decade. The incidence of black pod disease ranges from 6 to 51 per cent during monsoon season.

Zuidema *et al.* (2005) stated that over 70 per cent of the variation in simulated bean yield in cocoa could be explained by a combination of annual radiation and rainfall during the two driest months. Similar relations are observed in this study between cocoa dry bean yield and weather. The cumulative effect of changes in temperature, rainfall, humidity, evaporation, and sunshine hours has impact on the yield of cocoa.

The correlations between yield and weather parameters clearly indicate differential response of arecanut and cocoa (Table 4.3). The rainfall has negative impact on arecanut and positive impact on cocoa, which might be due to differences in yielding pattern. The nut development stage in arecanut invariably faces heavy rains resulting in yield loss due to water stagnation, pests and diseases, but cocoa escapes heavy monsoon rains during pod development stage. Another significant aspect of variability is sunshine hours showing negative impact on arecanut and positive impact on cocoa (Table 4.3). As cocoa is a shade crop in arecanut plantations with only 40 per cent of the incident radiation reaching the ground (Muralidharan, 1990), the positive relation between yield and sunshine hours indicates the need for higher sunlight availability to cocoa. Cocoa exhibits increased production under lowered light levels with optimal growth at 20to 30 per cent of full sunlight (Okali and Owusu, 1975; Galyuon *et al.*, 1996).Similarly, positive response of arecanut and negative response of cocoa to $RH_{forenoon}$ can be attributed to increased microclimatic humidity in arecanut plantation over atmospheric humidity.

Both arecanut and cocoa have similar evaporative demand in humid tropics (Abdul Haris *et al.*, 1999; Bhat *et al.*, 2007a), but the response of cocoa to pan evaporation was positive and significant. Reduced evaporation in recent years might impact the transpiration losses through metabolic activities of these crops and in turn the productivity. Model simulations of the potential yield of tea in north-east India predicted slight reduction in yield for each mm reduction in evapotranspiration (Panda *et al.*, 2003). The sensitivity of cocoa to T_{max} and positive response of arecanut clearly explains the shade requirement of cocoa ruling out the possibility of sole cropping of cocoa in humid tropics. Minimum temperature has shown positive impact on both crops suggesting clear adaptability of these crops. There is no influence of rainy days on both crops clearly indicating that the intensity and distribution of rainfall are important. The impact of weather variability on growth and yield would be different for dicot and monocot perennials.Even though simulation model in perennial crop like cocoa (Zuidema *et al.*, 2003) has been reported, the model is not yet validated for Indian conditions and suitability for climate change studies.

3. Adaptation Strategies

Crop management needs to fine-tune to weather changes as an adaptation strategy. Identification of genotypes tolerant to various biotic and abiotic stresses is need of the hour. During 2000-2015, successful technologies like nutrient and irrigation management (Bhat and Sujatha, 2004), drip fertigation (Bhat *et al.*, 2007a; Sujatha and Bhat, 2013a), cropping systems (Bhat and Sujatha, 2011; Sujatha *et al.*, 2011b; Sujatha *et al.*, 2016) and mixed farming approach (Sujatha and Bhat, 2015) in arecanut reduced the impact of weather changes (Table 4). Drip fertigation is a better adaptation strategy under changing climate scenario in humid tropics as it sustains yield levels in low rainfall years also as in 2002and 2012 (Bhat *et al.*, 2007a; Sujatha and Bhat, 2013a). In 2002, yield loss of 13 -14.5 per cent is reported in farmer's plantations due to less rainfall (Jose *et al.*, 2004). Adoption of farming systems is necessary to reduce income fluctuations due to weather changes (Sujatha and Bhat, 2015b). The predominant arecanut belt in India is laterite soil belt in humid tropics that receives higher average rainfall of above 3500 mm. Still water scarcity is noticed during summer months due to inherent soil constraints, higher evaporative demand of arecanut and faster depletion of ground water levels and less water harvesting possibilities (Mathew *et al.*, 2004 and 2008). The yield gap of 120-180 per cent between national/state average and on-station experiments clearly indicates that the adoption of these suitable technologies can be a better adaptation strategy under climate change scenario.The efficient arecanut based cropping system models for adaptation to climate change are given in Table 4.4. However, the scope for improving the productivity of arecanut by 200 to 300 per cent and profitability is demonstrated through different technologies at ICAR-CPCRI.

Coastal and hilly areas, where plantations are concentrated, are believed to be more vulnerable to climate change compared to other terrestrial areas. Thus, soil and water conservation incorporating concepts of water harvesting, *in situ* moisture conservation, life saving irrigation, mulching, permanent crop cover, ground water recharge through run off water ways, watershed management and erosion control measures attain utmost importance as climate change projects increase in frequency of extreme events. Adaption of cropping system and integrated farming approaches is very important for combating the risks of monocropping and climate change (Sujatha and Bhat, 2015; Sujatha *et al.*, 2016).

Table 4.4: Yield Gap between National Productivity and different Technologies

Suitable Adaptation Strategy	*Yield Level (kg ha^{-1})*	*Yield Gap (per cent) between Strategy and National Average Yield (1600 kg ha^{-1})*	*Reference*
Drip fertigation (2002-2006)	4017	151	Bhat *et al.*, 2007a;
Organic matter recycling (2003-2011)	2774	73	Sujatha and Bhat, 2013b; Sujatha and Bhat, 2016
Cropping system approach			
Arecanut+MAPs with sprinkler irrigation (2004-2007)	3010	88	Sujatha *et al.*, 2011a
Arecanut+vanilla with drip irrigation (2005-2008)	3114	95	Sujatha and Bhat 2010
Arecanut+cocoa with drip fertigation (2008-2011)	3117	95	Sujatha and Bhat 2013a
Mixed farming approach (2012-2014)	3418	114	Sujatha and Bhat, 2015

For reducing the climate change impact at regional levels, attempt should be made to adopt good agricultural practices like minimum tillage, drip fertigation, reduced input use, eco-friendly management of pests and diseases, and recycling of organic wastes to reduce emissions of GHG's especially CO_2.

4. Arecanut and Cocoa for Carbon Sequestration

Plantation crops have significant potential for offsetting and reducing the projected increases in green house gas (GHG) emissions and regarded as an important option for greenhouse gases mitigation. Plantation crops are generally cultivated in contiguous areas and can sequester considerable carbon due to higher growth and biomass increments. The plantation crops occupy the land for more than 3 decades and accumulate both above ground and below ground root stocks in this process. The reports indicate that the carbon sequestration potential of arecanut with and without the presence of inter/mixed crops is very high and are comparable to forests (Balasimha and Naresh Kumar, 2010). Annual increments in biomass or net primary productivity ranged from 3.34 -7.11 t ha^{-1} in arecanut. In arecanut, annual increment in carbon stock is 1.4-3.0 t ha^{-1}(Balasimha and Naresh Kumar, 2009). Arecanut-cocoa also is a good system for carbon sequestration with a potential to sequester 5 to 7 t CO_2/ha/year (Balasimha and Naresh Kumar, 2009). The standing biomass increased over time indicating accumulation of biomass in stem and also due to increase in yield by arecanut and cocoa plant with age up to 20th year of planting (Figure 4.1). Recent estimates indicated that the total carbon stocks vary from 129 to 169 t ha^{-1} in arecanut plantations of different ages (Sujatha and Bhat, 2015b). Further, soil carbon stocks (119-137 t ha^{-1}) are higher than standing above ground carbon stocks (10-21 t ha^{-1}) at 0-30 cm soil depth. The soil organic carbon levels are optimum in arecanut ecosystem Soil carbon stocks account for 80 per cent of carbon stocks in arecanut and mixed farming approach is the best

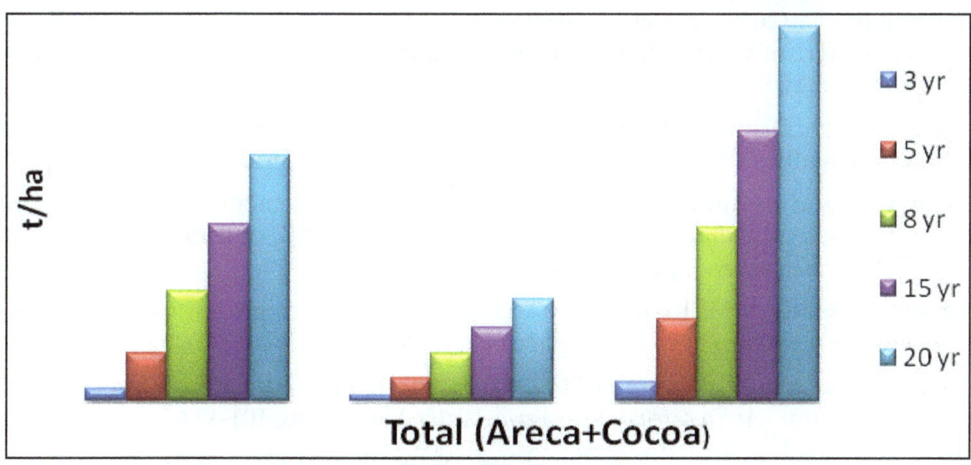

Figure 4.1: Carbon Sequestration in Areca and Cocoa System.

option for climate change scenario due to recycling of critical inputs among crop and livestock components (Sujatha and Bhat, 2015). Thus, carbon stocks in biomass and soil organic matter result in the net removal of CO_2 from the atmosphere.

5. Conclusions

The results imply that weather has a definite role in influencing the yield of arecanut and cocoa. A comprehensive analysis of 43-yr weather data from 1970-2012 revealed that humidity and temperature increase, while other variables like total rainfall, sunshine hours and evaporation decrease in humid tropics in India. The correlations between weather variables and yield was either positive or negative or without any relation. There was differential response of arecanut and cocoa to weather variability. The suitable adaptation strategies are also discussed for these two crops.The crop productivity remains highly dependent on weather, which can affect boththe quantity and quality,despite advances in technology and the widespread prevalence of irrigation facilities in arecanut belt due to climate change.

References

Abdul Haris, A., Balasimha, D., Sujatha, S., Bhat Ravi, Khader, K.B.A. (1999). The influence of drip irrigation and fertilizer on yield and photosynthetic characteristics of cocoa in mixed cropping system with arecanut. *J. Plantation Crops* 27, 131-135

Balasimha, D. and Naresh Kumar, S. (2013). Net primary productivity, carbon sequestration and carbon stocks in areca-cocoa mixed crop system. *Journal of Plantation Crops.* 41(1): 8-13.

Balasimha, D. (2007). Efficacy of pruning in enhancing bean yield of cocoa. *J. Plantation Crops* 35, 201-202.

Balasimha, D. (2009). Effect of spacing and pruning regimes on photosynthetic characteristics and yield of cocoa in mixed cropping with arecanut. *J. Plantation Crops* 37(1), 9-14.

Balasimha, D., Daniel, E.V., Bhat, P.G. (1991). Influence of environmental factors on photosynthesis in cocoa trees. *Agric. Forest Meteorol.* 55, 15-21.

Bhat, K. S. and Abdul Khader, K. B. (1982). Crop Management. B. Agronomy. In: *The Arecanut Palm.* (Eds. Bavappa, K.V.A., Nair, M.K. and Prem Kumar, T.). CPCRI, Kasaragod. pp 105-131.

Bhat Ravi, Reddy, V.M., Khader, K.B.A. (1999). Areca based high density multispecies cropping system in coastal Karnataka. *J. Plantation Crops*27(1), 22-26.

Bhat Ravi, Sujatha, S., Balasimha, D. (2007). Impact of drip fertigation on productivity of arecanut (*Areca catechu* L.). *Agric. Water Manage.* 90, 101-111.

Bhat, N. T., Mohapatra, A. R. (1989). Effect of supplying nutrients through organic manures, inorganic fertilisers and their combination on arecanut crop. *J. Plantation Crops*16 (Suppl), 443-447.

Bhat, Ravi and Sujatha, S. (2004). Crop Management. In: Balasimha, D., Rajagopal, V. (Eds.), Arecanut. CPCRI, Kasaragod. pp. 76-102.

Bhat, Ravi and Sujatha, S. (2011). Arecanut based high density multispecies cropping/ farming system. In: Thomas, G.V., Krishnakumar, V., Maheshwarappa, H.P., Bhat, R., Balasimha, D. (Eds.), Arecanut based cropping/farming systems. CPCRI, Kasaragod. pp. 27-44.

Burton, I., Lim, B. (2005). Achieving adequate adaptation in agriculture. *Clim. Change* 70, 191–200.

Chmielewski, F.-M., Rötzer, T. (2001). Response of tree phenology to climate change across Europe. *Agric. Forest Meteorol.* 108, 101–112,

Chuine, I., Cour, P., Rousseau, D.D. (1999). Selecting models to predict the timing of flowering of temperate trees: implication for tree phenology modelling. *Plant Cell Environ.* 22, 1–13.

CPCRI, 2015. Annual Report for 2014-2015. Central Plantation Crops Research Institute, Kasaragod, India.

CPCRI. (1996). Central Plantation Crops Research Institute, Annual Report for 1995-96, Kasaragod. pp. 220.

Galyuon, I.K.A., McDavid, F.B., Lopez, F.B., Spence, J.A. (1996). The effect of irradiance level on cocoa (*Theobroma cacao* L.): 1. Growth and leaf adaptations. Trop. Agric. (Trinidad) 73, 23–28.

GOI. (2015). Agricultural Statistics at a Glance, Directorate of Economics and Statistics, Ministry of Agriculture, Govt of India, New Delhi.

Joly, R. J., Hahn, D.T. (1989). Net CO2 assimilation of cocoa seedlings during periods of plant water deficit. *Photosynth.* Res. 21, 151-159.

Jose C.T., Balasimha, D., Kannan, C. (2009). Yield loss due to fruit rot disease in Karnataka. *Indian J. Arecanut, Spices and Med. Plants* 10, 45-51.

Jose, C. T. Jayashekar, S., Balasimha, D. (2004). Impact of drought on arecanut in Karnataka. *Indian J. Arecanut, Spices and Medicinal Plants* 5 (4), 144-149.

Josephrajkumar, A., Rajan, P., Chandrika Mohan, Nampoothiri, C.G.N. (2013). Distinguishing palm aphid and arecanut whitefly, two emerging pests in palms. *Indian J. Arecanut, Spices and Medicinal plants* 15 (2), 3-7.

Kramer, K., Leinonen, I., Loustau, D. (2000). The importance of phenology for the evaluation of impact of climate change on growth of boreal, temperate and Mediterranean forest ecosystems: an overview. *Int. J. Biometeorol.* 44, 67–75.

Lawal, J.O. and Emaku, L. A. (2007). Evaluation of the effect of climatic changes on cocoa production in Nigeria: Cocoa Research Institute of Nigeria (CRIN) as a case study. *African Crop Science Conference Proceedings,* 8, 423 – 426.

Mathew, A. C., Shajatnan, K. H., Sujatha, S. (2004). Water movement in the active root zone of arecanut under drip irrigation. *J. Plantation Crops* 32(Supplement), 248-252.

Mathew, A. C., Shajatnan, K. H., Sujatha, S. (2008). Development and Management of water harvesting structure in laterite soils for irrigating arecanut through conjunctive use of harvested and surface water. *J. Plantation Crops* 36(3), 304-309.

Mendelsohn, R., Dinar, A. (2003). Climate, water, and agriculture. Land Econ. 79:328–341.

Muralidharan, A. (1990). Intercropping in arecanut. *J. Plantation Crops*17, 25-38.

Nambiar, K. K. (1949). *A Survey of Arecanut Crops in Indian Union*. Indian Central Arecanut Committee, Calicut. pp 76.

Nellemann, C., MacDevette, M., Manders, T., Eickhout, B., Svihus, B., Prins, A. (2009). The environmental food crisis: the environment's role in averting future food crises. United Nations Environment Programme.

Nelson, G.C., Rosegrant, M.W., Koo, J., Robertson, R., Sulser, T., Zhu, T., Ringler, C., Msang, S., Palazzo, A., Batka, M., Magalhaes, M., Valmonte-Santos, R., Ewing, M., Lee, D. (2009). Climate change: impact on agriculture and costs of adaptation. Technical report, International Food Policy Research Institute.

Okali, D.H.H., Owusu, J.K. (1975). Growth analysis and photosynthetic rates of cocoa (*Theobroma cacao* L.) seedlings in relation to varying shade and nutrient regimes. *Ghana J. Agric.* Sci.8, 51-67.

Oredein O. (2011). Nigeria's cocoa output estimated 15 per cent -20 per cent lower. Market Watch –The Wall Street Journal. Internet file downloaded on 10thJanuary 2012 from http://www.marketwatch.com/story/nigerias-cocoa-output-estimated-15-20-lower-2011-07-29.

Oyekale A S. (2012). Impact of Climate Change on Cocoa Agriculture and Technical Efficiency of Cocoa Farmers in South-West Nigeria. *J Hum Ecol,* 40(2): 143-148.

Panda, R. K., Stephens, W. and Matthews, R. (2003). Modelling the influence of irrigation on the potential yield of tea (*Camellia sinensis*) in North-East India. *Experimental Agriculture.*

Rosenzweig, C., Hillel, D. (1998). Climate Change and the Global Harvest. Oxford University Press, New York, 324 pp.

Salvi, S. P., Mule, R. S., Agare, H. R.,(2015). Survey on yield of arecanut (*Areca catechu* L.) influence with weather parameters under South Konkan region of Maharashtra. *Indian J. Arecanut, Spices and Medicinal plants* 17(4), 38-40.

Smith, D. (1958). Cold tolerance of cultivated palms. *Principes.* **2**:120.

Sujatha, S. and Bhat Ravi, (2010). Response of vanilla (*Vanilla planifolia* A.) intercropped in arecanut to irrigation and nutrition in humid tropics of India. *Agric. Water Manage.* 97(7), 988 - 994.

Sujatha, S. and Bhat Ravi, (2013a). Impact of drip fertigation on arecanut–cocoa system in humid tropics of India. *Agroforest. Syst.*87(3): 643-656.

Sujatha, S. and Bhat Ravi, (2013b).Impact of vermicompost and NPK application on biomass partitioning, nutrient uptake and productivity of arecanut (*Areca catechu* L.). *J. Plant Nutr.* 36(06); 975-988.

Sujatha, S. and Bhat. Ravi. (2015). Resource use and benefits of mixed farming approach in arecanut ecosystem in India. *Agr. Syst.* 141, 126-137.

Sujatha, S., Bhat Ravi and Chowdappa, P. (2016). Cropping systems approach for improving resource use in arecanut (*Areca catechu*) plantation. *Indian Journal of Agricultural Sciences* **86** (9): 1113–20.

Sujatha, S., Bhat Ravi, Kannan, C., Balasimha, D. (2011a). Impact of intercropping of medicinal and aromatic plants with organic farming approach on resource use efficiency in arecanut (*Areca catechu* L.) plantation in India. *Ind. Crops Prod.* 33(1), 78-83.

Sujatha, S., Bhat Ravi, Reddy, V.M., Abdul Haris, A. (1999). Response of high yielding varieties of arecanut to fertilizer levels in coastal Karnataka. *J. Plantation Crops* 27, 187-192.

Sujatha, S., Bhat. Ravi, Balasimha, D. and Apshara, E. S. (2011b). Arecanut based inter/mixed cropping systems. In: Thomas, G.V., Krishnakumar, V., Maheshwarappa, H.P., Bhat, R., Balasimha, D. (Eds.), Arecanut based cropping/ farming systems. CPCRI, Kasaragod. pp. 6-26.

Sunil, K.M., Devadas, V.S. and George, P.S. (2011). Influence of Weather Parameters on Yield and Yield Attributes of Areca Nut (*Areca catechu* L.). *Journal of Agricultural Physics.* 11:88-90.

Tejaswini, A.B., Padmashri, H.S., Soumya, D.V., Kammardi, T.P., Hemalatha, H.G. (2014). Association studies of climate factors on arecanut yield in Western Ghats region Karnataka. *Environ. Ecol.*32(3A), 1085-1087.

van Ittersum, M.K., Rabbinge, R. (1997). Concepts in production ecology for analysis and quantification of agricultural input-output combinations. Field Crops Res. 52, 197-208.

Vijaya Kumar, B. G., Veerappadevaru, G., Balasimha, D., Abdul Khader, K. B. and Ranganna, G. (1991). Influence of weather on arecanut and cocoa yield. *J. Plantn. Crops.* **19**: 33-36

White, M.A., Running, S.W., Thornton, P.E. (1999). The impact of growing-season length variability on carbon assimilation and evapotranspiration over 88 years in the eastern US deciduous forest. *Int. J. Biometeorol.* 42, 139–145.

Zuidema, P.A., Leffelaar, P.A., Gerritsma, W., Mommer, L., Anten, N.P.R. (2005). A physiological production model for cocoa (Theobroma cacao): model presentation, validation and application. *Agric. Syst.* 84, 195–225.

2017, Impact of Climate Change on Plantation Crops Pages 75–86
Editors: K.B. Hebbar, S. Naresh Kumar & P. Chowdappa
Published by: ASTRAL INTERNATIONAL PVT. LTD., NEW DELHI

Chapter 5

Black Pepper and Cardamom

K.S. Krishnamurthy, K. Kandiannan,
S.J. Ankegowda and M. Anandaraj

1. Introduction

Black pepper (*Piper nigrum* L.), the king of spices and small cardamom or true cardamom (*Elettaria cardamomum* Maton), the queen of spices are native to south India. But the productivity of these two crops is low in India compared to many countries which grow these crops. Though India has gifted climate for cultivation of many spices, the changing climate is posing a threat to the production and productivity of many spices. One of the reasons for low productivity of spices in India is due to the fact that most of these crops are grown as rainfed crops and they suffer from lack of soil moisture during critical crop growth stages. Also, climate vagaries such as frequent flood, drought, cold waves, heat waves, shift in monsoon pattern, sun stroke etc. have increased in the recent past and are affecting the productivity of spices. In this chapter the effect of climate change variables on the production and quality of these two spices are discussed.

2. Black Pepper (*Piper nigrum* L.)

Black pepper belongs to the family Piperaceae. World production of black pepper mainly comes from Vietnam, Indonesia, India, Thailand, Sri Lanka, Brazil, Malaysia, China, Madagascar and Mexico. In India, it is generally grown in southern states *viz.*, Kerala, Karnataka, Tamilnadu, Maharashtra, Goa and is slowly spreading to non-traditional areas such as East and West Godavari districts of Andhra Pradesh, Odisha, West Bengal, Andaman and Nicobar islands and north eastern states. It requires a well distributed rainfall of 2000 - 3000 mm for better productivity. It is susceptible to excessive heat and dryness (Sivaraman *et al.*, 1999). The temperature range of 23- 32°C is optimum for pepper growth but it can be cultivated between

10–40°C. It can be grown from sea level to 1500 m above MSL (Radhakrishnan *et al.*, 2002). Hao *et al.* (2012) reported that the minimum temperature of the coldest month, the mean monthly temperature range, and the precipitation of the wettest month were identified as highly effective factors in the distribution of black pepper and could possibly account for the crop's distribution pattern. Such climatic requirements inhibited this species from dispersing and gaining a larger geographical range. This clearly suggests the comfort zone for the crop and hints on the specific climatic requirements of the crop. Black pepper annual cycle is broadly divided into i) lag period, the interval between harvest and subsequent spike initiation (March–May), ii) spike emergence and flower bud differentiation (June–July), new leaf accompanied by spike, iii) spike enlargement and berry formation (August–October), iv) berry development (November-December) and v) maturity and harvest (January–February). These phases may overlap and the duration may vary depending on rainfall distribution, variety and location (Kandiannan *et al.*, 2011).

2.1. Climate Change and Productivity

Weather parameters play a crucial role in production and productivity of black pepper. Pre-monsoon and early monsoon rains are very essential to induce new flush, flower initiation; berry setting as well as yield. Growth of fruit bearing lateral shoots and photosynthetic rate are maximum during peak monsoon in India (Ravindran *et al.*, 2000). Good soil moisture (16-18 per cent) is essential from flowering till fruit maturity to obtain better yields and any dry spell within this critical period of 16 weeks results in low yield. Kannan *et al.* (1988) noted that no rainfall in January–February and 40mm in March and good rainfall from third week of April–August resulted in good yield. This again highlights the importance of maintenance of sufficient soil moisture from flowering till berry filling to obtain higher yields. The rainfall pattern and pepper yields during two extremely adverse years (1980-81 and 1986-87) were compared to that of a favourable year (1981-82) by Pillay *et al.* (1988) and it was found that during both adverse years, there was a distinct break in the rainfall during critical period following flower initiation. The late commencement of south west monsoon delays flowering. Heavy north east monsoon showers after a spell of dry period after SW monsoon results in high spike drop. Rainfall after stress induced profuse flowering (Ridley, 1912; Pillay *et al.*, 1988). Heavy rains during flowering reduces the rate of pollination and continuous heavy rainfall promotes vegetative development and limits flowering (Pillay *et al.*, 1988).

Rainfall beyond normal during initial period of annual cycle (*i.e.*, 5 -11 March to 25 June – 1 July) was harmful or would reduce the yield in major black pepper growing regions of India (Kandiannan *et al.*, 2011a). A study on 140 years of climatic data of Kerala indicated cyclical rainfall pattern with a declining trend of annual as well as south west monsoon in the last 60 years and an increasing trend in post monsoon rains (Rao *et al.*, 2009). Climatic data of two decades (1984-2004) revealed a declining trend in rainfall and rainy days in major black pepper growing areas of the country (Krishnamurthy *et al.*, 2011). Increasing trend in rainfall during summer months was observed in black pepper growing regions of India (Parthasarathy *et al.*, 2010 and Kandiannan *et al.*, 2011b) that could affect the flowering pattern. In Idukki (a predominant black pepper growing region), the change in rainfall pattern during 1999-2000 crop season affected the flowering and yield (John *et al.*, 1999). Suparman (1998) reported similar observation in Bangka, Indonesia. In India, black pepper yield was affected during drought years (1987 and 2002). But the yield reduction was more dependant on the distribution of rainfall rather than the quantity of rainfall (Krishnamurthy *et al.*, 2015).

Meteorological parameters such as maximum temperature (°C) (TMAX) and minimum temperature (°C) (TMIN), maximum relative humidity (per cent) (RHMAX) and minimum relative humidity (per cent) (RHMIN), rainfall (mm) (RAIN), evaporation (mm) (EVPN), wind speed (WIND) (km h^{-1}) and bright sunshine hours (SUNS) were correlated with black pepper fresh yield and the. magnitude of their association was RHMAX > RAIN > TMIN > TMAX > SUNS > WIND > RHMIN > EVPN (Kandiannan *et al.*, 2011). Long term mean (1985–2004) for monthly rainfall in major black pepper growing regions show a major peak during June to September followed by a minor peak during October–November. Analysis of the climate of past 2-3 decades of black pepper growing regions of the country revealed that rainfall and rainy days are showing a decreasing trend and temperature (both TMAX and TMIN) is showing an increasing trend. Black pepper productivity shows decreasing trend/no change (Table 5.1). It is reported that rainfall during pre-monsoon period (March – April) is positively correlated and December rainfall is negatively correlated with pepper productivity (Krishnamurthy *et al.*, 2011). A study on 140 years of climatic data of Kerala indicated increase in day maximum by 0.64°C and night minimum temperature by 0.23°C (Rao *et al.*, 2009). Based on many studies covering a wide range of regions and crops, negative impacts of climate change on crop yields have been more common than positive impacts (IPCC, 2014).

Black pepper productivity is higher in higher elevations (400-1000 m above MSL) compared to plains. Diurnal temperature of higher elevations may be contributing to higher productivity as rainfall does not seem to play a role. It was revealed that the total annual rainfall was less in higher elevations compared to plains in most of the black pepper growing areas of Kerala. Studies revealed that Tmin of higher elevations had positive and both Tmax and Tmin in plains had negative correlation with pepper productivity. This suggests that increase in maximum temperature may reduce black pepper productivity especially in plains whereas the increase in minimum temperature could lead to higher productivity in higher elevations (Krishnamurthy *et al.*, 2015).

As per various emission scenarios, the temperature of the black pepper growing regions is likely to increase by around 1.5°C by 2100. It is reported that the rise in temperature would affect the native genotypes or wild types of black pepper that are endemic to Western Ghats region (Utpala Parthasarathy *et al.*, 2008). Controlled experiments in plant growth chamber under elevated temperature conditions (2.7 degrees higher than ambient) showed reductions in plant height, leaf area and photosynthetic rate of black pepper varieties grown under elevated temperatures during the initial growth period. (Krishnamurthy *et al.*, 2015). Hence, identification/ development of temperature insensitive varieties is essential to mitigate climate change effects.

Table 5.1: Trend Analysis of Climatic Variables and Black Pepper Productivity

Place	*Rainfall*	*Tmax*	*Tmin*	*Productivity*
KERALA				
Ambalavayal (Wynad) 1979–2004	Decreasing Y= -16.626X + 2196	Increasing Y= 0.0278X + 26.96	Increasing Y = 0.0501X + 17.08	No change Y= -0.410X + 406.8
Pampadumpara (Idukki) 1986–2004	Decreasing Y= -2.0596X + 1931.1	–	Increasing Y= 0.04X + 21.447	Increasing Y = 6.957X + 315.92
Panniyur (Cannanore) 1974–2004	Decreasing Y= -5.332X + 3518	Increasing Y= 0.01X + 32.85	Increasing Y= 0.0278X + 22.206	Decreasing Y= -5.022X + 291.3
Trichur1980–2004	Decreasing Y= -10.523X + 2868.5	Decreasing Y= -0.013X + 32.28	No trend Y= 0.004X + 23.29	Decreasing Y= -2.162X + 203.11
TAMILNADU				
Valparai (Coimbatore) 1976–2004	No trend Y= -0.261X + 5904	No trend Y= 0.0034X + 25.31	Decreasing -0.13X + 18.41	No change Y= 0.865X + 191.97
Nilgiris 1980–1992	Increasing Y= 28.56X + 1423.1	Increasing Y= 0.10X + 22.567	Decreasing Y= -0.05X + 14.11	No change Y= -0.824 + 202.6

Source: Krishnamurthy *et al.*, 2012.

Table 5.2: Relationship between Climatic Parameters, Altitude and Black Pepper Yield

Place	Temperature (°C)			Rainfall (mm)	Altitude (m above MSL)	Mean yield (kg/ha)
	Mean Tmax	Mean Tmin	Difference			
Wynad	27.3	17.6	9.7	1931	780	402
Idukki	27.5	15.6	11.9	1902	1200	327
Cannanore	33.1	22.6	10.5	3348	15	241
Trichur	32.1	23.4	8.7	2752	05	239

Our studies during 2013 and 2014 cropping seasons in Madikeri, Karnataka indicated that low rainfall, few rainy days and low temperature during pre-monsoon season and heavy rainfall, more number of rainy days, low light intensity, and low maximum and minimum temperatures during June to August (spike initiation period) in 2013 negatively influenced black pepper productivity. Low pre-monsoon rainfall, continuous and heavy rains leading to almost 100 per cent rainy days with a heavy cloud cover during the period and low temperatures delayed spike emergence and also reduced the number of spikes. Most of the spikes emerged in September which otherwise would have emerged in July as in 2014. These late emerged spikes had very low bisexual flower percentage. The flower emergence within the spike was also staggered. This low bisexual flower number together with staggered flowering resulted in very low number of berries per spike which ultimately lead to very low yield in 2013. Venugopal *et al.* (2013) also observed that high altitudes and heavily shaded conditions cause the proportion of female flowers to be higher compared to hermaphrodite flowers in the most popular variety Panniyur-1. This again indirectly suggests that low light and low temperature reduce bisexual flower production in black pepper as in high altitude temperature will be low. In 2014, there was good pre-monsoon showers (250-300mm spread over 15-20 rainy days during April and May) followed by normal monsoon season (well distributed rainfall of 1000-1200 mm during June and July), less cloud cover and slightly higher RH which resulted in early spike initiation (in July) and about 85-90 per cent of the flowers in the spikes were bisexual leading to good pollination and berry set.

2.2. Climate Change and Pathogens and Pests

Black pepper is affected by many diseases among which foot rot caused by *Phytophthora capsici*, followed by slow decline caused by nematodes, anthracnose caused by *Colletotrichum* and viral diseases are more common. Among insect pests, pollu beetle followed by mealy bugs and scales damage the crop. Weather parameters play a major role in the development of foot rot and anthracnose diseases. Anandaraj and Sarma (1994) reported that during peak monsoon period a daily rainfall of 15.8-23mm, temperature range of 22.7°C-29.6°C, 2.8-3.5 sunshine hours/day and 81-99 per cent RH favour disease development. Jayasekhar and Muthusamy (1999) reported that foot rot was negatively correlated with maximum and minimum temperatures and positively correlated with rainfall, number of rainy days and relative humidity. It is also noted that well distributed rainfall and high RH are congenial for the disease development. Maximum temperature and disease development were negatively correlated (Shamarao and Siddaramaiah, 2002). Similar results were also reported by Arasumallaiah *et al.* (2008) and Ramachandran *et al.* (1988). Climate change in cardamom hills of Kerala has resulted in increased incidence of anthracnose disease of black pepper (Murugan *et al.*, 2012). Maximum and minimum temperatures had negative correlation with anthracnose disease incidence while, rainfall and number of rainy days had positive correlation with disease initiation and subsequent spread (Biju and Praveena, 2015). Highest population densities of *M. incognita* on black pepper root were observed during the first half of the dry season (Thuy *et al.*, 2012) indicating that frequent drought may increase the population density of *M. incognita* on black pepper roots. Black pepper berry damage by pollu beetle was highest in the plains, and it was very low at higher elevations (300-900 m a.s.l.) and absent at > 900 m a.s.l. (Kumar and Nair, 1987). Our own studies (in press) suggests that temperature of 36°C and above trigger viral disease in black pepper under Kozhikode climatic condition. All these studies indicate that climate change in terms of increase in temperature, decrease in rainfall, frequent drought and floods *etc.* may bring about changes in disease incidence, population dynamics of pests, minor pests becoming major pests *etc.* which ultimately affects productivity.

2.3. Climate Change and Quality

Black pepper assumed king status among spices due to its inherent quality which is mainly contributed by the pungent principle, piperine. In general, quality in black pepper refers to physical quality constituents *viz.*, grade of berries, bulk density, test weight, fibre, starch and protein content of the berries and intrinsic quality constituents *viz.*, oil, oleoresin, piperine and oil constituents. Dry black pepper berries collected from low (10-200 m above MSL) and high elevations did not show differences in physical quality (Krishnamurthy *et al.*, 2015). But Sruthi *et al.* (2013) reported location wise variation for both primary and secondary metabolites such as essential oil, oleoresin, piperine, total phenol, crude fibre, starch, total fat and bulk density.

Intrinsic quality parameters *viz.*, piperine, oleoresin and oil also did not show variation between elevation groups. But oil components limonene and sabinene+myrcene showed positive correlation while β–caryophyllene showed negative correlation with elevation. Higher β-caryophyllene and lower limonene and sabinene+myrcene were observed under low elevation (warmer climate) in black pepper (Krishnamurthy *et al.*, 2015). Sruthi *et al.* (2013) also reported altitudinal variation in β-caryophyllene and total phenol contents. These two constituents were low at high elevations (>500 mean sea level (MSL) and high at plains. Similarly, monoterpenes like thujene, α-pinene, sabinene, limonene, α-phellandrene and linalool were relatively high at higher altitudes compared to plains. Telci *et al.* (2010) also observed higher trans-β-caryophyllene and germacrene D, in warmer climate and higher d-limonene and β-phellandrene in temperate climate in spearmint.Llusià *et al.* (2006) reported that maximum concentration of foliar volatile terpenes in four Mediterranean woody species were found in the coldest periods and minimum concentrations in the summer. In general, concentrations increased when soil moisture increased and decreased when air temperature increased. These studies show that climate change alters volatile oil constituents and hence the influence of climate in determining the quality of the produce.

3. Small Cardamom (*Eletteria cardamomum* Maton)

Small cardamom or the true cardamom belongs to the family Zingiberaceae. The crop is grown in a well distributed rainfall of 1500-2500 mm, temperature range of 15-25°C and an altitude of 600 -1200 m above MSL. Cardamom is a shade loving plant. In India, cultivation of cardamom is mainly restricted to Kerala, Karnataka and Tamil Nadu which is due to the very specific climatic requirement of the crop. It is very sensitive to climate change.

3.1. Climatic Influence on Cardamom Production

Most of the cardamom in India is grown in Indian cardamom hills of Kerala. Hence, studies on climate change in cardamom is generally based on this region. Spatial and temporal variations in air temperatures (maximum and minimum), rainfall and relative humidity were evident across stations in Indian cardamom hills. The mean air temperature increased significantly during the last 30 years. December and January showed greater warming across the stations. Rainfall during the main monsoon months (June–September) showed a downward trend (Murugan *et al.*, 2012a). The total number of rainy days has increased. The rainfall parameters had positive correlation with production of cardamom with significant relationship for number of rainy days (Murugan *et al.*, 2000). Cardamom productivity increased in the cardamom hills irrespective of the variety during the study period (1987 to 2007) indicating that warming may have positive influence on cardamom productivity (Murugan *et al.*, 2012). But apart from warming, crop management practices may also have influence on productivity.

Significant increasing trend was observed for minimum temperature than maximum temperature and this had caused decline in diurnal temperature. Both winter and summer monsoon rainfall as well as high relative humidity had a positive influence on the yield of cardamom. The variability of monthly mean precipitation is high for May, December and January under AR4 climate scenario (Murugan *et al.*, 2012b). But, the sustainable yield of cardamom may be possible only when the winter and summer rainfall variabilities are minimal. Increasing trend of soil temperature from 0-10 cm depth was recorded, which can cause considerable negative implications for sustainable cardamom production both in terms of reduced soil moisture availability and altered pest population dynamics (Murugan *et al.*, 2012b). Many studies have shown that phenology of crop is influenced by temperature and rainfall. Cardamom which was seasonal (June to December) in flowering up to early1990s is now flowering throughout the year which lead to increased harvests from five to nine per year and hence the productivity. This change from seasonal to year round flowering habit indicates the influence of climate change on cardamom phenology (Murugan *et al.*, 2012).

Rao *et al.* (2008) reported that Southwest monsoon and annual rainfall showed declining trends from 1951 onwards at rates of 5.2 and 5.6 mm/year, respectively in the humid tropics. However, the occurrence of floods and droughts, as evident in 2007 (floods due to a 41 per cent excess in monsoon rainfall) and the summer of 2004 (drought due to no significant rainfall from November 2003 to April 2004), is likely to increase and crop high losses are expected. Climate change in addition to deforestation will affect these thermo sensitive crops (cardamom, tea, black pepper etc.) as these are grown under the influence of typical forest and agricultural ecosystems. Deforestation, shift in cropping systems, decline in wetlands, and depletion of surface and groundwater resources may aggravate the adverse effects of floods and drought on crops.

The spatial and temporal distribution and proliferation of pests is determined, to a large extent, by climate, because temperature, light and water are the major factors controlling the growth and development of pests (Rosenzweig *et al.*, 2001). In cardamom hills, since 2000, the number of pesticide sprays has been significantly increased, and at present, 15–18 rounds of pesticide sprays are given (as against 7-8 rounds until 1990). But there was no great increase in the frequency of cardamom damage by major insect pests like thrips and borers indicating the involvement of more number of insect pests and diseases in damaging the crops. The incidence of many minor pests *viz.* insects and disease pathogens has increased in the recent years along with warming. Increased frequency of break period during monsoon seasons (wet and dry spells) as observed in Pampadumpara station might favour the development of dry rot during dry spell and wet rot during wet spell (Murugan *et al.*, 2012). The warming trend coupled with frequent wet and dry spells during the summer is likely to have a favorable effect on insect pests and disease causing organisms thereby pesticide consumption can go up both during excess rainfall and drought years. The incidence of many minor insect pest and disease pathogens has increased in the recent years along with warming (Murugan *et al.*, 2012a).

Differential climatic variability was observed between the two cardamom hot spots (Coban (Guatemala) and Pampadumpara (India). Indian cardamom hot spot showed higher variability than that of Guatemalan hot spot. The Indian spot had higher diurnal temperature range and had greater variability in rainfall amounts and pattern than the Guatemalan spot. Significant increases in major climatic elements were observed for both the spots. Insect pest and disease incidence levels were higher for the Indian hot spot compared to the Guatemalan region. Increased and higher productivity levels of cardamom were reported for Guatemala. Indian cardamom hot spot also showed increasing productivity (Murugan *et al.*, 2015)

4. Conclusions

Both black pepper and cardamom are generally grown under the cover of forest trees and are adapted to shade. Excess light is not beneficial for both the crops. Climatic parameters play a crucial role in production and productivity of these crops. Climate change is likely to bring down black pepper production while the available data supports that it may be an advantage in boosting cardamom production. Tmin is found to be more critical in affecting the production and quality as compared to the Tmax. Climate change is likely to alter both the physical and intrinsic quality of black pepper. Volatile oil constituents are more sensitive to climate change and the main pungent principle piperine is also reported to be affected. Increased frequency of climate extreme events such as drought, flood, high temperature and RH etc. may lead to increased disease incidence and severity and change in population dynamics of insect pests. Incidence of many minor pests and diseases with warming is also evident in cardamom.

References

Anandaraj, M. and Sarma., Y.R. (1994) Biological control of black pepper diseases. Indian Cocoa, *Arecanut and Spices Journal*, 18(1), 22-23.

Ankegowda, S.J., Venugopal, M.N., Krishnamurthy, K.S. and Anandaraj, M. (2011) Impact of basin irrigation on black pepper production in coffee based cropping system in Kodagu District, Karnataka. *Indian Journal of Horticulture*, 68(1), 71-74.

Arasumallaiah, L., Krishnamurthy, Y. L. and Krishnappa, M. (2008). Role of epidemiology on the incidence, development and spread of *Phytophthora capsici* Beon. causing foot rot disease of black pepper in malnad regions of Karnataka. *Environment and Ecology*, 26 (3A), 1427-1431.

Biju, C. N. and Praveena, R. (2015). The black pepper-*Colletotrichum* host-pathosystem: Biology, epidemiology and management. In: Krishnamurthy K S, Biju C N, Jayashree E, Prasath D, Dinesh R, Suresh J and Nirmal Babu K (Eds) 2015. Souvenir and Abstracts, National Symposium on Spices and Aromatic Crops (SYMSAC VIII): Towards 2050 - Strategies for sustainable spices production, Indian Society for Spices, Kozhikode, Kerala, India. 259p.

DaMatta, F.M., Cochicho, J.D. and Ramalho, G. (2006). Impacts of drought and temperature stress on coffee physiology and production: a review. Braz J Plant Physiol, 18(1),55–81.

De Waard, P. W. F. (1969). Foliar diagnosis, Nutrition and yield stability of black pepper (*Piper nigrum* L.) in Sarawak. Communication No. 58. Department of Agricultural Research, Koninklijk Instituut voor de Tropen, Amsterdam.

Jayasekhar, M. and Muthusamy, M. (1999). Influence of weather parameters on the incidence of foot rot of black pepper. *Madras Agricultural Journal*, 86 (4), 344-346.

John Koshy, Shankar, M. and Sudhakaran, K.V. (1999). Seasonal climatic influence in pepper production – Idukki district. Spice India, 12(12), 2-3.

Hao ChaoYun, Fan Rui, Ribeiro, M. C., Tan LeHe, Wu HuaSong, Yang JianFeng, Zheng WeiQuan and Yu Huan. (2012). Modeling the potential geographic distribution of black pepper (*Piper nigrum* L.) in Asia using GIS tools. *Journal of Integrative Agriculture*, 11 (4), 593-599.

Kandiannan, K., Utpala parthasarthy, Krishnamurthy, K. S., Thankmani, C. K., Srinivasan, V. and Aipe, K. C. (2011a). Modeling the association of weather and black pepper yield. *Indian Journal of Horticulture*, 68 (1), 96–102.

Kandiannan, K., Thankamani, C.K., Krishnamurthy, K.S. and Mathew, P.A. (2011b). Monthly rainfall trend at high rainfall tract of northern agro-climatic zone in Kerala. In: National Seminar on Recent Trends in Climate and Impact of Climate Change on South-West India. 11 October 2011. Department of Physics, St Joseph's College, Devagiri, Calicut.

Kandiannan, K., Thankamani, C. K. and Mathew, P. A. (2008). Analysis of rainfall of the high rainfall tract of northern agro-climatic zone of Kerala. *Journal of Spices and Aromatic Crops*, 17, 16–20.

Kannan, K., Devadas, V.S. and George Thomas, C. (1988). Effect of weather parameters on the productivity of coffee and pepper yield in Wynad. pp. 147–151. In: Agrometeorology of Plantation Crops, (Eds.) GSLVP Rao and RR Nair). Kerala Agricultural University, Thrissur.

Krishnamurthy, K.S., Kandiannan, K., Sibin, C., Chempakam, B. and Ankegowda, S.J. (2011) Trends in climate and productivity and relationship between climatic variables and productivity in black pepper (*Piper nigrum* L.). *Indian Journal of Agricultural Sciences*, 81 (8), 729–33.

Kumar, T. P. and Nair, M. R. G. K. (1987). Effect of some planting conditions on infestation of black pepper by Longitarsus nigripennis Mots. *Indian Cocoa, Arecanut and Spices Journal*, 10 (4), 83-84.

Llusia, J., Peñuelas, J., Alessio, G.A. and Estiarte, M. (2006). Seasonal contrasting changes of foliar concentrations of terpenes and other volatile organic compounds in four dominant species of a Mediterranean shrubland submitted to a field experimental drought and warming. *Physiologia Plantarum*, 127(4), 632-649.

Murugan, M., Raj, N. M. and Joseph, C. R. (2000). Changes in climatic elements and their impact on production of cardamom (*Elettaria cardamomum* Maton) in the Cardamom Hills of Kerala, *India. Journal of Spices and Aromatic Crops*,9 (2), 157-160.

Murugan, M.,Shetty, P. K., Raju Ravi, Aavudai Anandhi and Rajkumar, A. J. (2012). Climate change and crop yields in the Indian Cardamom Hills, 1978-2007 CE. Climatic Change, 110 (3), 737-753.

Murugan, M.,Shetty, P. K., Anandhi, A. and Ravi, R. (2012b) Present and Future Climate Change in Indian Cardamom Hills: Implications for Cardamom Production and Sustainability. *British Journal of Environment and Climate Change*, 2(4), 368-390.

Murugan, M.,Anandhi, A., Ravi, R.Dhanya, M. K. and Deepthy,K. B. (2015). Climate change in the cloud forest cardamom hot spots in relation to cardamom productivity in Guatemala and India. In: Krishnamurthy K S, Biju C N, Jayashree E, Prasath D, Dinesh R, Suresh J and Nirmal Babu K (Eds) 2015. Souvenir and Abstracts, *National Symposium on Spices and Aromatic Crops* (SYMSAC VIII): Towards 2050 - Strategies for sustainable spices production, Indian Society for Spices, Kozhikode, Kerala, India. 259p.

Parthasarathy, V.A., Kandiannan, K., Utpala Parthasarathy and Ankegowda, S.J. (2010). Climate change and spices production. **In:** *19th Biennial Symposium on Plantation Crops (Placrosym XIX), 7-10, December 2010. Rubber Research Institute of India*, Kottayam.

Pillay, V. S., Sasikumaran, S. and Ibrahim, K.K. (1988). Effect of rainfall pattern on the yield of black pepper. **In** : *Agrometeorology of Plantation Crops*. pp. 152-159. Kerala Agricultural University, Trichur.

Radhakrishnan, V.V., Madhusoodanan, K.J., Kuruvilla, K.M. and Vadivel, V. (2002). Production technology for black pepper. *Indian Journal of Arecanut, Spices and Medicinal Plants*, 4(2), 76-80.

Ramachandran, N., Sarma, Y.R. and Anandaraj, M. (1998). Effect of climatic factors on Phytophthora leaf infection in black pepper grown in Arecanut- Black pepper mixed cropping system. *Journal of Plantation Crops*, 16(2), 110-118.

Ridley, H.N. (1912). Pepper. In : Spices pp. 239 - 312. Macmillan and Co. Ltd. London.

Rao, G.S.L.H.V.P., Mohan, H.S.R., Gopakumar, C.S. and Krishnakumar, K.N. (2008). Climate change and cropping systems over Kerala in the humid tropics. *Journal of Agrometeorology*, 10 (Special Issue 2), 286-291.

Rao, G.S.L.H.V.P., Kesava Rao, A.V.R., Krishnakumar, K.N. and Gopakumar, C.S. (2009). Impact of climate change on food and plantation crops in the humid tropics of India. In : ISPRS Archives XXXVIII-8/W3 *Workshop Proceedings: Impact of Climate Change on Agriculture*. Space Applications Centre (ISRO), Ahmedabad, India.

Rosenzweig, C., Iglisias, A., Yang, X.B., Epstein, P.R. and Chivian, C. (2001). Climate change and extreme weather events—implications for food production, plant diseases, and pests. *Glo Change Hum Health*, 2(2),90–104.

Shamarao, J. andSiddaramiah, A. L. (2002). Influence of weather forecasters on the epidemiology of foot rot of black pepper in Karnataka. *Indian Journal of Agricultural Research*, 36 (1), 49-52.

Sivaraman, K., Kandiannan, K., Peter, K.V. and Thankamani, C.K. (1999). Agronomy of black pepper (*Piper nigrum*L.) - a review. *Journal of Spices and Aromatic Crops*, 8, 1-18.

Sruthi, D., John Zachariah, T., Leela, N. K. and Jayarajan, K. (2013). Correlation between chemical profiles of black pepper (*Piper nigrum* L.) var. Panniyur-1 collected from different locations. *Journal of Medicinal Plants Research*, 7(31), 2349-2357.

Sunil, K. M., Devadas, V. S., Sreelatha, A. K. and George, S. P. (2010). Effect of weather parameters on the yield of small cardamom (*Elettaria cardamomum* Maton.). *Indian Journal of Arecanut, Spices and Medicinal Plants*, 12 (3), 6-8.

Suparman, U. (1998). The effect of El-Nino and La-Nina on the production of white pepper in Bangka, Indonesia. *International Pepper News Bulletin*, XXII (3 and 4), 44-45.

Telci, I., Demirtas, I., Bayram, E., Arabaci, O. and Kacar, O. (2010). Environmental variation on aroma components of pulegone/piperitone rich spearmint (*Mentha Spicata* L.) *Industrial Crops and Products*, 32(3), 588-592.

Thuy, T. T. T., Yen, N. T., Tuyet, N. T. A., Te, L. L. and Waele, D. de. (2012). Population dynamics of *Meloidogyne incognita* on black pepper plants in two agro-ecological regions in Vietnam. *Archives of Phytopathology and Plant Protection*, 45 (13), 1527-1537.

Utpala Parthasarathy, Parthasarathy, V.A. and Jayarajan, K. (2008). A temperature sensitivity analysis on plantation crops: A GIS approach. *Journal of Plantation Crops*, 36, 372-374.

Wahid, P. and Sitepu, D. (1987). Current status and future prospect of pepper development in Indonesia. Food and Agricultural Organization, Regional Office for Asia and Pacific, Bangkok, Thailand.

Sruthi, D., John Zachariah, T., Leela, N. K. and Jayarajan, K. (2013). Correlation between chemical profiles of black pepper (*Piper nigrum* L.) var. Panniyur-1 collected from different locations. *Journal of Medicinal Plants Research*, 7(31), 2349-2357.

2017, Impact of Climate Change on Plantation Crops Pages 87–99
Editors: **K.B. Hebbar, S. Naresh Kumar & P. Chowdappa**
Published by: **ASTRAL INTERNATIONAL PVT. LTD., NEW DELHI**

Chapter 6

Cashew

T.R. Rupa

1. Introduction

Climate change that is defined by high atmospheric carbon dioxide (CO_2) concentrations (>400 ppm), increasing air temperatures (2-4 °C or greater), significant or abrupt changes in daily seasonal, and interannual temperature, changes in the wet/dry cycles, intensive rainfall and/or heavy storms, extended periods of drought, extreme frost, and heat waves and increased fire frequency, is expected to significantly impact terrestrial systems, soil properties, surface water and, stream flow (Patterson *et al.*, 2013), ground water quality, water supplies, and terrestrial hydrologic cycle (Pangle *et al.*, 2014). Agricultural productivity is sensitive to two broad classes of climate-induced effects: (i) direct effects from changes in temperature, precipitation, or carbon dioxide (CO_2) concentrations, and (ii) indirect effects through changes in soil moisture and the distribution and frequency of infestation by pests and diseases (IPCC 1996; 2001). Climate change impacts on agriculture are being witnessed all over the world, but countries like India are more vulnerable in view of the high population depending on agriculture and excessive pressure on natural resources. The warming trend in India over the past 100 years (1901 to 2007) was observed to be 0.51°C with accelerated warming of 0.21°C per every 10 years since 1970 (Krishna Kumar, 2009). Climate change projections made up to 2100 for India signify an overall increase in temperature by 2-4 °C with no substantial change in precipitation quantity. However, different regions are expected to experience differential change in the amount and distribution of rainfall that is likely to be received in the coming decades. It is projected that some parts of country will receive higher amount of rainfall. Another significant aspect of climate change is the increase in the frequency of occurrence of extreme events such as droughts, floods and cyclones. Climate change impacts are likely to vary in different parts of the country. Parts of western Rajasthan, Southern Gujarat, Madhya Pradesh,

Maharashtra, Northern Karnataka, Northern Andhra Pradesh, and Southern Bihar are likely to be more vulnerable in terms of extreme events.

Climate change is a result from emission of greenhouse gases (GHGs), for example, carbon dioxide (CO_2), methane (CH_4), and nitrous oxide (N_2O,) etc. in the past century that will cause atmospheric warming (IPCC, 2007). Agriculture is considered to be one of the major anthropogenic sources of atmospheric greenhouse gases (Lal, 2000). The atmospheric concentration of CO_2, CH_4 and N_2O and other GHGs have increased since the industrial revolution (1750 A.D.) due to natural and anthropogenic activities. The atmospheric concentration of CO_2 has increased from 280 parts per million by volume (ppmv) in 1750 to 394 ppmv in 2012 and is currently increasing at the rate of 1.9 ppmv year^{-1} (NOAA, 2012). Atmospheric CH_4 concentration has increased from about 715 to 1826 parts per billion by volume (Ppbv) in 2012 over the same period and is increasing at the rate of 7 ppbv year^{-1}. Similarly, the atmospheric concentration of N_2O has increased from about 270 ppbv in 1750 to 325 ppbv and is increasing at the rate of 0.8 ppbv year-1 (lPCC, 2007; EPA, 2013). In order to compare emissions from different sources, the global warming potential (GWP) of each gas is equated to the GWP of CO_2, For example, the GWP of 1 tonne of CH_4 is 24.5 times more potent than 1 tonne of CO_2 over a 100 year period (IPCC, 2007).

All these expected changes will have adverse impacts on climate sensitive sectors such as cashew. In India, cashew is habitually grown as a rainfed crop in ecologically sensitive areas for example coastal belts, hilly areas and areas with high rainfall and humidity, and therefore its performance primarily be influenced by climate. Cashew adapted well in west and east coast regions and subsequently spread to hilly and plain regions of Karnataka, Tamil Nadu, Gujarat, Chhattisgarh, and NEH States. There are both threats and opportunities for cashew sector with climate change, the key is in understanding the likely impacts and then managing the risks. Any variability in climate has direct impact on reproductive phase of cashew.With climate variability, there will be various impacts on cashew production such as reduction in yields, variation in flowering, fruit setting, nut development and kernel quality, incidence of pests and diseases and water stress. The increase in weather extremes like torrential rains, heat waves, mid waves and floods, besides year-to-year variability in rainfall affects cashew productivity considerably and leads to stagnation/decline in production across various agro-climatic zones. The challenge for cashew crop will be to improve yields in marginal lands under rainfed conditions where the harsh environment strongly limits crop growth, productivity and quality of the produce. Hence, adaptation and mitigation strategies are required to be formulated to minimize the climate change related impacts on cashew.

2. Climatic Requirement

Cashew, a tropical nut crop can be found growing between 28°N and 28°S latitude. In India, cashew is grown in the coastal belt between 8°N and 28°N. Flowering time depends on the latitude. In Brazil and Tanzania, peak flowering is between August and September. The highest flowering occurs in October in Mozambique, while in Philippines it is March. In the west coast of India, flowering is from October to March. Whereas in the east coast of India, flowering is delayed by about 2-3 months. However, the crop is ready for harvest in summer, both in the north and south of the equator. As cashew grows at an elevation ranging from 0 to 1000 m above mean sea level (MSL) the The phenology of cashew is very much influenced by the altitude of the region. However, the productivity is the highest up to the altitude of 750 m above MSL. Low temperatures at higher altitudes have adverse effect on the crop. Flowering and fruiting are delayed irrespective of latitude at higher altitude. Low altitude areas with a mean rainfall of 1500 to 2000 mm is excellent for cashew. Damage to young trees or flowers occurs below the minimum temperature of 7° C and above the maximum of 45° C. Only prolonged cool temperatures will damage mature trees; cashew can survive temperatures of about 0° C for a short time (Ohler, 1979). Environments with maximum temperature ranging from 28 to 32 °C, minimum winter temperature around 19 °C and 70 to 80 per cent relative humidity are satisfactory for better output. Frost is detrimental to the crop Mandal (1992) (Table 6.1).

Table 6.1: Environmental Rating for Growing Cashew

Parameter	*Very Good*		*Good*	*Fair*	*Poor*
	Class I	*Class II*	*Class III*	*Class IV*	*Class V*
Altitude (m)	20	20-120	120-450	450-750	
Rainfall (mm/year)	1500-2000		1300-1500	1100-1300	900-1100
Proximity to sea (km)	<80	80-160	160-240	240-320	
Maximum temperature (°C)	28-32	32-33	33-34	34-35	
Minimum temperature (°C)	19	18-19	17-18		15-17
Humidity (per cent)	70-80	65-70	60-65		50-60
Occurrence of frost	None	None	Very rare	Once in 5 years	

Though cashew can tolerate wide range of temperature but the optimum monthly temperature is between 24 °C and 28 °C. Cashew grows in the semi-arid regions like northern Mozambique where daily maximum temperatures exceeds 40 °C and in Assam, cashew survives up to 7 °C. It has been reported that cashew cultivation is not economical in regions where annual temperature falls below 20 °C for prolonged periods. In major cashew growing regions, the mean daily maximum temperature vary between 25 °C and 35 °C and the mean daily minimum temperature vary between 15 °C and 25 °C. The productivity of cashew is higher in regions where the mean annual temperature ranged from 22.5 °C to 27.5 °C and the minimum temperature ranges from 10 °C to 22 °C. The productivity is lower in regions where the minimum temperature drops below 10 °C. For flowering, cashew requires mild winter, that is low minimum surface air temperature ranging between 16 °C and 20 °C coupled with more dew nights. Cashew is highly sensitive to light and produces more foliage, flowers and fruits on braches exposed to sunlight than on the shaded branches. Bright sunshine (>9 h/day) with moderate dry weather is good for flowering. In Togo the optimum sunshine is held to be 2464 h/year with 1285 h in the flowering/fruit set period (November-March) which is only found in the north and centre of the country. In Brazil the optimum sunshine lies between 1500 and 2000 h/year, while the precise sunshine for Venezuela is considered to be an average of 2000-2400 h/year.

Cashew can survive under very high and low rainfall conditions. Although cashew can tolerate drought conditions but for proper vegetative development and regular fruit-setting, the average annual rainfall of 1,300 to 2,500 mm is considered to be suitable. The average annual rainfall distribution in cashew areas ranged from low rainfall (300-600 mm in Gujarat) to high rainfall (2700-3000 mm in west coast and NEH region) but the productivity is highest in regions with a mean annual rainfall distribution of 600-1500 mm. Cashew needs a clearly defined dry season of at least 4 to 5 months. A dry spell from January to May with occasional light summer rains ensures better cashew production. A well distributed North-East monsoon rainfall of about 500 mm during September to December and about 100 mm during February to April are good for successful production of cashew. Light rains during flowering do not affect the production but heavy rains during flowering affects the yield. Year to year variation in time of flowering of a variety is common even under uniform cultural and management practices. It signifies the influence of weather factors on flowering behavior of cashew. Unusual heavy rainfall between January and March may encourage high incidence of pest like tea mosquito bug (TMB) and reduce yield and quality. Cashew is highly sensitive to high relative humidity (more than 80 per cent) during the harvesting season. High relative humidity adversely affects the nut quality.

3. Climate Change Impacts

The rainfed cashew crop is highly sensitive to changes in climate and weather, especially during reproductive phase. High temperature (>34.4 °C) and low RH (<20 per cent) during afternoon cause drying of flowers. The maximum temperature, humidity and rainfall are the major climatic factors that influence the productivity of cashew. According to Prasada Rao *et al.* (2010), the maximum temperature plays a crucial role on nut size and kernel weight of cashew during the nut development stage. Haldankar *et al.* (2003) reported that the relative humidity during pre-flowering phase is the main factor which explains yield variation in cashew plantations. The unusual rains between November and December inordinately delay reproductive phase of the late flowering varieties. Unseasonal rainfall and heavy dew during flowering and fruiting intensify the incidence of pests and diseases as well as deterioration of nut quality. For example, a rainfall of about 201 mm received at ICAR-DCR experimental farm during 15-25 March, 2008 adversely affected the yield and quality of nuts. Large quantity of nuts which could not be picked up in time germinated in the field itself. The nuts could not be dried in time due to non-availability of good sunshine hours resulting in poor quality of nuts. The surplus moisture in raw cashewnut damages the kernel inside by changing its colour from white to cream. The nut yield losses due to unseasonal rains were estimated to be between 50 and 65 per cent in March, 2008.

Cashew genotypes vary noticeably in their heat units (day °C) requirement. Early variety (Anakkayam-1) requires only 1953 heat units for reproductive phase, while late variety (Madakkathara-2) requires 2483 heat units. Cashew kernel weight is positively correlated with heat units especially for mid and late varieties. Continuous rains without critical dry spells and late winter rains delay the bud break in cashew. A dry spell of 7 days is usually necessary 30 days prior to the bud break. Late and extended winter rains reduce the number of bright sunshine hours invariably which results in delaying of bud break and better availability of soil moisture during flowering (December and January). For example incessant rains received at Pilicode (Kerala) until November in 1998 delayed the bud break. The delay in bud break was prominent in early varieties. In late varieties, bud break was normal, because the required dry spell of 7 days was met 22-26 days before bud breaking (Prasada Rao *et al.*, 2001).

Cashew is infested by numerous insect pests, thereby limiting the production considerably. One of the main reasons for reduction of cashew nut yield is the occurrence of an important sucking insect pest, tea mosquito bug (TMB) (*Helopeltis antonii* Signoret) during the cropping season. The production loss from the TMB alone is estimated to be about 30 per cent.This pest has got potential to cause 100 per cent loss in yield in severe cases. The incidence and severity of the pest is highly dependent on climate and weather factors. The minimum temperature plays a vital role in the incidence of pest population and is negatively correlated with the TMB pest incidence (Godse *et al.*, 2005). The favourable minimum temperature for TMB incidence ranges between 13-18 °C. Low temperature (12 °C) is antagonistic for pest build up.

Apart from the damage caused by the infestation of TMB, infection of the panicles by the fungal pathogens, *viz.*, *Colletotrichum gleosporoides* and *Gleosporium mangiferae* have been reported to cause drying up of young shoots, inflorescence and immature nuts in cashew. The characteristic symptom is the drying of floral branches. The symptoms appear as minute water soaked lesions on the main rachis and secondary rachis. The incidence is very severe when cloudy weather prevails. The incidence of this disease is being reported from different new locations in which it was not prevalent earlier. High relative humidity in forenoon during December - February both in 1997 and 1998 and the minimum temperature of 18-20 °C were found to be favourable for sporulation by fungi. A significant increase in dewfall was one of the most important factors which favoured the growth, sporulation and spread of fungi. Cloudiness leading to low bright sunshine hours (2 h/day) followed by dewfall triggered the growth, sporulation and spread of fungal pathogens causing inflorescence blight during 1998-99 in Kannur and Kasaragod districts (Prasada Rao, 2002).

4. Extreme Events of Weather Impacts

Changes in temperature and precipitation patterns together with occurrence of extreme events due to climate change are a major threat to future cashew sector. The Tsunami in coastal Tamil Nadu on 26 December, 2004 harshly affected cashew crop. Most of the standing crop was inundated with salty sea water ingressing in the mainland. Salinity is a major environmental stress and is a substantial constraint to crop production. Salt affected soils are the soils that contain considerable amounts of soluble salts and or sodium on the exchange complex. The high amount of soluble salts (in saline soils) and of sodium on the exchange complex (in sodic soils) hinder crop growth and have rendered them barren (Gupta and Abrol, 1990). In saline soil the available moisture range is low and crop has to spend extra energy to extract water from the soil because of high osmotic potential of the soil solution. These also adversely affect water and nutrient availability. Most of the cashew varieties are sensitive to salinity. Rise in sea water level due to climate change conditions may adversely affect the cashew plantation. Electrical conductivity of 1.48 dS m^{-1} in irrigation water is a threshold tolerance for precocious cashew during the initial growth (Carneiro *et al.*, 2002).

'Thane' cyclone with a very high wind speed created havoc in Cuddalore district of Tamil Nadu and Pondicherry on 30 December 2011. The entire cashew area of Panruti, Cuddalore and Kurinjipadi taluks of Cuddalore district were totally devastated. The extent of damage to cashew trees in Vridhachalam taluk was 60-70 per cent due to decrease in the wind speed away from the coast. A team of scientists of AICRP-Cashew Centre at Regional Research Station (Tamil Nadu Agricultural University), Vridhachalam visited and surveyed the 'Thane' cyclone affected area and the recommendations of the team to cope up with the situation include i) Removal of uprooted trees for new planting, ii) Mass replanting programme, iii) Government support for laying out a borewell for every 10 acres for establishment of the new plants, iv) Government support programmes for raising annual crops like pearl millet, blackgram, horsegram, groundnut and kodo millet upto five years etc.

5. Adaptation and Mitigation Strategies

Adaptation strategies can work in two ways, by reducing vulnerability (susceptibility) to changing condition, or by increasing resiliency (to recovery) by reducing suffering during and immediately after the events (Bedsworth and Hanak, 2010). There are two basic counter measures regarding climate change, mitigation and adaptation (Mimura, 2010). Mitigation includes different measures, undertaken to reduce GHGs emission and increasing absorption of GHGs, which already emitted, and adaptation indicates different potential measures to protect the adverse effects of climate change. Some of the adaptation measures in common that the cashew sector can undertake to cope with future climate change include: Changing planting dates, planting different varieties, developing new drought and heat resistant varieties, salt tolerant varieties, adoption of intercropping, using sustainable fertilizer, weed management, more use of water harvesting techniques, supplemental irrigation, drip irrigation, fertigation, better pest and disease management etc. Climate change mitigation refers to efforts to reduce/ prevent emission of GHGs. Protecting natural carbon sinks like forests and oceans, or creating new sinks through silviculture or green agriculture are also elements of mitigation. In order to mitigate the ill effects of climate change on soil quality and to protect the soil and land resource, it is important to give more focus on adoption of soil and water conservation practices, mulching, cashew leaf litter retention and recycling, addition of animal based manures etc.

5.1. Soil and Water Conservation Practices

Major cashew growing area in India is under undulated topography (steep slopes). As a rainfed crop, cashew experiences severe moisture stress from January to May. It adversely affects flowering and fruit set. The main aim of soil and water conservation practices is to reduce or prevent either water erosion or wind erosion, while providing the desired moisture for crop growth/production. Erosion by water is the most serious soil degradation problem in the humid tropical and sub tropical India. In India, soil erosion has been taking place at an average rate of 16.35 t ha^{-1} $year^{-1}$ (Dhruvanarayana and Rambabu, 1983), totaling an annual loss of 5334 million tonnes, nearly 29 per cent of the total eroded soil is permanently lost into the sea, and nearly 10 per cent is deposited in reservoirs. Remaining 61 per cent of the eroded soil is transferrable from one place to another. The soil erosion by water and wind result in loss of top soil and depletion of soil organic matter. Depletion of organic matter in soil discourages the activity of soil microflora responsible for decomposition of organic matter to enrich soil fertility. It becomes imperative for the land use planners to adopt appropriate conservative measures.

In order to harvest the rainwater and to make it available to the cashew plant during critical period, an *in situ* soil and water conservation and rainwater harvesting are very important. Research work carried out at ICAR-DCR indicated that soil and water conservation measures such as modified crescent bund (at 2 m radius having a crescent shaped bund of 6 m length, 1m width and 0.5 m height on the upstream of the plant so that a trench of 6 m length and 50-75 cm deep will be formed while making the bund) and coconut husk burial (Trenches of size 5 m length, 1 m width and 0.5 m depth in the middle of 4 plants with coconut husk buried) were found effective in terms of higher yield, reduction in annual runoff and conserving moisture (Rejani and Yadukumar, 2010). It was shown that reduction in peak runoff and increase in recession time and groundwater recharge due to soil and water conservation practices (Deshmukh *et al.*, 1992). According to Badhe and Magar (2004), trapezoidal shaped staggered trenches (230/ha) having dimensions of 4.5 m length, 0.60 m top width, 0.30 m bottom width and 0.30 m depth were effective for reducing runoff and conservation of soil and nutrients. Mane *et al.* (2009) demonstrated that continuous contour trench (0.50 m x 0.60 m) is the best soil conservation practice for cashew on areas having 7 to 8 per cent slope.

5.2. Mulching

Cashew is often planted in areas which are totally dry and less fertile. Under such situations mulching is highly useful to conserve the soil moisture for a long period, protecting the soil from erosion and maintaining, restoring and improving soil organic carbon status. The basin area of cashew plants can be mulched either with green leaves or dry leaves and weeds soon after planting. Black polythene mulch was helpful to conserve soil moisture (Nawale *et al.*, 1985). Using coconut coir pith as soil mulch in cashew plantations resulted in 14.15 per cent more water retention and suppression of weeds to an extent of 73.52 per cent (Kumar *et al.*, 1989). Formation of terrace and crescent bund and mulching the base area with cashew leaf litter and other jungle growth available in the garden are helpful.

5.3. Green Manuring

Growing green manure crops like *glyricidia*, *sesbania*, sunhemp and cover crops between two rows of cashew have potential to improve soil moisture retention capacity and nutrient content. According to research studies at ICAR-Directorate of Cashew Research, Puttur, Karnataka, higher soil moisture content was observed in cashew orchard with *glyricidia* (17.0 to 18.6 per cent dry basis), sunhemp (17.8 to 18.3 per cent dry basis), sesbania (15.5 to 18.2 per cent dry basis) compared to control (15.5 to 17.0 per cent dry basis). The nutrient addition to soil was about 186 kg N, 23.6 kg P_2O_5 and 126.2 kg K_2O and 141 kg N, 17.9 kg P_2O_5 and 162.3 kg K_2O/ha through *glyricidia* and *sesbania*, respectively.

5.4. Site Specific Nutrient Management (SSNM)

It is another approach with potential to mitigate effects of climate change. The SSNM is critical for GHGs mitigation so as to reduce input cost and enhance nutrient use efficiency considerably. This approach facilitates cashew grower to invest only on deficient nutrients and omit nutrient application which was in sufficient range in soils. Various benefits of SSNM practice include lowering in input cost, higher nutrient use efficiency, and reducing GHGs particularly N_2O. Application of fertilizers particularly N during water stress escalates water stress problem further as higher N improves leaf canopy which results higher evapotranspiration. Using nitrification inhibitors, and fertilizer placement practices need further consideration for GHGs mitigation. Management of soil in combination with optimum soil moisture is essential to protect the plants during weather aberrations and overall reduction of CO_2, N_2O and CH_4 from soil. For example, increased rainfall in regions that are already moist could lead to increased leaching of minerals, especially nitrates. Increases in fertilizer applications would be necessary to restore productivity levels. Placement of fertilizer materials and split application of nutrients into soil will substantially improve both nutrient and water use efficiency. However, KNO_3 foliar sprays during intermittent droughts results in the adaptation of plants by closing stomata.

5.5. Intercropping

Cashew takes 8 to 10 years for the canopy to cover the entire area under normal spacing. In hilly regions there is a possibility of soil erosion and weed growth during initial years of planting. Under such situations, intercropping not only minimizes erosion and conserves soil and moisture and also to realize higher returns from unit area during the early stages of cashew plantation. Growing field crops (groundnut, black gram and green gram), vegetables (cucumber and bottlegourd), tuber crops and fruit crops (pineapple), spices (turmeric, ginger and pepper) are suitable and profitable intercrops in cashew plantations. Ginger was recommended as an intercrop in the initial 3-4 years of cashew plantation in the west coast region. In east coast, crops like groundnut and cowpea are grown profitably under rainfed situations. Intercropped pineapple in cashew plantations has been adopted by farmers and resulted in 1.5 fold higher yield and net profit. Pepper vine and kokum are the other popular intercrops for enhancing the farm profitably as mixed crops grown in cashew orchards by trailing on to the stem of cashew in case of pepper and planting kokum in the middle of 4 cashew plants in west coast where rainfall is high.

5.6. Contingency Plan for Rainfall Deficit Management for Cashew

Cashew is planted under normal onset of monsoon during June to September. However, dry spells occur at various stages during the planting season resulting in early, mid-season and terminal droughts adversely impacting performance of the plants. The effect of drought is pronounced due to delays in onset of monsoon, prolonged breaks in monsoon and deficit rainfall. Contingent planning for rainfed cashew is important when the onset of monsoon is delayed. Small changes in climatic parameters can often be managed reasonably well by altering dates of planting. In case of normal onset followed by early season dry spell immediately after planting leading to poor crop establishment, gap filling, mulching, nutrient management etc. may be necessary. In addition, moisture conservation measures and life saving irrigation wherever possible are suggested. The irrigation through pitcher (hold pots) is recommended in dry land situations. In case of terminal drought in September to October due to dry spells or early withdrawal of the south west monsoon, crop management measures such as life saving or supplemental irrigation, if available, from harvested pond water or other sources are suggested as in the case of mid season drought wherever feasible.

Established plants survive even in adverse soil moisture conditions. The rainfall deficit or cessation of rains at early stage also adversely affects the nut yield particularly in late maturing varieties. In order to minimize the yield loss under such situations, one or two protective irrigations are highly beneficial. Supplemental irrigation of 200 L of water/plant once in 15 days during January to March from water collected in ponds through rain harvesting helps in flowering and nut development by improving the microclimate with increased humidity. It also leads to increased nut and kernel weight by reducing flower and nut drying to some extent. Drip irrigation during fruit development stages wherever water is available is helpful under drought situations to rainfed cashew crop. Normally for west coast of Dakshina Kannada, irrigation by drip at 20 L/tree/day for mature cashew plantations (10 to 15 years) is recommended. Research results indicated that yield can be doubled, if irrigated. Irrigation can be started after the commencement of flowering for better nut set, filling and yield. It has been reported that fertigation saved 50 per cent in the fertilizer requirement and doubled the cashew yield (Richards, 1993; Yadukumar and Mandal, 1994; Mishra *et al.*, 2008). To maintain the proper soil moisture regime, the harvesting of rain water and recycling them during deficit period is suggested. Moreover, adoption of soil conservation measures and installation of drip wherever water source is available will be helpful. Developing heat and drought tolerant cashew varieties by utilizing genetic resources that may be better adapted to new climatic and atmospheric conditions should be the long-term strategy.

In the recent past, continuous high rainfall in a short span leading to water logging. Heavy rainfall coupled with high speed winds in a short span are being experienced at various growth stages, of which the 2008 one was the most recent, significantly affecting the production of cashew. In addition to that, the amount and distribution of rainfall is becoming more and more erratic, causing greater incidence of drought and flood. The increase in frequency of heavy rainfall events in last 50 years over Central India points towards a significant change in climate pattern in India. Unseasonal rains received during harvesting season in Dakshina Kannada district of Karnataka caused considerable yield loss and deterioration of nut quality both in trees and on the ground. The contingency measures suggested are drying of collected matured fallen nuts immediately by sun drying/artificial drying in case there is not sufficient bright sun light.

The climatic change threats include likely increase of temperature, extreme weather conditions, increased water stress and drought, and desertification. All these may seriously increase the vulnerability of resource poor cashew farmers to global climate change. Policies that support crop insurance can provide protection to the farmers in the event of crop failure or lower production due to natural calamities. Some of the measures to cope up with climate change effects include weather based crop and an early warning system of environmental changes.

5.7. Pest and Disease Management

Tea mosquito bug (TMB) pest and inflorescence blight disease cause considerable damage to cashew. Incidence and severity of both are dependent on climate and weather factors. Need based sprays are recommended for TMB during most vulnerable periods of crops such as flushing, flowering and fruiting stage of the crop. Forewarning the outbreak of the dreaded inflorescence blight disease in west coast region of India is of utmost importance for its effective management. Proper training of cashew trees from early stages and timely pruning of side and criss-cross branches may ensure that microclimate is not congenial for build-up of sucking pests and inflorescence blight disease.

5.8. Carbon Sequestration

Cashew has substantial carbon sequestration potential for carbon sequester for mitigation of climate change and can also be grown in vast degraded/wasteland existing in cashew-growing regions. Soil carbon (C) has gained increased interest in the recent past owing to its importance in C sequestration studies and its potential impact on sustainable crop production. Carbon sequestration implies removing atmosphere carbon and storing it in natural reservoirs for extended periods (Lal, 2004; Srinivasarao *et al.*, 2013). Soil carbon sequestration is the process of transferring carbon dioxide from the atmosphere into the soil through crop residues and other organic solids, and in a form that is not immediately remitted. This transfer or "sequestering" of carbon helps off-set emissions from fossil fuel combustion and other carbon-emitting activities while enhancing soil quality and long-term agronomic productivity.

The C fixed in plants by photosynthesis and added to soil as above and below ground litter is the primary source of C in ecosystems (Warembourg and Paul, 1977). Although most carbon enters ecosystems via leaves, and carbon accumulation is most obvious when it occurs in above ground biomass, more than half of the assimilated carbon is eventually transported below ground via root growth and turnover and exudation of organic substances from roots. Numerous studies have convincingly proved that the inclusion of trees in the agricultural landscapes often improves the productivity of systems while providing opportunities to create carbon sinks.

Since C sequestration is an essential component of mitigation of green house gases, it is important to assess the sequestration potential of crops. It has been estimated that soil organic carbon sequestration potentially could offset about 15 per cent of the global CO_2 emission. Cashew has dense green leaves with good photosynthetic capacity and it can also be grown in high-density planting system. Cashew is suitable crop for C sequestration. It can be grown in vast degraded/ wasteland existing in cashew growing regions. Based on research undertaken at ICAR-DCR, it was found that, 7 years old trees of cashew genotype VTH-174 sequestered about 2.2 fold higher C under high density planting system (625 trees/ ha) as compared to normal density planting system (156 trees/ha). Carbon storage by cashew has been estimated as 32.25 and 59.22 t CO_2/ha at 5th and 7th years of growth, respectively under high density planting. The extent of C sequestered will depend on the amounts of C in standing biomass, age of the crop, tree density, variety etc.

6. Future Line of Work

- ☆ Development of cashew varieties resistant to abiotic stresses like temperature, drought, salinity, floods etc.
- ☆ Development of climate resilient agro-techniques in order to suit unfavourable abiotic stresses.
- ☆ Development of crop simulation models for cashew.
- ☆ Assessment of new pest and diseases in the scenario of climate change and development of cost effective, eco-friendly approaches for management of emerging pests.
- ☆ Mitigation and adaptation strategies to cope up with expected climate change impacts on cashew required to be streng

References

Badhe, V.T. and Magar, S.S. (2004). Influence of different conservation measures on runoff, soil and nutrient loss under cashew nut in lateritic soils of south Konkan region. *Indian Journal of Soil Conservation*, 32(2), 143-147.

Bedsworth, L.W. and Hanak, E. (2010). Adaptation to climate change: a review of challenges and tradeoffs in six areas. *Journal of the American Planning Association*, 76, 477-495.

Carneiro P.T., Fernandes, P.D. and Gheyi, H.R. (2002). Germination and initial growth of precocious dwarf cashew genotypes under saline conditions. Revista *Brasileira De Engenharia Agricola E Ambiental*,6(2), 199-206.

Deshmukh, M.T., Sawke, D.P., Borude, S.G., Hurni, H. and Tato, K. (1992). Effects of platform bench terraces on the growth and yield of mango and cashew grafts. Erosion, conservation and small scale farming report. Department of Agricultural Engineering, Konkan Agriculture University, Ratnagiri, India, pp. 477-482.

Dhruvanarayana, V.V. and Rambabu, (1983). Estimation of soil erosion in India. *Journal of Irrigation and Drainage Engineering*, 109, 419-434.

Godse, S.K., Bhosle, S.R. and Patil, B.P. (2005). Population fluctuation studies of tea mosquito bug on cashew and its relation with weather parameters. *Journal of Agrometeorology*, 7(1), 107-109.

Gupta, R.K. and Abrol, I.P. (1990). Salt affected soils - Their reclamation and management for crop production. *Advances in Soil Science*, 12, 223-275.

Haldankar, P,M., Deshpande, S,B., Chavan, V.G. and Rao, E.V.V.B.(2003). Weather associated yield variability in cashewnut. *Journal of Agrometeorology*,5(2), 73-76.

IPCC, (2007). Agriculture. In: Metz, B., Davidson, OR, Bosch PRo (Eds.), Climate Change 2007: Mitigation. Contribution of Working Group III to the Fourth Assessment Report of the Intergovernmental Panel on Climate Change. Cambridge University Press, Cambridge, United Kingdom/New York, NY, USA.

IPCC, (2001). Climate Change 2001. Impacts, Adaptation, and Vulnerability, Contribution of Working Group II to the Third Intergovernmental Panel on Climate Change. In: McCarthy, J.J; Canziani, O.F; Leary, N.A; Dokken, D.J. and White, K.S. (Eds), Cambridge University Press, Cambridge, UK.

IPCC, (1996). Climate Change 1995: Impacts, Adaptations and Mitigation of Climate Change: Scientific- Technical Analyses.

Krishna Kumar, (2009). Impact of climate change on India's monsoon climate and development of high resolution climate change scenarios for India. Presented at MoEF, New Delhi on October 14, 2009 (http: moef.nic.in).

Kumar, D.P., Subbarayappa, A., Hiremath, I.G., Khan, M.M. and Sadashivaiah (1989). Use of coconut coir-pith: a biowaste as soil mulch in cashew plantations. *The Cashew*, 3(3), 23-24.

Lal, R. (2000). World cropland soils as a source or sink for atmospheric carbon. *Advances in Agronomy*,71, 145-191.

Lal, R. (2004). Soil carbon sequestration impacts on global climate change and food security. Science, 304, 1623-1627.

Mandal, R.C. (1992). Cashew Production and Processing Technology. Agro Botanical Publishers (India), Bikaner, pp. 195.

Mane, M.S., Mahadkar, U.V., Ayare, B.L. and Thorat, T.N. (2009). Performance of mechanical soil conservation measures in cashew plantation grown on steep slopes of Konkan. *Indian Journal of Soil Conservation*, 37(3), 181-184.

Mimura, N. (2010). Scope and roles of adaptation to climate change. In: A. Sumi, K. Fukushi, and A. Hiramatsu, eds., Adaptation and Mitigation Strategies for Climate Change. Springer Tokyo Berlin Heidelberg New York. Ch.9.

Mishra, J.N., Paul, J.C. and Pradhan, P.C. (2008). Response of cashew to drip irrigation and mulching in coastal Odisha. *Journal of Soil and Water Conservation,* 7(3), 36-40.

Nawale, R.N., Sawke, D.P. and Salvi, M.J. (1985). Effect of black polyethylene mulch and supplemental irrigation on fruit retention in cashewnut. Cashew Causerie, 7(3), 8-9.

NOAA/ESRL. (2012). Available at http://www.esrl.noaa.gov/gmd/ccgg/trends.

Pangle, L.A., Gregg, J.W., McDonnell, J.J. (2014). Rainfall seasonality and an ecohydrological feedback offset the potential impact of climate warming on evapotranspiration and groundwater recharge. *Water Resources Research,*50, 1308-1321.

Prasada Rao, GS.L.H.V., Giridharan, M.P., Jayaprakash Naik, B., Gopakumar, C.S., Krishnakumar, K.N. and Xavier Tony (2001). Influence of weather factors on bud break, flowering and nut quality of cashew (*Anacardium occidentale* L.). *Indian Journal of Agricultural Sciences,*71(6), 399-402.

Prasada Rao, G.S.L.H.V. (2002). Climate and Cashew. AICRP on Agro Meteorology, Department of Agricultural Meteorology, College of Horticulture, Kerala Agricultural University, Kerala, India, pp. 1-100.

Prasada Rao, G.S.L.H.V., Rao, G.G.S.N. and Rao, V.U.M. (2010). Climate Change and Agriculture over India. PHI Learning Private Limited, New Delhi, India. 328 p.

Patterson, L.A., Lutz, B., Doyle, M.W. (2013). Climate and direct human contributions to changes in mean annual streamflow in the South Atlantic, USA. *Water Resources Research,*49, 7278-7291.

Rejani, R. and Yadukumar, N. (2010). Soil and water conservation techniques in cashew grown along steep hill slopes. *Scientia Horticulturae,* 126, 371-378.

Richards, N.K. (1993). Cashew Research in Northern Territory, Australia, 1987-1991. Tech. Bull. No. 202.

Srinivasarao, Ch., Venkateswarlu, B., Rattan Lal., Singh, A.K. and Sumantha Kundu. (2013). Sustainable management of soils of dryland ecosystems of India for enhancing agronomic productivity and sequestering carbon. *Advances in Agronomy,* 121, 253-329.

Warembourg, F.R. and Paul, E.A. (1977). Seasonal transfers of assimilated ^{14}C in grassland: plant production and turnover, soil and plant respiration. *Soil Biology and Biochemistry,* 9, 295-301.

Yadukumar, N. and Mandal, R.C. (1994). Effect of supplementary irrigation on cashewnut yield. Water Management for Plantation Crops - Problems and Prospects. Centre for Water Resource Development and Management, Calicut, Kerala, India, pp. 79-84

2017, Impact of Climate Change on Plantation Crops *Pages* **101–122**
Editors: **K.B. Hebbar, S. Naresh Kumar & P. Chowdappa**
Published by: **ASTRAL INTERNATIONAL PVT. LTD., NEW DELHI**

Chapter 7

Oil Palm

K. Suresh, S.K. Behera, K. Manorama and B.N. Rao

1. Introduction

Oil palm is known to be the highest edible oil yielding perennial crop and produces 4 to 6 tonnes of crude palm oil ha^{-1} yr^{-1} and 0.4 to 0.6 tonnes of palm kernel oil ha^{-1} yr^{-1} from 4th to 30th year of its productive life span. Palm oil is an important source of vegetable oil (Corley 2009) and also used in a wide range of products. It is increasingly being used for production of biofuel (Basiron 2007; Henderson and Osborne 2000; Koh 2007). Oil palm is one of the most important agricultural crops cultivated in the tropics. Indonesia and Malaysia are the leading producers of palm oil (Basiron 2007), which are also located in global biodiversity hotspots (Myers *et al.*, 2000). Hence, bringing more area under oil palm in these countries may negatively impact the biodiversity at the global level (Sodhi *et al.*, 2010). In India, an area of 19.33 lakh ha, spread over 18 states, is identified as potential for oil palm cultivation by various Expert Committees constituted by Ministry of Agriculture and Farmers Welfare, Government of India. About 80 per cent of the potential area is identified in the states of Andhra Pradesh, Karnataka and Tamil Nadu, under irrigated conditions. Till 2014, an area of 2.69 lakh ha has been brought under oil palm cultivation in different states. Oil palm has established as a successful crop in a number of states in the countryand productivity levels up to 6-8 tonnes oil per ha could be achieved. The FFB yields obtained by progressive farmers of Andhra Pradesh and Karnataka, under optimum cultural and irrigated conditions, are between 20 and 25 tonnes of FFB ha^{-1} y^{-1} *i.e.*,4-5 tonnes of oil ha^{-1} y^{-1} from fourth year onwards. If efforts were made to bring 1 M ha under oil palm in the country, it would be possible to get 3 - 4 MT of palm oil and 0.3 - 0.4 MT of palm kernel oil within two decades.

As reported in the Third Assessment Report of Inter-Governmental Panel on Climate Change (IPCC), average surface temperature of the earth has increased by 0.6°C over the 20th century. It was also estimated that during the last century, the sea level rise has been at a rate of 1 - 2 mm annually. The report also forecasts that globally averaged surface temperature would rise by 1.4 to 5.8°C and the global mean sea level may rise by 0.09 to 0 88 mm during 1990-2100 (UNIES 2007). It has already been forecasted by IPCC that, in the coming decades, agriculture worldwide will have to face the negative effects of the changing climate. It is anticipated that the crop yields in the mid and high latitude regions would be less affected than in low latitude regions. Shortening of crop growth period, decrease in water availability due to higher rates of evapo-transpiration and poor vernalization (Chattopadhyay 2005) may be the likely reasons for the decrease of potential yields in warmer areas.

Climate change may lead to increase in temperature and variability in rainfall pattern which may further lead to development of abiotic stresses like heat, drought and flooding. These stresses could occur in varying degrees and may in turn affect the productivity and quality of the crops by coinciding with different growth and phenological phases. To overcome the negative impacts of climate change timely and appropriate measures are to be initiated. In this paper, the effect of high temperature, water stress and elevated CO_2 on oil palm and possible adaptation measures to overcome them are briefly discussed.

2. Climate Change and Oil Palm

The ideal climatic requirements for oil palm are:

a. Annual rainfall of 2000 mm or greater, evenly distributed, preferably at least 100 mm in each month and without a marked dry season,.

b. Mean maximum temperature of 29 – 33°C.

c. Mean minimum temperature of 22 – 24°C.

d. Relative humidity above 45 per cent.

e. Low vapor pressure deficit.

f. Sunshine of 5 – 7 h day^{-1} in all months and

g. Solar radiation of 15 MJ m^{-2} day^{-1}.

The effect of climate change on oil palm plantations will be in many ways, either beneficial or negative, dominating in different agricultural agro climatic regions. As gradual and possibly abrupt climate change develop in this century, the factors that prevail regionally may change over time. The factors that signicantly affect oil palm production with associated consequences for water resources and pest/disease distributions are rising atmospheric CO_2 concentration, higher temperature, changing patterns of precipitation, and altered frequencies of extreme events.

2.1. Temperature

Oil palm is an equatorial crop and demands high temperature for its growth and yield. A general increase in temperatures in the order of 3-6°C over the base period average is indicated by the climate projection studies, with more warming in northern parts than southern parts, depending on the scenario. Based on Phytotron studies, best growth of oil palm was at 32/22 (mean 27°C) and the next level, with a mean temperature of 22°C, gave only slightly slower growth. At a mean temperature of 17°C, only half of the best was observed and at a mean temperature of 12°C (17/7) very little growth occurred. However, it is difficult to separate the effect of minimum and maximum temperature on growth and yield of oil palm. Although palms may be growing at high elevation or at the geographical limit of about 15°N with mean minimum temperatures of less than 20°C for part of the year, the best mean temperature range is 24 – 28°C.

Higher temperatures can reduce or even halt photosynthesis, prevent pollination and anthesis, decrease the production of female inflorescences, lead to abortion of bunches and thus bunch failure. In oil palm, as temperature increases, photosynthetic activity also increases until the temperature reaches 20°C. The photosynthetic rate then plateaus until temperature reaches 35°C, then it begins to decline and at 40°C, photosynthetic activity completely ceases. High temperatures can also cause dehydration in oil palm. The leaflets are folded to reduce exposure to the sun and thereby photosynthesis is also reduced. At high temperatures, inorder to reduce the transpiration load the stomata on abaxial side of leaflets are closed, hence CO_2 uptake is also reduced and thereby photosynthesis. Diurnal variations in oil palm under irrigated conditions also explains the sensitivity of stomata leading to decreased photosynthetic rates after 10.00 AM, which could also be due to increased ambient and leaf temperatures. During summer (1-2 weeks) early ripening of bunches in the same whorl of palm (pseudo-ripening) may occur due to higher temperature (> 38°C) thus leading to drastic reduction in oil extraction rates (< 8 per cent). In oil palm, the effect of any kind of stress (temperature/water deficits) is revealed only after 1.5 to 2 years in the form of reduced bunch yield. In oil palm there is limited scope for scheduling crop calendar, as the crop is perennial.

2.2. Precipitation

A change in the precipitation by 5-25 per cent over India by the end of the century is predicted by the climatic models, according to which, more reductions in winter rainfall than summer monsoon are expected leading to droughts during summer months. Due to more erratic rainfall patterns and unpredictable high temperature spells the major effects of global climate change are likely to occur on crop water relations. The effect of drought on oil palm growth can be divided into five stages (Luibis *et al.*, 1993). During the first stage (when water deficit is less than 200 mm y^{-1}), palms do not show any serious problem. The second stage occurs when water deficit is 200-300 mm y^{-1} and symptoms like sticking of frond and immature leaves together are observed. Leaves may not open and the old fronds may also become defective. The third stage occurs when water deficit is 300-400 mm y^{-1} During this stage, the sticked and unopened leaves number increases to 4-5 and defective old fronds will be seen in 1-1.5 spirals and fronds become dry. Subsequently, when the water deficit increases to 500 mm y^{-1}, young fronds will not open, leaf bud cracks, becomes defective and breaks.

The palms of Deli origin showed more tolerance to drought when crossed with palms from Tanzania, Yangambi and La Mé. It has also been observed in many oil palm families that tolerant crosses have a better-developed root system than susceptible crosses. Stomatal opening, leaf water potential and membrane breakdown can be considered as possible selection criteria for drought tolerance in oil palm (Cornaire *et al.*, 1989). During dry season leaf starch is hydrolyzed and there is an increase in soluble sugar concentration in oil palm (Adjahossou and Silva 1978). Accumulation of large amounts of proline contributes to osmotic adjustment and serves as cytoplasmic osmotic balance for potassium accumulation in the vacuole (Harun 1997).

Due to moisture stress, female inflorescence formation will be suppressed and abortion of female inflorescences also increases. During dry period the evapo-transpiration would be high and inadequate water supply to meet this demand would severely affect sex differentiation, subsequent inflorescence development process and eventually reduce the yield. As evidenced from an irrigation trial conducted on a semi-commercial scale in a dry area, 38.8 T FFB ha^{-1} y^{-1} could be achieved during third year of irrigation as compared to 14.1 T FFB ha^{-1} y^{-1} without irrigation. Under irrigation, 8-10 year old palms could sustain a yield of 30-35 T FFB ha^{-1} y^{-1} as compared with 16-18 T FFB ha^{-1} y^{-1} of non-irrigated palms (Foo, 1998).

In India, African dura germplasm was screened for drought tolerance based on physiological markers (Mathur *et al.*, 2001; Suresh *et al.*, 2004; Suresh *et al.*, 2008; Suresh *et al.*, 2010) and superior drought tolerant duras have been identified. When compared to other duras ZS-1 was the most drought tolerant dura, while TS-9 was the most susceptible. Based on membrane susceptibility indices (MSI), ZS-2 recorded highest MSI closely followed by TS-9 indicating their better tolerance to drought (Suresh, 2010). Studies were conducted in oil palm nursery on membrane stability indices (MSI) and fourteen dura (Africa) x pisifera crosses were screened for drought tolerance. 34CD X 110P cross recorded highest MSI closely followed by 124CD X 17P and 66CD X 129P, indicating their better tolerance to drought. 254CD X 14P and 435CD X 14P crosses are less tolerant to drought due to MSI values recorded by these crosses.

Another study was taken up in oil palm nursery with five tenera crosses to observe the genotypic variations in leaf water potential, gas exchange parameters and chlorophyll fluorescence (Suresh *et al.*, 2012). 913X1988 cross recorded highest leaf water potential after 24 days of imposition of water stress which did not differ significantly with that of 1425X2277 and 7418X1988. A significant decrease in photosynthetic rate and stomatal conductance was observed in all hybrids due to water stress. At 12 days of rehydration reduced apparent electron transport rate was observed in 7418X1988 and 1425X2277 indicating its lesser tolerance to water stress. 7418X1988 exhibited highest non photochemical quenching after rehydration indicating its better adaptation to counter excess light energy produced due to low photosynthesis.

Sap flow studies were taken up to understand plant stress responses in the form of sap flux and transpirational adjustments made by oil palm under Indian conditions (Suresh *et al.*, 2006; Suresh and Nagamani, 2007). Results revealed that sap flux gradually increased from 9.00 AM onwards, recorded a peak during 1.00 to 2.00 A.M and decreased as day progressed. A similar trend was observed with Evapo-transpiration and vapor pressure deficit as that of sap flux. Studies on seasonal variations in sap flux indicated that the flux is higher during February and March and lower during May and June. The lower flux during dry months could be due to stomata closure after mid day with increased atmospheric vapor pressure deficit.

2.3. Atmospheric Carbon Dioxide Concentration

In the arid regions of India increased CO_2 enhances the productivity of the C_3 plants, but the high temperatures may offset the beneficial effects of CO_2. Studies conducted on CO_2 enriched oil palm seedlings indicated increase in photosynthetic rate (12 folds) due to higher intercellular CO_2 level, increase in stomatal conductance as well as transpiration (3 folds), and water-use-efficiency (4 folds) as compared to the control. These studies indicate that in the tropical lowlands, CO_2 enrichment technique under the growth house prototype controlled environment is technically feasible. It also has a great application potential in the seedling/nursery (seedlings and advanced planting materials), production research and development study, and in the climate change impact analyses for possible enhanced bio-productivity and income generation (Hawa, 2006).

Oil palm when exposed to 800 and 1200 µmol.mol of CO_2^{-1}, the CO_2 imposes a marked effect on growth and leaf gas exchange parameters although all the variables measured did not differ significantly. Oil palm seedlings exposed to higher (800 µmol mol^{-1}) CO_2 concentration, recorded higher total biomass, net assimilation rate, relative growth rate, plant height, frond number, basal diameter and total leaf area compared to the controlled seedlings. Further increase in CO_2 concentration (1200 µmol mol^{-1}) resulted in seedlings becoming acclimatized to increased quantum efficiency of PSII. However total chlorophyll content and stomata density (pores. mm^{-2}) reduced. With CO_2 enrichment, net photosynthesis and water use efficiency increased, but a reduction in stomata conductance and evapo-transpiration rate has been observed. Increase in water use efficiency with increased CO_2 concentration indicates that plant could utilize water per unit carbohydrate produced especially under stress. Seedlings treated with high CO_2 have recorded an increase in apparent quantum yield and maximum photosynthetic capacity but a reduction in light compensation point.

Studies indicated that leaves of many C_3 plants respond to atmospheric CO_2 enrichment to a greater extent than the leaves atambient CO_2 as a result of increased quantum yields (Hanstein and Felle 2002). Under low light conditions, a small increase in quantum yield may increase daily carbon gain (Kiirats *et al.*, 2002). In oil palm, elevated CO_2 resulted in an increase in apparent quantum yields in 800 and 1200 µmolmol^{-1} CO_2 treatments. In oil palm seedlings, light response curve analysis indicated that CO_2 enrichment had reduced the dark respiration rate by 43 to 70 per cent through enhancement of saturated photosynthetic capacity and apparent quantum yield by 52 to78 per cent and 15 to 62 per cent, respectively. The enhancement of light compensation point and quantum yield signifies direc tinhibition of the activity of key respiratory enzymes under elevated CO_2 (Drake *et al.*, 1997). This result is in conformity with that of Henson and Haniff (2005) who reported that productivity or dry matter production of plants would increase if respiration could be minimized without affecting gross assimilation, or if gross assimilation could be increased without increasing respiration.This indicates that increased CO_2would enhance gross assimilation and reduce respiration by compensating respiration rate with high carbon gain. Usually, in plants under elevated CO_2, compensation irradiance is reduced while quantum efficiency is increased (Vavinetal,1995). The same result was observed with redoak seedlings grown atelevated CO_2 in shaded open top chamber (Kubiske and Preigitzer, 1996).

In oil palm seedlings, enhancement of the leaf gas exchange was observed with enrichment of high levels of CO_2. The up-regulation of photosynthetic rate might be due to increased leaf inter cellular CO_2concentration (Ci) which could also be related to increased leaf thickness obtained under elevated CO_2 that contains high photosynthetic protein especially Rubisco (Ramachandra and Das, 1986). The latter might also up–regulate several enzymes related to carbon metabolism which simultaneously increase the Ci (Anderson*et al.* 2001). During 2002, Lawson *et al.*, explained that in oil palm, high photosynthetic rate under elevated CO_2 could be due to more efficient net assimilation resulting from extra carbon fixation as exhibited by high Ci per unit area which is related to increased thickness of mesophyll layer, mainly due to increased palisade layer.

Oil palm enriched with high CO_2 concentrations (possessing higher apparent quantum yield and net assimilation rates), may record an increase in photosynthesis which might be justified by reduced light compensation point and dark respiration rate (Kubiske and Preigitzer, 1996). It has also been observed that in oil palm seedlings enriched with high levels of CO_2, stomatal conductance decreased, as levels of CO_2 increased, despite increases in A and WUE. The decreased gs simultaneously decreased the transpiration rate under elevated CO_2. This phenomenon is usually reported in plants treated with high CO_2 than ambient (Rashke 1986; Lodge *et al.*, 2001; Lawson *et al.*, 2002). It was believed that reduced gs might contribute to plant acclimation to high Ci (Morrison 1987). Application of CO_2 levels influences the nitrogen content of the seedlings. As the CO_2 levels increased from 400 to 1200 µmol mol-1 CO_2, nitrogen content was found to be decreased (Porteus *et al.*, 2009). This phenomenon has been explained by the researchers that under elevated CO_2 level there is a reduction in stomata conductance and with decreased transpiration rate there is decreasing uptake of nitrogen (Conroy and Hawking, 1993).

2.4. Variations in Climate

Climate change is characterized by an increase in climate variability (IPCC 2001a), which may increase the risks of crop failures, often connected to specic extreme events like heat waves or late frosts during critical crop phases like flowering. In addition, crops grown on their marginal climate ranges would be put under pressure with increase variability in temperature and precipitation. Extremes in precipitation like foods or droughts are also detrimental to crop productivity. Heavy precipitation and fooding could increase crop damage in some areas, due to soil water-logging, physical plant damage, and pest infestation (Rosenzweig *et al.*,

2002a, b). On the other hand, higher drought frequency and increased evaporation may increase the need for irrigation in specicfic regions, further straining competition for water with other sectors (Rosenzweig *et al.*, 2004). In regions where additional water resources are not available, entire cropping systems may go out of production. In coastal agricultural regions, sea-level rise and associated saltwater intrusion and storm-surge fooding can harm crops through poor soil aeration and salinity.

2.5. Pests and Diseases

Under climate change, pests associated with specic crops may become more active (Coakley *et al.*, 1999; IPCC 1996). Increased use of agricultural chemicals might become necessary, with consequent increase in health, ecological, and economic costs (Rosenzweig *et al.*, 2002a, b; Chen and McCarl 2001). Development rates of some insect species may speed up under warmer temperatures and result in shortened times between generations and improved capacity for over-wintering at northern latitudes. Some insects species might become more established earlier in the growing season and their number might increase during more vulnerable crop stages.

In oil palm warm winter temperatures favor the increase of insect populations. Crop pathogen interactions may also be influenced by overall temperature rise thus speeding up the pathogen growth rates, increasing reproductive generations per cycle, decreasing pathogen mortality rate and by making the crop more vulnerable. In Andhra Pradesh, India incidence of slug caterpillar occurs during the summer months (April – May), causing severe defoliation and reduction in yields. Incidence of bunch end rot is in observed during summer, wherein a group of distal fruits abort and are delineated from the basal portion of the bunch and 25-50 per cent of the bunch gets affected. Manifestation of spear rot disease in oil palm indicated a positive correlation of the disease with rainfall and number of rainy days and a negative relationship with average maximum temperature. During rainy season higher incidences of spear rot and bud rot diseases were noticed. The insect-crop relations are also indirectly modified with elevated $CO_{2;}$ increased C:N ratio in crop leaves renders them less nutritious per unit mass and further stimulates increased feeding by insects, leading to more plant damage (Lincoln *et al.*, 1984; Salt *et al.*, 1995).

2.6. Greenhouse Gas (GHG) Emissions

The total green house gas (GHG) emissions from oil palm plantations in Indonesia were found to be 119 kg COeq/t FFB, where major GHG contributors were due to application of pesticides, fungicides and nitrogenous fertilizers (Yew *et al.*, 2012). The CO_2 emissions of oil palm plantations in peat soil also differ as many factors are involved in the process. For example, a joint research involving IOPRI/BAU/IAERI showed that CO_2 emission was around 25.48 – 45.45 t CO/ha/year at a depth of 23 – 63 cm, while it was lower compared to the study by Agus *et al.* (2007), where CO_2 emission was about 73 t/ha/year at a water table of 80 cm.

3. Impacts of Climate Change: Case Studies

Research on *El Nino* – Southern Oscillation (ENSO) phenomenon has found that ocean atmosphere interaction especially in tropical Pacific Ocean dramatically affects weather and climate in the nations bordering Pacific Ocean and even beyond the sub tropics. Due to this phenomenon, *El Nino* and *La Nina* phases are observed in these regions. The Southern Oscillation Index (SOI) is calculated from monthly or seasonal fluctuations in the air pressure difference between Tahiti and Darwin and numerical index is a good approximation of occurrences of *El Nino* and *La Nina*. SOI values exceeding (+10) for several months indicate *La Nina* events, while SOI of more than (-10) show *El Nino* episodes (Samah and Sootyanararayana 2000).

3.1. India

In India, most of the potential areas identified for oil palm cultivation lie on the eastern and western coast of India. During 1901 to 2006, a mean temperature increase of 0.8°C and 1.2°C was recorded in coastal Andhra Pradesh (East Coast) and Kerala (West Coast) respectively. The rainfall pattern shows an increase of 3.47 per cent in the coastal Andhra Pradesh and 2.58 per cent in Kerala from 1901 to 2006. In East and West Godavari districts of Andhra Pradesh an increase in the mean temperature was recorded to the extent of 1.6°C and 1.59°C respectively during 1971-2002 (Suresh and Kochu Babu, 2009).

In the major oil palm growing areas *viz.*, East and West Godavari districts of Andhra Pradesh and OPIL plantations of Kerala, correlations between FFB yields and mean temperatures were worked out to design crop models for oil palm under Indian conditions. The increase in mean temperature during 1998 to 2007 of East Godavari district of Andhra was 0.3°C. However, FFB yields during the period did not vary among the different years, which indicate that mean air temperature had little effect on yield. As oil palm cultivation in Andhra Pradesh is mostly taken up under irrigated conditions, nutrients and water management would be the key factors in determining FFB yields (Suresh and Arulraj, 2010).

A study on the mean temperatures and FFB yields of oil palm plantations in West Godavari district of Andhra Pradesh during 1999 to 2007 indicates that there has been an increase of 0.61°C during the period and a weak correlation between mean temperature and FFB yields was observed. However, an increase in FFB yields over the years has been observed which might be due to better water and nutrient management. In the oil palm plantations, which are cultivated under rainfed conditions of Kerala (OPIL), the relationship between mean temperature and average FFB yields (1988–2005) is very weak and a probable dependence on rainfall (Suresh and Kochu Babu 2009; Suresh and Arulraj, 2010).

3.2. Indonesia

The impact of climate change on oil palm in Indonesia is manifold. Friedrich (2011) reported that climate change causes variations in higher rainfall and less reliable rainfall due to extreme precipitation and extended drought periods. The productivity of oil palm is affected by drought and high rainfall. The long dry season (which periodically occurs every 3 – 5 years) as occurred on 1972, 1977, 1982, 1987, 1991, 1994, 1997, 2002, 2004, and 2006 causes significant effect on oil palm in the southern equator. Siregar *et al.*, 2007 also reported that the periodicity of the long dry season that used to be every 3-5 years has become shorter, occurring every 2-3 years.

Growing of oil palm on high altitudes used to be major problem for better growth and development. The major limiting factor for reduced growth and development under such conditions is minimum temperature as the minimum temperature in such altitudes would be around <18ºC. The base temperature of oil palm is 15ºC. The increase of annual minimum air temperature (above 18ºC) after 1990 at 850 m above sea level has a vital role in the development of oil palm plantation at higher altitudes (600 – 850 m asl) and has become possible to grow oil palm at higher altitudes. However, Siregar *et al.* (2006) reported that low temperature stress will be observed as minimum temperature go down to <18ºC causing abortion of bunches, rotten bunches and long inflorescences. In high altitudes of Indonesia, air temperature has been increasing in the morning as per the data collected from several observation stations by Indonesian Meteorological and Geophysical Service (Badan Meteorologi dan Geofisika, BMG). Similar observations were seen from the data of air temperature collected in several altitudes in North Sumatra during 1970 – 2005. Siregar *et al,* (2006) also reported that the largest increment of minimum air temperature happened at an altitude of 850 m above sea level.

The impact of climate change on pests and diseases in oil palm in Indonesia is the change of their status from minor to major or vice versa (Chamhuri *et al.,* 2009). In many instances, changes of previously minor pests become major pests often occur. Schowalte (2006) has observed changes on life cycle, fecundity and distribution of insect pests as well as suppressing the natural enemies due to increased temperatures. Also the economic losses caused by these pests were very prominent, as shown on the outbreak of the leaf miner *Coelaenomenodera elaedis* in Nigeria (Aneni *et al.,* 2012). Over the last few decades, outbreaks of leaf eater caterpillar species, often occurred in Indonesia replacing the nettle caterpillar. Implementation of sound integrated pest management strategies on the basis of biological control including introduction and conservation of predators and parasitoids could give far reaching solutions.

3.3. Malaysia

The root system of oil palm extends deep into the soil and cover a wide area laterally. It has also been documented that the affect of soil moisture deficit is revealed only after 1.5 to 2 years by increased production of male inflorescences compared to that of female inflorescences and by inducing abortion of female inflorescences, thereby decreasing the yield. Oil palm Fresh Fruit Bunches (FFB) yield residuals when plotted against average derived Southern Oscillation Index showed a random scatter indicating no relationship between oil palm FFB yield and the assumed climate water availability parameter SOI. The derived SOI indicator was imperfect as it did not match the particular yield sensitive development phase of oil palm (Samah and Sootyanararayana, 2000).

The episodes of *El Nino* in Malaysia during 1981 and 1982 have resulted in 21 per cent loss during the following year, while the 1990 episode had resulted in more than 10 per cent yield loss in 1991. A yield loss of 16.8 per cent was recorded with the most severe *El Nino* episode of 1997. The resultant water stress due to the *El Nino*, especially during 1997, resulted in a significant reduction in productivity.

(CGPRT Monograph 2002). The oil extraction rates varied greatly due to climate change.During the late 1990s the worst effects of the *El Niño* reduced the OER in both Indonesia and Malaysia (Carter and Jackson, 2007).

La Niña causes increased rainfall in northern Australia, Kalimantan, Indonesia, Papua New Guinea and Malaysia. Oil palm requires at least 2,000 mm of rain a year or more and hence the increased rainfall is beneficial for oil palm. This would also help oil palm trees to recover from the after-effects of *El Niño*. It has also been observed that due to *La Niña* the palm oil yields across Indonesia and Malaysia recovered and production started to equal pre-*El Niño*levels. However, excessive rainfall is also detrimental and yield levels are significantly affected. In southern Malaysia flood related problems had decreased the crude palm oil production to 1.1 million metric ton or 26.3 per cent during December 2006 (Greenall, 2008).

Models by Sung and MOSTE (2000) have predicted the following in oil palm under Malaysian conditions: On an average, for every:

- ☆ 100 ppmv increase in CO_2, oil palm yield increases by 40 per cent,
- ☆ 1 degree celsius increase in air temperature, oil palm yield declines by 27 per cent,
- ☆ 1 meter per second increase in wind speed, oil palm yield increases by 32 per cent, and
- ☆ 1 hour increase in sunshine hour, oil palm yield increases by 3 per cent.

3.4. Columbia

The unfavorable climatic conditions will have a negative impact on oil palm *viz.*, reduction in evapo-transpiration, delay in leaf opening (first stage) and sexual differentiation is affected (reducing the proportion of female-productive vs. male non-productive inflorescences). Water is fundamental in sexual differentiation, development of central arrows, flowering and fruit production. More female inflorescences are formed when the level of solar radiation is good. And if any of these requirements are not met the yield will be reduced during the first and consecutive harvests. In oil palm, higher (lower) FFB production was observed during and after 6 months of enhanced (reduced) precipitation due to the fruit's accelerated (slowed) maturity (final stages). During the period that corresponds to the initial stages of bunch formation (27 months) and central arrow development (17 months), the same pattern was observed, that is when the water supply in the soil became the most important factor for future production (Cadena *et al.*, 2006).

Since rainfall exceeded evapo-transpiration, the region of Tumaco, in general, had a suitable HAI. However, decrease in rainfall during the second semester is emphasized for the years with higher drought probabilities (*La Niña* events). This deficit could be reflected in the next 3 year's production, which corresponds to bunch development. High (low) HAI values are vital to oil palm 17 to 27 months later, owing to a preservation of rainwater in soils. This is the period in which the spear develops, which means that low production in oil palm will be observed two year later due to inadequate water supply. Climatic changes had important influence on oil palm production during various stages of the oil palm life cycle. The *El Nino* which occurred during 1997/98 was favorable to oil palm production during all its development stages, and had a maximum cross-correlation with production 2.6 years later (2000).The 1999/2000 *La Nina* which occurred during the second half of the year had the highest negative impact, causing droughts and oil palm production is reduced during 2002 (Cadena *et al.,* 2006).

4. Adaptation Strategies

A certain degree of climate change and associated impacts during the coming decades has already took place on the earth due to the cumulative past emissions, regardless of local, regional and global actions planned and policy recommendations that are framed to be adopted to slow anthropogenic the greenhouse gases emissions and thus to reduce the magnitude of climate change. The effect of actions taken today to combat climate change will further evolve only in the second half of this century. Worldwide, the recent observations of increased frequency of climate extremes, as well as shifts in eco-zones, might be an indication of global warming-related changes already under way (IPCC 2001a, b, c; Milly *et al.,* 2002; Root *et al.,* 2003). Thus, in the future, sectoral adaptation is very likely and integral to the study of impact of climate change on agriculture (IPCC 2001a, b, c; Smit and Skinner 2002; Smith *et al.,* 2003).

In agriculture, adaptation is the norm rather than the exception. Farmers always had to adapt to the vagaries of weather, in addition to changes driven by several socio-economic factors like policy frameworks and market conditions, on weekly, seasonal, annual and longer timescales. The degree and nature of climate change compared to the adaptation capacity of farmers will be the real issue in the coming decades. The farmers may successfully adapt to changing climates if the future changes are relatively smooth. They can apply a variety of agronomic techniques that already work well under current climates, like adjusting the time of planting and harvesting operations, substituting cultivars and modifying or changing their cropping systems. Adaptation strategies also will vary with a change in agricultural systems, location, and scenarios of climate change.

4.1. Timely Weather Forecasts

Adaptation of oil palm plantations to climate change requires the knowledge and technology to understand the patterns of variability of current and projected climate, seasonal forecasts, hazard impact mitigation methods, land use planning, risk management, and resource management. To assess the adaptation practices in oil palm, extensive high quality data and information is required on climate and agricultural, environmental and social systems affected by climate, in order to carry out realistic vulnerability assessments. In oil palm vulnerability assessment can oversee the impacts of variability and changes in mean climate (inter-annual and intra-seasonal variability).However, in oil palm production system adaptation has a particular emphasis on future agriculture and the plantations have their own dynamics. To reduce the risk of climate change on farmers, early warning and risk management systems are the efficient contributors that can facilitate adaptation to climate variability and change. These include a historical climate data archive, monitoring tools using systematic meteorological observations, climate data analysis, information on the characteristics of system vulnerability and adaptation effectiveness and crop weather insurance indices.Adaptation strategies should focus on securing oil palm productivity in a sustainable manner. The short- and long-term impact of (extreme) events can be assessed with the improved use of Early Warning and Information Systems and Disaster Information Management Systems, while contributing to disaster preparedness and mitigation of potential risks.

4.2. Improved Water Use Efficiency (WUE)

Land conservation techniques help significantly in enhancing the residual soil moisture at the margin of dry periods while in areas where rainfall intensities increase buffer strips, mulching and zero tillage help to mitigate soil erosion risk.

4.3. Development of Drought Tolerant Oil Palm Hybrids

By utilizing the genetic variability existing in new crop varieties, crops and cultivars tolerant to abiotic stresses like, drought, flooding, high salt content in soil, high temperature, pest and disease resistance can be selected, however, the national programmes should have the required capacity and long-term support to use them. During June 2007, FAO and other like-minded institutions have started Global Initiative on Plant Breeding Capacity Build initiative (for implementation of Article 6 of the Treaty for supporting the development of capacities in plant breeding) to strengthen capacity of developing countries to implement plant breeding programmes and thereby develop locally-adapted crops. The Initiative emphasizes - conserving diversity, adapting varieties to diverse and marginal conditions, broadening the genetic base of crops, promoting locally adapted crops and underutilized species and reviewing breeding strategies and regulations concerning variety release and seed distribution. FAO's work on adapted crops includes decision-support tools such as EcoCrop to identify alternative crops for specific ecologies is in this direction.

4.4. Soil and Land Management

The oil palm cropping systems require a higher resilience against both excess of water (due to high intensity rainfall) and lack of water (due to extended drought periods) for adaptation to climate change. Soil organic matter is the key element to respond to both problems. It improves and stabilizes the soil structure, because of which soils can absorb higher amounts of water without causing surface run off, which otherwise leads to soil erosion and, further downstream, to flooding. Soil organic matter also improves the water holding capacity of the soil during extended drought.

Management practices like low tillage, maintenance of permanent soil cover are to be promoted which will increase soil organic matter and reduce the impacts of flooding, erosion, drought, heavy rain and winds. Conservation agriculture, organic agriculture and risk-coping production systems are to be explored. Soil organic matter content will be reduced due to intensive soil tillage (through aerobic mineralization), low tillage and maintenance of a permanent soil cover increases soil organic matter through crop residues or cover crops. Conservation agriculture and organic agriculture have the ability to increase soil organic carbon, reduce the use of mineral fertilizers and reduce on-farm energy costs and hence are promising adaptation options in oil palm.

5. Mitigation Strategies

Oil palm production, though stands to be affected by projected climate change, cultivation of the crop itself has been a source of greenhouse gases to the atmosphere, thus contributing to climate change. Over the past century, clearing and management of land for food and livestock production was responsible for cumulative carbon emissions of about 150 GT C, compared to 300 GT C from fossil fuels (LULUCF 2000). At present, about a quarter of the carbon dioxide is emitted by agriculture and associated land use changes (through deforestation and soil organic carbon depletion, machine and fertilizer use), half of the methane (via livestock and rice cultivation), and three-fourths of the nitrous oxide (through fertilizer applications and manure management) are released annually into the atmosphere by human activities. The global anthropogenic emissions could be mitigated by modifying the current agricultural systems management. In the coming decades, such activities could be visualized as new forms of environmental services to be provided by the farmers to society, who in turn could sell the carbon-emission credits to other carbon-emitting sectors and get additional income.

5.1. Carbon Sequestration

The only practical way of removing large volumes of green house gases from the atmosphere is to absorb CO_2 from air and inject it into the biomass. Carbon Sequestration is the process of removal of carbon from the biosphere. Carbon emissions, arising mainly from fossil fuel consumption are offset by standing crops like oil palm which serve as net accumulators of carbon. Under the terms of Kyoto Protocol, developing countries like India can gain financially through carbon trading by net fixation or sequestration of carbon. In India, oil palm seems to be a ideal crop for storing carbon with more than 2.0 lakh hectares of area under oil palm plantations and is also eligible for Clean Development Mechanism (CDM). Large projects which are expected to save millions of tons of greenhouse gases per year, qualify as a CDM project. About 1,000 hectares of land would need to be assembled for a project. In India, a large number of plantations would need to be assembled to make it large, as the where land holdings are small.

It is estimated that the global carbon storage by oil palm is 74 Mt C y^{-1} for 12 m ha due to its high annual biomass productivity (50 t DM ha^{-1}) compared to that of coconut (28 t DM ha^{-1} y^{-1}). Under irrigated conditions of Andhra Pradesh, India, the amount of carbon sequestered by eleven year old oil palm hybrids ranged between 17.98 and 35.44 T C ha^{-1} with Papua New Guinea hybrids sequestering the highest carbon content. The carbon sequestered by the ASD Costa Rican hybrids was between 20.54 (ASD Deli X Ghana) and 35.44 T C ha^{-1} (ASD Deli X Ekona), while the carbon sequestered by Palode hybrids ranged from 20.26 (P12 X 313) to 26.05 T C ha^{-1} (P12 X 266). The Ivory Coast hybrids sequestered carbon between 17.98 and 28.73 T C ha^{-1} (Suresh and Kochu Babu, 2008). Another study indicated that the carbon contents ranged from 0.413 to 1.314 kg in different fronds of a mature palm.

In the younger leaves the carbon contents were low and no pattern was observed among the middle and lower whorls (Suresh *et al.*, 2008).

5.2. Zero Burning

Zero burning technique mitigates carbon dioxide emissions which otherwise would have been released from burning the vegetation to clear land for cultivation. In this technique, the material cleared from the site will be chipped or stacked between the palm rows and left to decay. However, a significant portion of the biomass from land clearing should be used to manufacture wood products which have a long useful life (instead of simply decaying), otherwise mechanical land clearing methods will only have a short term effect on carbon emissions.Zero burning is now being adopted by a majority of estate companies in Indonesia to clear their land. However, a clear gap exists between the company policies stated about zero burning and the interpretation by middle management on the ground and contractors engaged on behalf of oil palm companies to clear land (Sargent 2001). As viewed by the satellite imagery, large scale estates continue to use fire to clear land rather than dong it manually or adopting zero burning. Many companies use fire to clear the land because it is easier and also thought to be a cheaper method of clearing land. Guyon and Simongkir (2002) undertook extensive economic analysis on the costs and benefits of zero burning versus burning in commercial oil palm plantations and confirmed the same.

5.3. Water Management

In the existing oil palm plantations with heavy soils, appropriate drainage system must be designed to remove excessive water but maintain the water table at 0.5-0.7 m from the soil surface. This helps to prevent excessively rapid depletion of the soil layer. A good drainage system retards oxidation, and also improves the yields and performance of the palms because palms tend to fall over and the stem becomes bent as the palm reestablishes growth in a vertical plane. This results in an uneven canopy and reduced yield. It also makes harvesting and other field operations difficult (Kee, Uexkull, Hardter, 2003).

5.4. Reduced Usage of Chemical Inputs

A host of greenhouse gases (carbon dioxide, nitrous oxide and methane) are released into the atmosphere by the chemical inputs, such as fertilizers, pesticides and herbicides. The global community ultimately aims to identify strategies that can mitigate global warming, hence emission of other greenhouse gases resulting from chemical inputs applied to oil palm plantations should also be considered. Greenhouse gases (GHGs) are atmospheric compounds that store energy, thus influencing the climate. The global warming potential of each of the GHGs is different which takes into account the effectiveness of each gas to trap heat radiation and its longevity in the atmosphere. Oil palm is one of the largest mineral fertilizer consumers in Southeast Asia (Hardter and Fairhurst, 2003). During the first 5 years of planting, a typical oil palm plantation on mineral soil requires around 1114 kg/ ha of nutrient inputs. From the climate point of view nitrogen is the most relevant nutrient input. Oil palm plantations planted on both mineral and peat soils requires around 354 kg/ha of nitrogen during the first five years. This is usually applied in the form of nitrogen based fertilizers like ammonium nitrate, ammonium sulphate and urea. In oil palm plantations, in addition to nitrogenous fertilizers, pesticides and insecticides are also used liberally to control weeds, diseases and other pests. It has also been observed that the pests and diseases problem is more prevalent in oil palm plantations that use zero burning to clear land rather than fire. Zero burning can also require higher amount of fertilizers during the initial years (0-2) because fire is an efficient means to convert the standing biomass into nutrients. However, oil palm plantations require lower fertilizer application when planted on land cleared by mechanical means because the nutrients from decaying wood debris left after zero burning are released very slowly into the soil (Guyon and Simorngkir 2002).

Integrated pest management which involves the use of cultural, biological and physical methods may be practiced, as it minimizes the pesticide requirement. Rhinoceros beetle, a major pest in oil palm plantations can be managed by using a combination of cropping practices (pulverization, shredding the vegetation debris and covering it with leguminous cover crops) or through biological control, such as pheromone traps. However, Integrated Pest Management practices are considered to be more expensive than traditional pest management methods using pesticides. Though cheap, excessive use of pesticides alone for managing pests and diseases of oil palm can have significant environmental impacts, like incidental killing of non target species due to run off of pesticides into water sourses.

5.5. Composting

In oil palm mills, zero waste management could be achieved by composting the Empty Fruit Bunches (EFB) and Palm Oil Mill Effluent (POME) while at the same time organic compost produced could successfully substitute chemical fertilizer (partially) in the plantations. This leads to substantial savings and develops a sustainable business model.

5.6. Biogas Production from POME

According to Gold Standard CDM projects production of biogas is not sustainable but in future, it may become essential in order to ensure premium carbon credits for project developers.

5.7. Conversion of Biomass to Energy

To replace fossil fuel and thereby reduce carbon dioxide generation in the mills, palm oil can be used either as palm bio-diesel or can be burnt directly. To generate power oil palm biomass is normally used as fuels for boilers. Thus, palm oil would displace fossil fuel based energy in mills.

6. Future Research Strategies

Oil palm has been introduced into India in the late eighties and ICAR-Indian Institute of Oil Palm Research was established only during 1995, hence, research on oil palm on the aspects of climate change is very scanty. Under the present scenario of the importance of climate change there is urgent need to focus more on its impact on the basic physiological aspects of the crop. The specific research priorities for oil palm crop are as follows:

6.1. Quantification of CO_2 Flux, Energy Budget and Water Transfer in Oil Palm

As per the provisions proposed under Clean Development Mechanism (CDM) there is scope for establishment of carbon sinks, through afforestation or reforestation. In India, due to its high biomass production and its recent dynamic area expansion, oil palm is of particular interest making thus making it a potentially important carbon sink. Cultivation of oil palm has many advantages and also the crop has the potential to mitigate global climate change, but under Indian conditions, information available on carbon storage potential of oil palm in biomass and soil is very scarce. Hence, data available on carbon content in oil palm plantations and soil needs to be collected to understand the carbon sequestration in soil and plants with careful attention to the Kyoto protocol for getting carbon permits.

At IIOPR several studies were conducted on oil palm photosynthesis and gas exchange, but have been restricted to leaf level measurements only. Though by using mechanistic models, predicting the response of whole canopy based on single leaf photosynthetic parameters is possible, validating these models becomes very difficult. In this context, Eddy Covariance studies will aid in studying CO_2 flux, energy budget and water transfer within a plantation. The validated CO_2 and H_2O fluxes will be used to model gas exchanges of the oil palm plantations according to climate and other environmental parameters as well as crop management.

6.2. Assessment and Quantification of Impact of CO_2 and Tmperature in Oil Palm using CO_2 Enrichment Studies

The ameliorating capacity to withstand adverse drought could be better understood by studying the effects of elevated CO_2 and changes in soil water supply on gas exchange processes and plant water relations (that affect net primary production). Experiments can be planned to understand the effect of elevated CO_2 on oil palm growth and other physiological processes; to investigate the effect of different concentrations of CO_2 on seedling growth, leaf gas exchange characteristics and nutrient status of oil palm hybrids belonging to different sources.

6.3. Understanding of Seasonal Variations in Vegetative and Reproductive Growth, Phenology and Yield Components to Climatic Variables

Physiological causes for general seasonal variation are largely unknown and hence Intra and Inter annual yield variations, which are very high in oil palm crop, are difficult to interpret. Though fresh fruit bunches production in oil palm is continuous, marked seasonal peaks are generally observed which cannot be explained either by carbon assimilation or by phenology. Information available on endogenous or environmental control of periodic events in oil palm is limited. In oil palm, the effect of climatic variability on yield components is complex because at a given point of time, large number of inflorescences having different developmental stage are available and also due to longer developmental duration of female inflorescences. Drought sensitive processes like sex determination and abortion takes place months or years before maturity of given bunch and hence the effect of drought on yield components occur with long time lags.

6.4. Phenotyping of Drought, Salinity and High Temperature Tolerance Traits in Oil Palm

In crops like oil palm phenomics platform would help in understanding the physiological basis for tolerance to different stresses like drought, salinity and high temperature. Rapid screening of large populations can be done at very short time for different stresses as it provides high throughput imaging using visible, near infra-red, far infra-red and fluorescence imaging. This would also enable the development of capabilities and facilities to provide a comprehensive, continuous analysis of plant growth and performance using modern technologies.

7. Conclusions

In order to address the impact of climate change, the possible regional impacts of climate change on oil palm needs to be investigated. The impact of increasing temperature and precipitation patterns is projected to be different over the various agro-ecological regions of the country. Hence the effect of climate change will be different in different oil palm growing regions. Crop based and agro-ecological region based adaptation strategies need to developed, based on vulnerability of individual crop and the region by integrating all the measures available for sustainable productivity.

References

Adiahossou, F, Silva, V.D. (1978). Soluble glucide and starch contents and resistance to drought in the oil palm. *Oleagineux*, 33(12), 603-604.

Agus, F., Suyanto, W. and Van Noordwijk, M. (2007). Reducing Emission from Peatland Deforestation and Degradation: Carbon Emission and Opportunity cost. Paper presented in: International Symposium and Workshop on Tropical Peatland "Carbon-Climate-Human Interaction-Carbon Pools, Fire, Mitigation, and Wise Use". Yogyakarta, Indonesia, 27-29 of August, 2007.

Aneni, T.I., Aisagbonhi, C.I., Iloba. B.N. and Ogbebor, C.O. (2012). Empirical assessment of climate change impacts on the leaf miner, a major pest of the oil palm in nifor, Nigeria. *Proceedings of The International Biometric Conference,* Kobe, Japan.

Anderson, J.M. (2000). Strategies of Photosynthetic Adaptations and Acclimation." In Probing Photosynthesis:Mechanisms, Regulationand Adaptation. M.Yunus,U. Pathre, P. Mohanty,eds. London.

Basiron, Y. (2007). Palm oil production through sustainable plantations. *Eur. J. Lipid Sci. Tech.* 109 (4), 289-295.

Cadena, M.C., Devis-Morales, A., Pabon, J.D., Reyna-Moreno, J.A. and Ortiz, J.R. (2006). Relationship between the 1997/98 El Nino and 1999/2001 La Nina events and oil palm tree production in Tunaco, South Western Columbia. *Adv. in Geosciences*, 6, 195-199.

Carter, C. and Jackson, D. (2007). Secret to survival. Global oils and fats business magazine 4(2). 37.

CGPRT Centre monograph No. 43 (2002). Coping against El Nino for stabilizing rainfed agriculture. Lessons from Asia and the Pacific. *Proceedings of a joint workshop held in Cebu*, Philippines, September 17-19, 130-131 pp.

Chamhuri, S., Mahmudul, A. and Wahid M, Al- Amin. (2009). Climate change, agricultural sustainability, food security and poverty in Malaysia. *IRBRP J*, 5(6),309-321.

Chattopadhyay, N. (2005). Climate change and its implication to Indian agriculture. Needs for Agrometeorological solutions to farming problems.

Chen, C.C. McCarl, B.A. (2001). An investigation of the relationship between pesticide usage and climate change. *Clim Change*, 50, 475–487.

Coakley, S.M., Scherm, H. and Chakraborty, S. (1999). Climate change and plant disease management. *Ann Rev Phytopathol*, 37, 399–426.

Conroy, J. and Hocking, P. (1993). Nitrogen nutritions of C3 plants at elevated carbon dioxide concentration.*Plantphysiol*,89,570–576.

Corley, R.H.V. (2009). How much palm oil do we need? *Env. Sci. Policy*, 12, 134-139.

Cornaire, B., Houssou, M.S. and Meunier, J. (1989). Breeding of drought resistance in oil palm. 2. Kinetics of stomatal opening and protoplasmic resistance. Paper presented at Int. Conf. Palms and palm products, 21-25 Nov., *Nigerian Inst. Oil Palm Res.*, Benin City.

Drake,B.G.,Gonzalez-Meler,M.A. and Long,S.P.(1997).More efficient plants: a consequence of rising atmospheric CO_2? *Ann. Rev. Pl. Physiol. Pl. Mol. Biol*,48, 609–639.

Foo, S.F. (1998). Impact of moisture on oil palm yield. Kemajuan Penyelidikan, 32, 5-17.

Freidrich, T. (2011). Conservation Agriculture for Climate Change Adaptation in East Asia and the Pacific. Climate Change and Adaptation in Agriculture for East Asia and the Pacific Region: Issues and Options. FAO-WB Expert Group Meeting, Rome. http://www.fao.org/ag/ca.

Greenall, M. (2008). *La Nina* is good for oil palms but bad for soybeans. *Asia Plantation*, pp. 6-7.

Guyon, A. and Simorangkir, D. (2002). The Economics of Fire Use in Agriculture and Forestry: A preliminary review for Indonesia, Project Fire Fight South East Asia, Jakarta.

Hanstein, S.M. and Felle, H.H. (2002). CO_2 triggered chloride release from guard cells inintact fava bean leaves. Kinetics of the onset of stomatal closure. *Plant Physiol*, 130: 940–950.

Hardter, R. and Fairhurst, T. (2003). Introduction: Oil Palm: Management for Large and Sustainable Yields, Hardter R and Fairhurst, Oxford Graphic Printers Pte Ltd.

Hawa, Z.E.J. (2006). CO_2 enrichment technique for lowland controlled environment system. Geneva during 23-26th Feb, 2006.

Henderson, J. Osborne, D.J. (2000). The oil palm in our lives: how this came about. *Endeavour*, 24(2) p. 63.

Henson, I.E, Haniff.M.H.(2006).Carbondioxideenrichmentinoilpalmcanopiesand itspossible influenceonphotosynthesis.*Oil Palm Bul*, 51, 1–10.

IPCC, (1996). Watson, R.T., Zinyowera, M.C. and Moss, R.H. (eds) Climate change (1995). impacts, adaptations and mitigation of climate change: Scientic-Technical Analyses. Chaps. 13 and 23. The Intergovernmental Panel on Climate Change second assessment Report, vol. 2. Cambridge University Press, Cambridge UK, 878 pp.

IPCC, (2001a). Climate change 2001: The scientic assessment. Intergovernmental Panel on Climate Change. Cambridge University Press, Cambridge UK, 845 pp.

IPCC, (2001b). Climate change 2001: impacts, adaptation, and vulnerability. Intergovernmental Panel on Climate Change. Cambridge University Press. Cambridge, UK, 1032 pp.

IPCC, (2001c). Climate change 2001: mitigation. Intergovernmental Panel on Climate Change. Cambridge University Press. Cambridge, UK, 1076 pp.

Kee, N. and Uexkull, Hardter. (2003). Botanical Aspects of the Oil palm Relevant to Crop Management, in Hardter R and Fairhurst, Oil Palm: Management for Large and Sustainable Yields, Oxford Graphic Printers Pte Ltd.

Kiirats, O., Lea, P.F., Franceschi, V.R. Edwards,G.E. (2002). Bundle sheath diffusive resistance toCO_2 and effectiveness of C4 photosynthesis and refixation of photorespiredCO_2in a C_4cyclemutantandwild-type *Amaranthusedulis.Plant Physiol*,130, 964–976.

Koh, L.P. (2007). Potential habitat and biodiversity losses from intensified biodiesel feedstock production. *Conserv. Biol*, 21, 1373–1375.

Kubiske,M.E. Pregitzer, K.S. (1996). EffectsofelevatedCO_2andlightavailabilityonthe photosyntheticresponseoftreesofcontrastingshadetolerance.*TreePhysiol*,16, 351-358.

Lawson, T., Carigon, J., Black, C.R., Colls, J.J., Landon, G. and Wayers. J.D.B. (2002). Impact of elevated CO_2 and O_3on gas exchange parameters and epidermal characteristics of potato (*Solanum tuberosum* L.). *J. Exp. Bot*, 53 (369), 737-746.

Lincoln, D.E., Sionit. N. and Strain. B.R. (1984). Growth and feeding response of *Pseudoplusia includens* (Lepidoptera: Noctuidae) to host plants grown in controlled carbon dioxide atmospheres. *Env. Ento*, 13,1527–1530.

Lodge, R.J., Dijkstra, P., Drake, B.G. and Morrison, J.I.L. (2001). Stomatal acclimation to increased level of carbon dioxide in a Florida scrub oak species *Quercus myritifolia*. *Plant Cell Env*, 14, 729–739.

Lubis. A.U., Syamsuddin. E. and Pamin, K. (1993). Effect of long dry season on oil palm yield at some plantations in Indonesia. Paper presented at PORIM International palm oil congress, 20-25 Sept., Kuala Lumpur, Malaysia.

LULUCF, (2000). IPCC special reports: land use, land-use change, and forestry. Watson, R.T, Noble, I.R., Bolin, B., Ravindranath, N.H., Verardo, D.J., Dokken, D.J. (eds) (2000) Cambridge University Press, Cambridge, 324 pp.

Mathur, R.K., Suresh, K., Nair, S., Parimala, K. and Sivaramakrishna, V.N.P. (2001). Evaluation of exotic dura germplasm for water use efficiency in oil palm (*Elaeis guineensis* Jacq.) *Ind. J. Plant Gen. Res*, 14, 257-259.

Milly, P.C.D., Wetherald, R.T., Dunne, K.A. and Delworth, T.L. (2002). Increasing risk of great oods in a changing climate. Nature, 415,514–517.

Morison,J.I.L.(1987).Intercellular carbon dioxide concentration and stomatal responses to carbon dioxide. In Stomatal Function, ed. Zeiger,E,Farquhar, G.D., Cowan, I.R., 229–251 pp, Stanford,California,StanfordUniversityPress.

MOSTE, (2000). Malaysia. Initial National Communication. Report submitted to the United Nations Framework Convention on Climate Change. Ministry of Science, Technology and Innovation, Putrajaya.

Myers, N., Mittermeier, R.A., Mittermeier, C.G., de Fonseca, G.A.B. and Kent, J. (2000). Biodiversity hotspots for conservation priorities. *Nature,* 403(6772), 853-858.

Porteaus,F.,Hill,J.,Ball,A.S.,Pinter,P.J.,Kimbal,B.A.,Wall,G.W., Ademsen,F.J. and Morris,C.F.(2009). Effects of free air carbon dioxide enrichment (FACE) on the chemical composition and nutritiv evalue of wheat grain straw. *Ani .Feed Sci. Tech,* 149, 322– 332.

Ramachandra, A.R. and Das, V.S.R. (1986).Correlation between biomass production and net photosynthetic rates and kinetic properties of RuBP carboxylase in certain C_3 plants. *Biomass*,10, 157–164.

Raschke, K. (1986). The influence of carbon dioxide content of the ambient air on stomatal conductance and the carbondioxide concentration in leaves. In: Enoch H Z and Kimball B.A. [eds] Carbon dioxide enrichment of greenhousecrops, Vol. 2,87-102 pp,*Boca Raton*, CRCPress.

Root, T.L., Price, J.T., Hall, K.R., Schneider, S.H., Rosenzweig, C. and Pounds, J.A. (2003). Fingerprints of global warming on wild animals and plants. *Nature,* 421,57–60.

Rosenzweig, C., Iglesias, A., Yang, X.B., Epstein, P.R. and Chivian. E. (2002a). Climate change and extreme weather events: implications for food production, plant diseases, and pests. *Global Change Human Health,* 2(2), 90–104.

Rosenzweig, C., Tubiello, F.N., Goldberg, R., Mills, E. and Bloomeld, J. (2002b). Increased crop damage in the US from excess precipitation under climate change. *Global Environ Change*, 12,197–202.

Rosenzweig, C., Strzepek, K.M., Major, D.C., Iglesias, A., Yates, D.N., McCluskey, A. and Hillel. D. (2004). Water resources for agriculture in a changing climate: international case studies. *Global Environ Change*, 14, 345–360.

Salt, D.T., Brooks, B.L. and Whitaker, J.B. (1995). Elevated carbon dioxide affects leaf-miner performance and plant growth in docks (Rumex spp.). *Global Change Biol*, 1,153–156.

Samah, A.A. and Sootyanarayana, V. (2000). Possible impact of climate change on agriculture in Malaysia – A Report.

Sargeant, H. (2001). Vegetation Fires in Sumatra Indonesia. Oil palm Agriculture in the wetlands of Sumatra: destruction or development?, European Union Forest Fire Prevention and Control Project, Government of Indonesia and the European Union.

Schowalte, T.D. (2006). *Insect Ecology: An Ecosystem Approach, Second Edition.* Academic Press, Elsevier. USA.

Siregar, H.H., Sutarta, E.S. and Darlan, N.H. (2006). Implication of Climate Change on Oil Palm Plantation in Higher Altitude (Case of North Sumatra Province). Paper on International Oil Palm Conference 2006, Bali, 19-23 June 2006.

Siregar, H.H., Darlan, N.H. and dan Sutarta, E.S. (2007). Dampak Kemarau Panjang dan Kekeringan Terhadap Pertanaman Kelapa Sawit. Seminar GAPKI Sumatera Selatan, 2 Agustus 2007. 15p.

Smit, B. and Skinner, M.W. (2002). Adaptation options in agriculture to climate change: a typology. *Mit Adapt Strategies Glob Ch*, 7, 85–114.

Smith, J.B., Klein, R.J.T. and Huq, S. (2003). Climate change, adaptive capacity, and development. Imperial College Press, London, 347 pp.

Sodhi, N.S., Posa, M.R.C., Lee, T.M., Bickford, D., Koh, L.P. and Brook, B.W. (2010). The state and conservation of Southeast Asian biodiversity. *Biodiversity and Conservation*, 19(2): 317-328.

Suresh, K. and Arulraj, S. (2010). Climate change and oil palm productivity. Invited paper presented in Plantation Crops seminar PLACROSYM XIX hosted by Rubber Research Institute of India, Kottayam from 7 - 10th December 2010. 28 pp.

Suresh, K. and Kochu Babu, M. (2008). Carbon sequestration in oil palm under irrigated conditions. Poster presented during Golden Jubilee Conference on Challenges and Emerging strategies for improving plant productivity held during 12-14th Nov, 2008 at IARI, New Delhi.

Suresh, K. and Kochu Babu, M. (2010). Impact of climate change on oil palm in India. In Challenges of climate change – Indian Horticulture(HP Singh, JP Singh and SS Lal Eds.), Westville Publishing House, New Delhi 36-41 pp.

Suresh, K., Mathur, R.K. and Kochu Babu, M. (2008). Screening of oil palm duras for drought tolerance – Stomatal Responses, gas exchange and water relations. *J. Plant. Crops*, 36(3), 270-275.

Suresh, K. and Nagamani, C.h. (2007). Partitioning of water flux in oil palm plantations – Seasonal variations in sap flow under irrigated conditions. Paper published in Proceedings of International workshop on advanced flux nut and flux evaluation (Asiaflux workshop, 2007) held during 19-21 October, 2007 at Taoyuam, Taiwan.

Suresh, K., Nagamani, C., Lakshmi Kantha, D. and Kiran Kumar, M. (2012). Changes in photosynthetic activity in five common hybrids of oil palm (*Elaeis guineensis* Jacq.) seedlings under water deficit. *Photosynthetica* (In Press).

Suresh, K., Nagamani, C., Ramachandrudu, K. and Mathur, R.K. (2010). Gas exchange characteristics, leaf water potential and chlorophyll a fluorescence in oil palm (*Elaeis guineensis* Jacq.) seedlings under water stress and recovery. *Photosynthetica*, 48 (3), 430 – 436.

Suresh, K., Nagamani, C.H. and Reddy, V.M. (2006). Photosynthesis and chlorophyll fluorescence in oil palm grown under different levels and methods of irrigation. *J. Plant. Crops*, 34(3), 621-624.

Suresh, K., Nagamani, C.H., Sivasankar kumar, K.M. and Vinod Kumar. P. (2004). Variations in photosynthetic rate and associated parameters in the different dura oil palm germplasm. *J. Plant. Crops*, 32 (suppl.), 67-69.

Suresh, K., Reddy, V.M. and Kochu Babu, M. (2008). Biomass, Carbon and Nitrogen Distribution with Leaf Age in Oil Palm. Poster presented in the third Indian Horticultural Congress held at Bhubaneswar during 6th - 9th Nov, 2008.

United Nations UNIES Report (2007). Climate change – An Overview. Paper presented by the secretariat of the United Nations Permanent Forum on Indigenous Issues, November, 2007.

Vivin, P.,Gross, P.,Aussenac, G. and Guehl,J.M.(1995).Wholeplant CO_2exchange, carbonpartitioning andgrowthin*Quercus robur* seedlings exposedto elevatedCO_2. *Plant Physiol. Bio*, 33, 201–211 pp.

Yew, A.I.Tan., Muhamad, H., Hashim, Z., Subramaniam, V., Wei, P.C. and May, C.Y. (2012). GHG Emission inventories and mitigation stategis in the palm oil sector. 3 ICOPE, Bali 22-24 February 2012.

2017, Impact of Climate Change on Plantation Crops *Pages* **123–143**
Editors: **K.B. Hebbar, S. Naresh Kumar & P. Chowdappa**
Published by: **ASTRAL INTERNATIONAL PVT. LTD., NEW DELHI**

Chapater 8

Tea

E.Edwin Raj, K.V. Ramesh, B. Radhakrishnan and R. Raj Kumar

1. Introduction

Climate variability and change is the most important factor that causes year-to-year uncertainty in crop productivity in general especially rainfed plantation crops like tea. Change in local climate is due to changes associated with large-scale features like the changes in the interactions between land, atmosphere and oceanic changes, and also changes in local land use and land cover changes. However, the observed rainfall trends show reduction in land rainfall during southwest monsoon season over India (Ramesh and Goswami, 2007) as well as tea growing regions (Raj Kumar *et al.,* 2012). However, model projections to date have not provided any conclusive evidence of the patterns of trends (Ramesh and Goswami, 2014) and inter-annual variation in terms of intensity, frequency and distribution of extreme ecological events that is expected in the near future (IPCC, 2007). The impact of local change in climate further hampers global crop productivity and thus affects the food security. Recent simulation models developed have the effect of region specific crop production (Horie, 1993; Horie *et al.*, 1995; Matthews *et al.*, 1995; Rosenzweig *et al.*, 1995), which predict global climate change that may have either positive or negative influence on crop production. However, these modelling studies pointed out the need for further studies in order to evolve and improve our understanding of environmental effects on crop production. Specifically, further research is required to quantify the interactive effects of CO_2, air temperature and other environmental variables on crop production.

Commercial tea manufactured from cultivated *Camellia* sp. is one of the oldest beverages consumed almost all over the world. Its ideal growing conditions are at high risk and expected to change significantly under the local changes in climate.

As evidenced, most of the tea growing countries are experiencing drastic reduction in tea production, which are associated with significant changes in local climatic variables (Raj Kumar and Mohan Kumar, 2010). The distribution of tea yield during a particular season is dependent on many environmental factors, in which local climate are of great importance (Saharia and Bezbaruah, 1984). Variation in yield between regions shows the profound influence of soil health and inputs besides climatic variables on yield (Chanda, 1957). Studies have also demonstrated that black tea quality changes are due to soil type, altitude, seasons, weather factors, geographical areas of production, agronomic inputs, processing technologies and management (Owuor *et al.*, 2011). As these relationships between tea production and environmental factors are not uniform in nature, but highly variant both in space and time, in which climate change is a spatio-temporal phenomenon. Adoption of new varieties, a commonly applicable option for enduring climate change which is much more slower than any other annual/seasonal crops (Lobell *et al.*, 2006). To develop mitigation strategies to adopt to the impact of climate change in tea crops, it is important to understand and quantify the crop response to climate change in tea. This will provide valuable inputs in developing methods to retain the productivity and also to maintain/improve the quality of tea. The present chapter deals with the impact of climate change on tea crop productivity and the reliability of climate models in simulating the regional climate.

2. Tea Production

2.1. Global Tea Production

Tea is grown in a wide range of climates from Mediterranean to warm, humid tropical regions that include countries such as Russia and Georgia in the Northern latitudes, Argentina and Australia in the southern latitudes. South-East Asia and Eastern and Southern Africa are the major producers of black tea while China and Japan are the major green tea producers. World tea production reached 5.3 million tons (Mt) in 2013 (FAOSTAT, 2015) in which China shared 36.1 per cent. World tea productivity significantly is increasing ($R^2 = 0.9247$, $P<0.001$) at the rate of 16.72 kg/ha/yr (1961-2013; Figure 8.1). Tea growing countries are disturbed in terms of tea yield, primarily due to the biotic and abiotic factors. The possible fallouts of the climate change impacts are already being witnessed and have escalating management costs (Raj Kumar and Mohan Kumar, 2009).

2.2. Indian Tea Production

India is the second largest producer of tea and manufactures 1.2 Mt of tea, accounting for 22.7 per cent of global tea production. The Indian tea industry includes small growers, corporate and government plantations. Though major portion of tea production in India is highly concentrated in some specific regions, it is scattered in sixteen states. India holds around 16 per cent (~5,67,000 ha) of total tea area in the world. The increasing trend of tea production in India (Figure 8.2a) since 1951 is similar to that of north (Figure 8.2b) and south India (Figure 8.2c). Though the increasing trend in yield fluctuates from year to year, the production

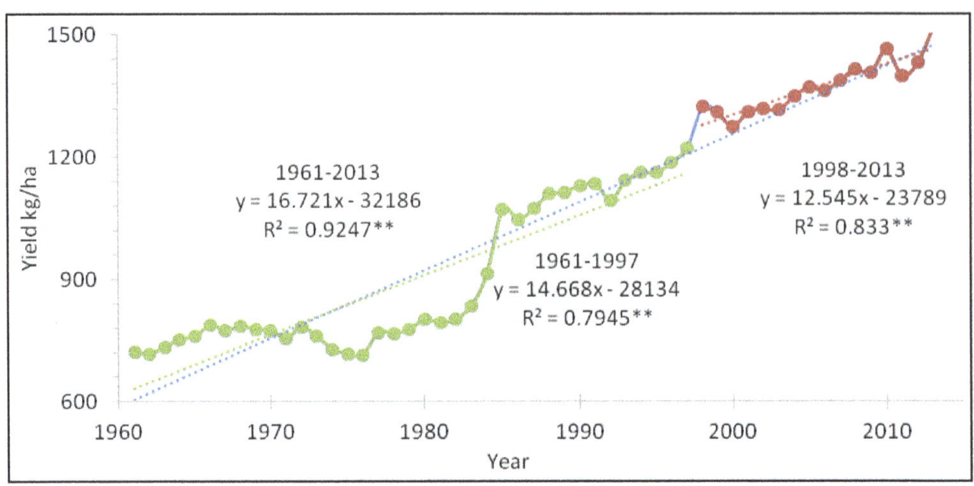

Figure 2.1: Average Annual Tea Productivity of the World from 1961 to 2013 (FAO, 2015).

has moved positively upward. Increase after the mid-1960s is attributed to the green revolution during which numerous changes in agricultural technologies and agro inputs occurred. These include development in crop varieties, improved pest, disease and weed control measures, enhanced use of fertilizers and follow-up irrigation (Evans, 1997). Indian tea production sharply declined from 9.81 to 9.66 Mt between 2007 and 2010. The recent crisis in the Indian tea industry has, however challenged the plantation model economy during which around 36 plantations were closed down or abandoned which directly affected thousands of workers and the economy of states. It also resulted in the drastic expansion of the small tea-growers in India. The Indian tea industry has overcome and has been increasing productivity since 2004, even though the plantations experienced drastic cultivation practices and pest and disease menace.

2.3. Tea Producing Regions of India

The tea production areas can be broadly classified into two regions, north and south India. North Indian tea growing states includes Assam, West Bengal, Tripura, Bihar, Uttaranchal, Himachal Pradesh, Manipur, Sikkim, Arunachal Pradesh, Nagaland, Odisha, Meghalaya and Mizoram (Figure 8.3a). The tea growing states of south India comprises of Tamil Nadu, Kerala and Karnataka (Figure 8.3b). Yields began to increase by about 1980 and reached 2000 kg/ha in the 1990s (Figure 8.2d). The yield per hectare in south India (2004 kg/ha) is higher than that of the north India (1601 kg/ha). It has to be noted however that unlike in the southern gardens, plucking does not take place throughout the year in north Indian tea gardens and hence the volume is low. But the yield started to decline after 1997 and reached ~1832 kg/ha due to increase in sizable small growers with poor crop management.

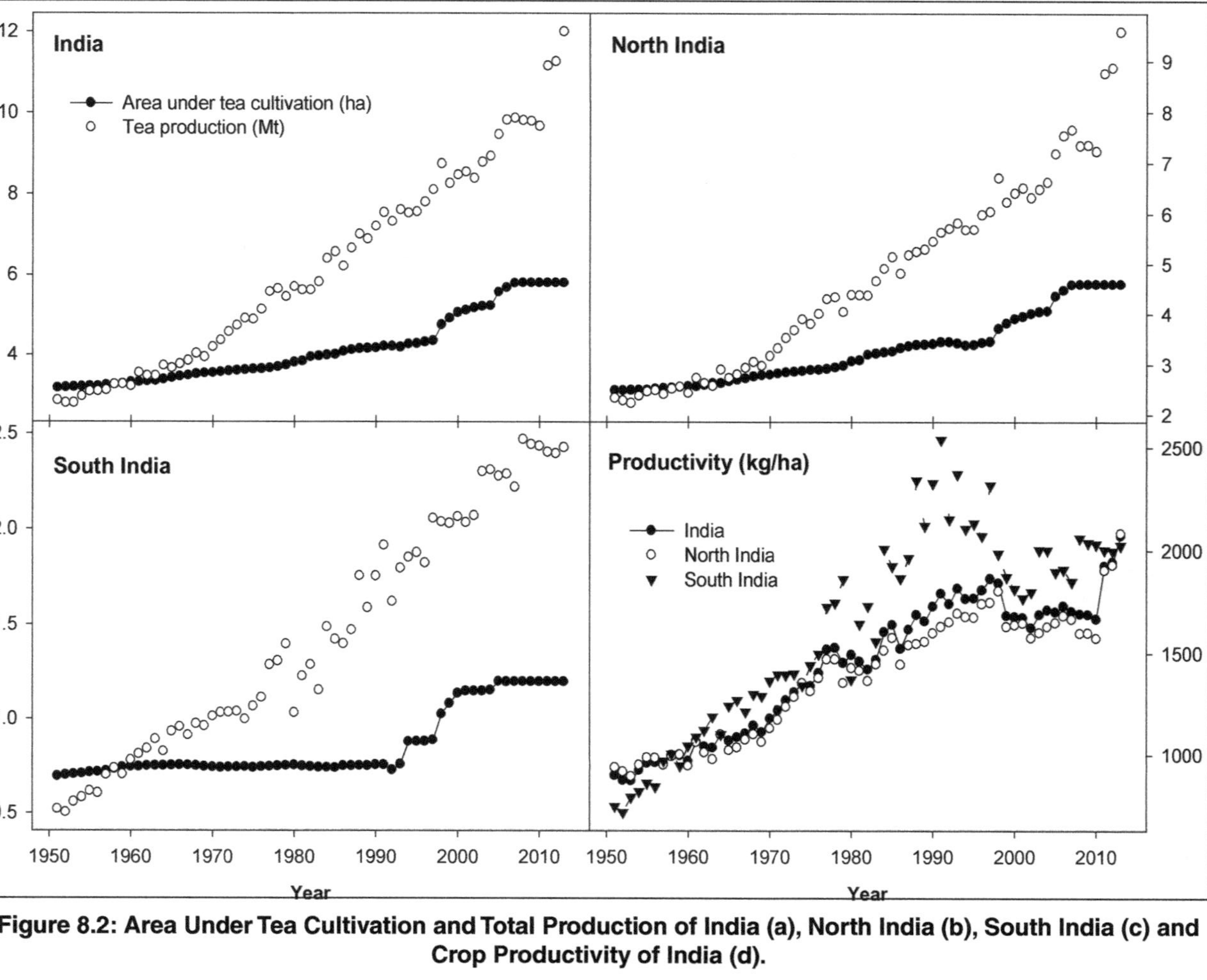

Figure 8.2: Area Under Tea Cultivation and Total Production of India (a), North India (b), South India (c) and Crop Productivity of India (d).

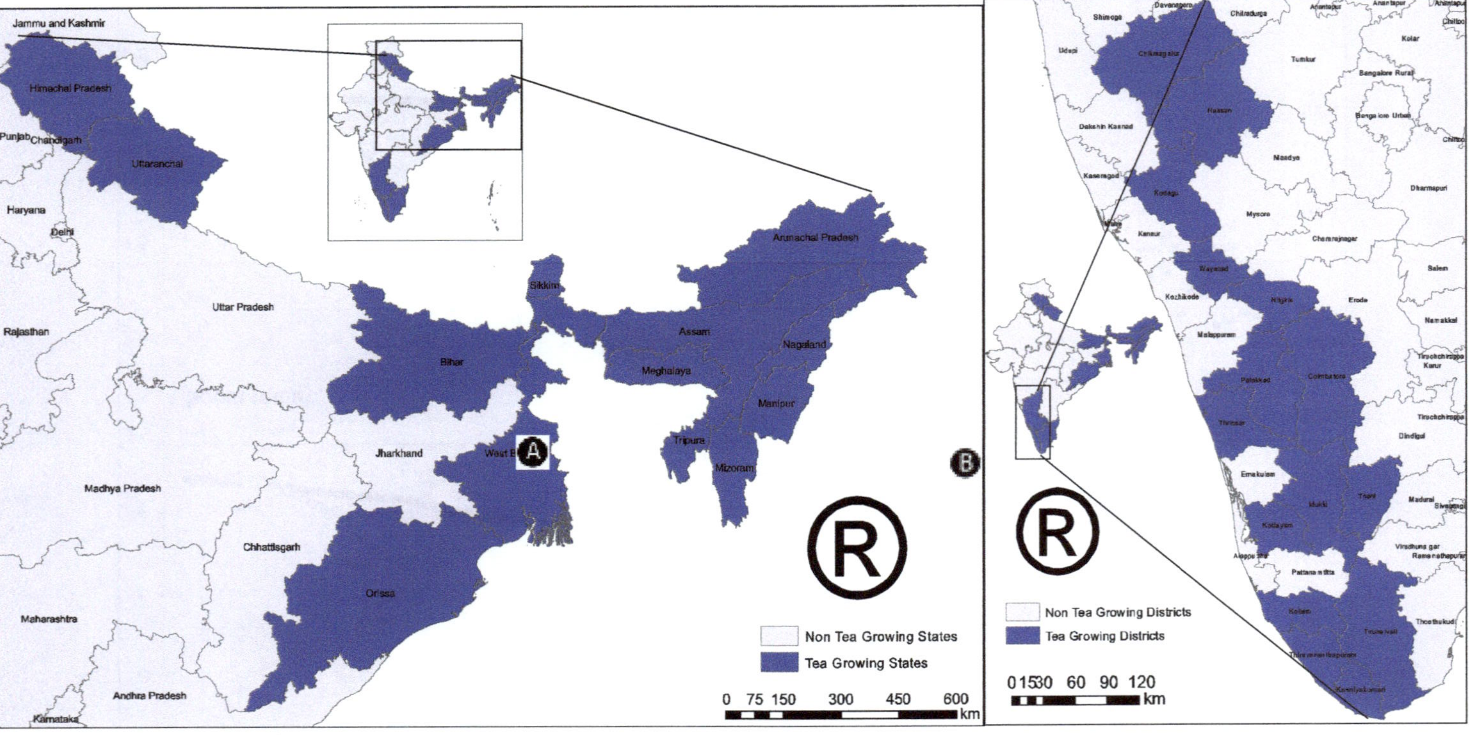

Figure 8.3: Tea Producing States of North India (a), and Districts of South India (b).

3. Favourable Growth Conditions for Tea

3.1. Climatic Requirements of Tea

Tea crop has rather specific agro-climatic requirements that are only available in tropical and sub-tropical climates, while some varieties can tolerate marine climates of British mainland and Washington area of the Unites States. Tea plant needs a hot, moist climate. Its specific requirements for growth and development are temperature ranging from 10 to 30 °C, minimum annual precipitation of 1250 mm, preferably acidic deep and well-drained soil with pH of 4.5-5.5 and adequate organic matter content. The botanical classification of tea of commercial plantations depends on the hybrids of three distinct 'ecotypes', the Assam-type (*Camellia assamica* sp. *assamica* (Masters) White), China-type (*Camellia sinensis* L. (O.) Kuntze) and Cambod type (*Camellia assamica* ssp. Lasiocalyx (Planch ex. Watt) (Wight, 1959). The Assam-type is believed to have originated under the shade of humid, tropical forests. China-type is thought to have originated under cool open conditions, humid tropics (Carr and Stephens, 1992). Because of the distinct difference in the geographical and ecological origins of the ecotypes, they exhibit considerable variations in their physiological, yield and quality traits (De Costa *et al.*, 2007).

4. Relationship between Yield and Climate

The impact of climate change on tea production of south Indian tea growing regions was established in the section using observed values. Preliminary investigation on mean yield of all the study area showed that in the last two decades, the yield recorded negative trend (Figure 8.4). Therefore, twenty years of climatic and productivity data (1991 – 2010) have been used in the present investigation.

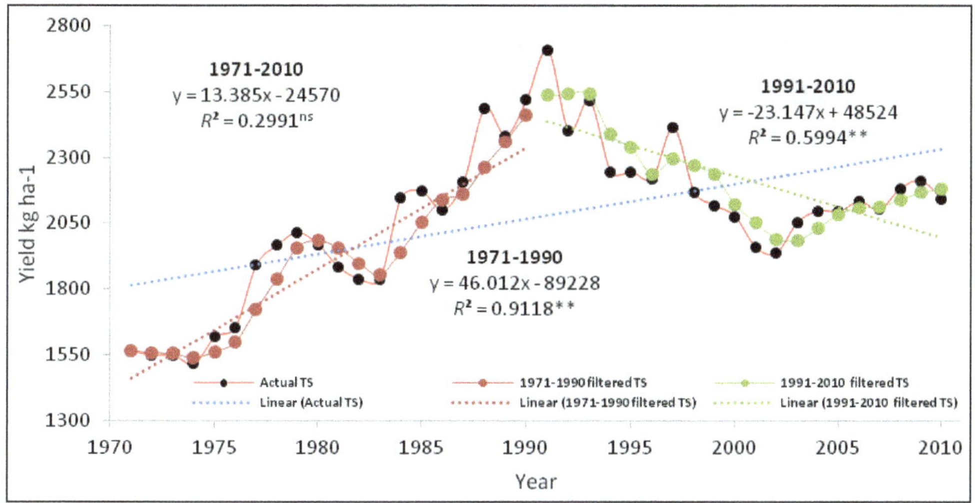

Figure 8.4: Time Series of Mean Yield of Four Tea Growing Regions of South India.

4.1. Yield

The mean yield recorded was maximum in the Anamalais (2832 kg/ha) and minimum in Vandiperiyar (1933 kg/ha) (Table 8.1). Based on the observed values, it is understood that if the climatic conditions are favourable, the Nilgiris could be the most benefiting region and if not, the Anamalais will be the most damaging region followed by Vandiperiyar and Munnar. Among the tea growing regions, the Anamalais, the Nilgiris and Vandiperiyar illustrated decreasing trend in productivity than that of Munnar, which demonstrated a positive increasing yield trend corresponding to time series. Nevertheless, it is highly significant at the Nilgiris ($r=-0.52$, $p<0.005$) and Vandiperiyar ($r=-0.68$, $p<0.001$).

4.2. Temperature

The minimum and maximum temperature recorded in these regions fluctuated from 12.72 to 18.05 °C and 22.20 (the Nilgiris) to 27.04 °C at Munnar and Vandiperiyar, respectively. It is astonishing to note that the mean deviation of annual minimum temperature of the study area oscillated from -2.3 (Munnar) to 2.2 °C (the Anamalais) and the maximum temperature exhibited -2.0 (Munnar) and 2.5 °C (the Nilgiris). Temperature minimum of the all tea growing regions are invariably decreasing and has significantly positive correlation with the yield. The increasing trend of maximum temperature had a negative relationship to crop production but not in Vandiperiyar, which may be due to some other biotic variables.

4.3. Relative Humidity

Relative humidity (RH) 14.30 h of all tea growing regions is consistently declining and RH 08.30 h exhibited an increasing trend except at Vandiperiyar. However, the decreasing trend of RH recorded during 14.30 h in the Anamalais and the Nilgiris showed a positively weak and non-significant correlation, while it has anegative trend at Munnar and Vandiperiyar.

4.4. Rainfall

The statistical analysis associated with the rainfall distribution is important aspects of rainfall climatology and integrated with crop production. It is well known that the bulk of the monthly, seasonal and annual rainfall at a station is contributed by a small percentage of the total number of rainy days with huge rainfall. On an average, the number of rain days (*i.e.*, ≥0.1 mm of rainfall) varied from 91.5 (The Nilgiris) to 156 mm (The Anamalais). Figure 8.5 showed the yearly frequency of number of rain days. A variety of form of relationship between monthly totals (T), number of rainy days (RD) and mean daily rainfall intensity (MDIs) were investigated on monthly and seasonal scale, wherein MDI was obtained by dividing T by RD. The lowest intensity (*i.e.*, mean rainfall/day) was recorded at Vandiperiyar (15.1 mm/day) and the moderate rainfall intensity at the Nilgiris (20.10 mm/day). These regions are situated in the rain-shadow region of the Western Ghats receiving highest rainfall during southwest rainfall. The highest rainfall intensity occurred over Munnar region (27 mm/day) followed by the Anamalais (25.23 mm/day) situated at the centre of Western Ghats which receives downpour during both the rainy seasons.

Table 8.1: Descriptive Statistics and Relationship between Yield and Climatic Bariables of the Study Locations

Region	*YPH (kg ha^{-1})*	*Tmin (°C)*	*Tmax (°C)*	*RH830 (%)*	*RH1430 (%)*	*Rainfall (mm)*	*Rain Days*	*Sunshine Hours*	*Sunny Days*
The Anamalais									
Mean	2832	15.43	25.38	90.12	72.89	3928	156.0	4.949	289.1
SD	206	0.572	0.874	3.509	5.518	941	19.85	0.452	14.38
MaxD	388	2.243	1.764	2.165	1.455	2955	54.05	1.041	25.90
MinD	–443	–1.342	–1.887	–0.965	–2.126	–1265	–32.95	–1.029	–29.10
Correlation	–0.14^{ns}	0.07^{ns}	–0.26^{ns}	–0.26^{ns}	0.10^{ns}	–0.31^{ns}	–0.07^{ns}	0.19^{ns}	0.20^{ns}
Trend	↓	↓	↑	↓	↑	↓	↓	↓	↓
The Nilgiris									
Mean	2350	13.79	22.20	78.89	75.78	1880	91.50	5.648	304.1
SD	247	0.439	0.344	3.231	2.895	300	9.758	0.498	20.46
MaxD	578	1.595	2.468	1.836	2.112	763	17.50	0.926	31.91
MinD	–304	–1.706	–1.257	–1.669	–1.740	–466	–18.50	–0.884	–27.58
Correlation	–0.52*	0.40^{ns}	–0.02^{ns}	0.75**	0.19^{ns}	–0.19^{ns}	0.55*	–0.31^{ns}	–0.39^{ns}
Trend	↓	↓	↓	↓	↑	↑	↓	↑	↑
Munnar									
Mean	2393	12.72	22.38	90.39	77.25	4038	149.9	4.645	280.2
SD	161	0.449	0.527	2.301	2.827	576	13.47	0.565	34.19
MaxD	274	2.187	1.410	1.568	2.212	1036	29.10	1.048	64.80
MinD	–395	–2.268	–1.990	–1.836	–1.944	–1162	–26.90	–1.486	–66.20
Correlation	0.03^{ns}	0.45*	–0.23^{ns}	–0.49*	–0.02^{ns}	–0.42^{ns}	–0.08^{ns}	0.17^{ns}	0.22^{ns}
Trend	↑	↓	↑	↓	↑	↓	↓	↓	↓
Vandiperiyar									
Mean	1933	18.05	27.04	89.85	68.74	2026	133.7	4.324	268.4
SD	235	0.576	0.458	1.371	2.993	370	17.30	0.558	22.40
MaxD	278	2.164	2.454	3.389	1.870	751	25.35	1.770	51.58
MinD	–440	–1.708	–1.898	–1.353	–1.416	–595	–31.65	–0.900	–38.42
Correlation	–0.68**	0.30^{ns}	0.13^{ns}	–0.36^{ns}	–0.34^{ns}	–0.07^{ns}	0.05^{ns}	0.47*	0.42^{ns}
Trend	↓	↓	↑	↑	↑	↓	↓	↓	↓

YPH: Yield per hectare; Tmin: Temperature minimum; Tmax: Temperature maximum; RH830 and RH1430: relative humidity respectively at 8:30 and 14:30 hours; SD: standard deviation; MaxD: Deviation maximum; Deviation minimum.

* Correlation is significant at the 0.05 level (2-tailed).

** Correlation is significant at the 0.01 level (2-tailed).

ns: Correlation is not significant at the 0.05 and 0.01 levels (2-tailed).

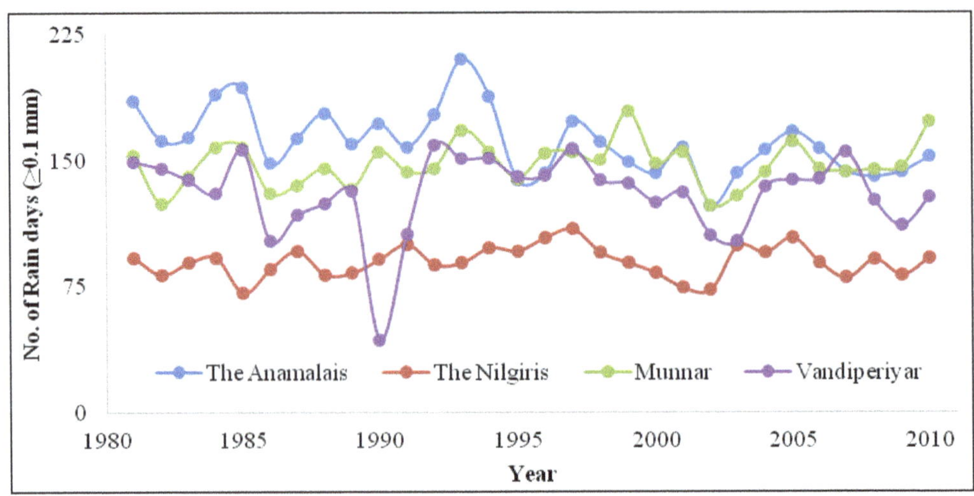

Figure 8.5: Yearly Frequency of Number of Rain Days at Four Tea Growing Regions of South India.

5. Impact of Carbon Dioxide and Technological Advancements on Tea Production

Carbon dioxide, most abundant greenhouse gases has increased from 280 parts per million (ppm) in the pre-industrial period to 398 ppm in 2014, with an average annual increase of 2.1 ppm per year (2005-2014, Figure 8.6a). The average for the previous decade (1995-2004) is 1.9 ppm per year (NOAA, 2015a). Sixty years trend of tea production in India has correlated with that of the increased area of cultivation (Figures 8.2b, c and d), however, it did not increase the yield significantly. To accurately calculate the yield change in future, it is necessary to consider technology development *e.g.* fertilizer (Figure 8.6b) and cultivar (Reynolds *et al.*, 1999) in addition to changes in the climatic variables and CO_2 effects (Ewert *et al.*, 2005). The main aim of the present exploration is to develop certain models for future changes in tea crop productivity. The four scenario families describe future worlds that may be global economic (A1F1), global environmental (B1), regional economic (A2) and region specific environment (B2). Productivity changes were calculated for the time period from 2010 (baseline year) until 2080 with an emphasis on the years 2020, 2050 and 2080 using IMAGE 2.2 implementation of the SRES scenarios (IMAGE-team, 2001).

5.1. Impact of Carbon Dioxide on Tea Production

The effect of higher CO_2 concentration on crop yield differed among years and scenarios from baseline (2010) but not among locations (Figure 8.7). This might be due to the fact that change in CO_2 concentration is global and not regional. The highest yield change by increasing CO_2 concentration was under scenario A1F1 in 2080, at Coimbatore district and the Nilgiris seem to be lowest under scenario B1 in 2020 of Idukki district. The calculated crop yield increases were the smallest for the regional environmental scenario (B2) and largest increases were the global

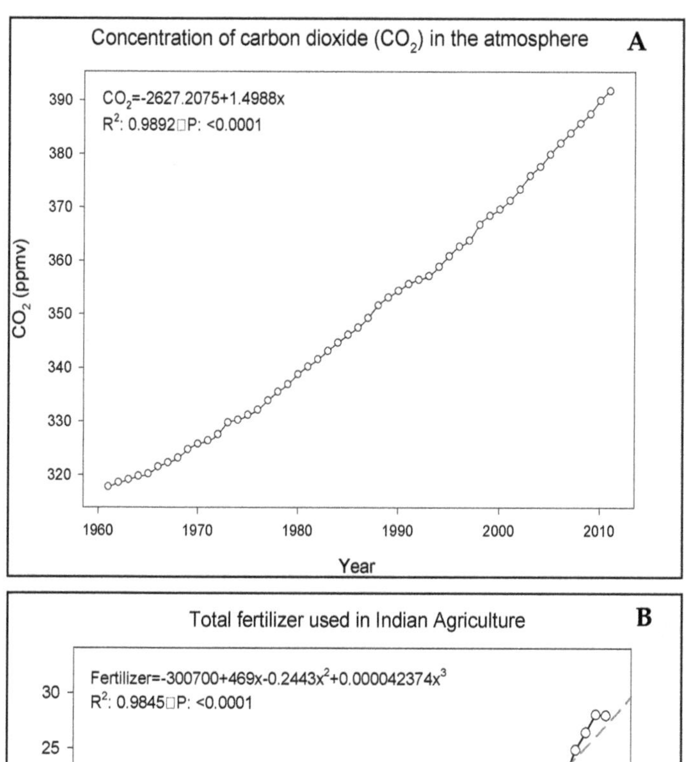

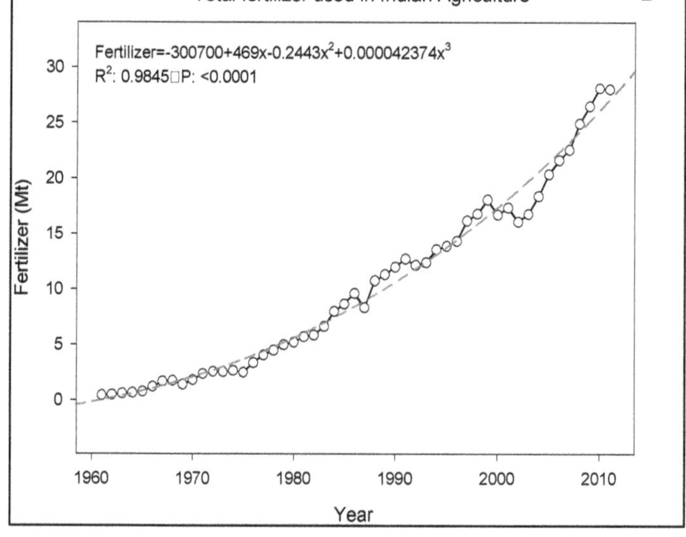

Figure 8.6: Concentration of Carbon Dioxide in the Atmosphere (a); and Total Fertilizer Used in Indian Agriculture (b).

economic scenario (A1FI) for 2020, 2050 and 2080, respectively due to the lower increase in CO_2 concentration under scenario. Differences among scenarios were relatively low in 2020, but increased with time indicating higher uncertainties for productivity change estimates at more distant futures. Elevated CO_2 concentration stimulates the rate of photosynthesis and causes higher biomass and crop yield (Bannayan *et al.*, 2009; De Costa *et al.*, 2006).

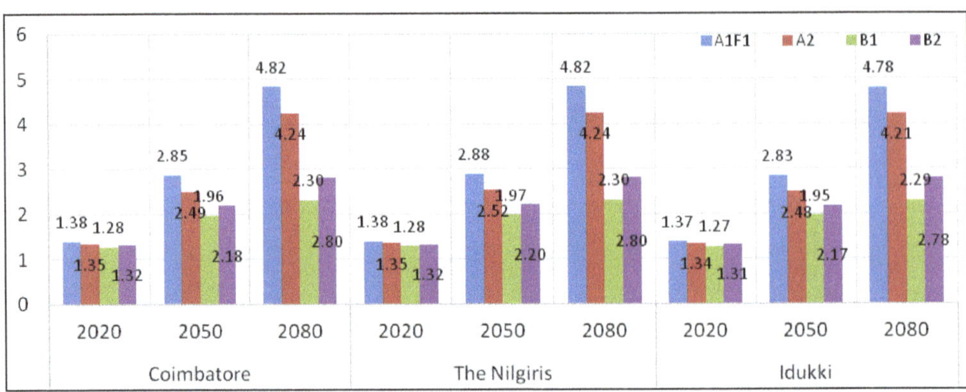

Figure 8.7: Estimated Relative Change in Crop Production as Affected by Carbon Dioxide for different Scenario IPCC Special Report on Emission Scenario.

5.2. Impact of Technological Advancements on Tea Production

Change in crop yield in tea production affected by technological development varied from location to location, scenarios and among the years (Figure 8.8). The effect of technology development on crop yield was in the range of 1.0 (Idukki, B2, 2050 and 2080) to 4.57 (Coonoor, A1F1, 2080) compared to the baseline year of 2010. Yield change under scenario A1F1 was higher because of higher values of yield potential and gap between actual and potential yields under scenario A1F1. Calculation of future yields from estimated relative productivity changes indicated that productivity will increase from baseline from 1.02 to 4.57 t/ha for the B2 and A1FI scenarios, respectively in 2080.

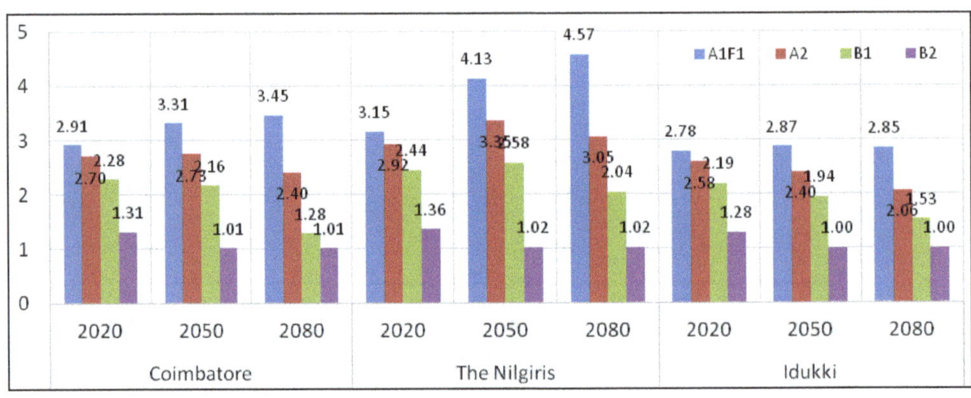

Figure 8.8: Estimated Relative Change in Crop Production as Affected by Technological Development for different Scenario IPCC Special Report on Emission Scenario.

5.3. Interactive Effects of CO_2 and Technology on Tea Production

Considering the factors together, which include climate change, CO_2 and technology development on yield is significantly important. As far as the data generated, when compared to baseline data, tea production was highest in the Nilgiris followed by the Anamalais and Idukki districts, which ranged from 0.59 to 7.08 tonnes per hectare (Figure 8.9). Estimated yield change as affected by integrated management system indicated either negative or positive impacts on yield.

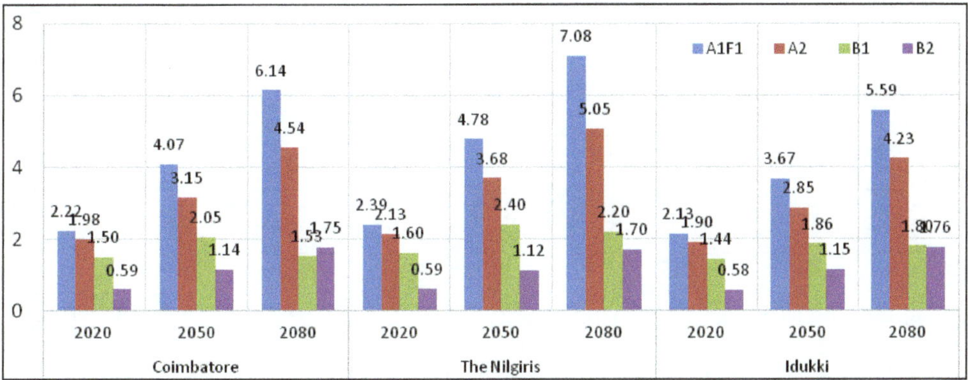

Figure 8.9: Estimated Relative Change in Crop Production as Affected by CO_2, Technological Development and Climate Change for different Scenario IPCC Special Report on Emission Scenario.

6. Impact of ENSO and SST on Crop Yield

6.1. *El Niño* Southern Oscillation

El Niño Southern Oscillation (ENSO), which can roughly be characterised as an irregular oscillation of the coupled atmosphere/ocean system in the tropical Pacific, which is associated with significant extreme weather conditions in the tropics and beyond (Kiladis and Diaz, 1989). ENSO events are the most important source of inter-annual climatic variability with a marked effect on rainfall patterns (IPCC, 2007). The NINO3 SST index is based on recorded data from 1981 to 2010 (NOAA, 2015b). Out of thirty years, five years were considered as warm (1982, 1986, 1987, 1991 and 1997), another five years as cold (1984, 1985, 1988, 1998 and 1999) and remaining 20 years were classified as neutral.

6.2. Impact of ENSO on Crop Yield

Crop yield and rainfall statistics during neutral and ENSO years are presented in Table 8.2. ENSO has both positive and negative influence on rainfall and crop productivity across the tea growing regions. Mean change either C-N or W-N in annual yield was positive in the Anamalais and Munnar and negative in the Nilgiris and Vandiperiyar. Typically, the regions showing positive trend characterised by mid elevation region except for the regions of high and mid elevation which showed a negative impact. The results demonstrated that both cold and warm ENSO event will not affect the crop productivity of mid elevation but it is not significant whereas, high and low elevation tea growing regions will be vulnerable to ENSO events.

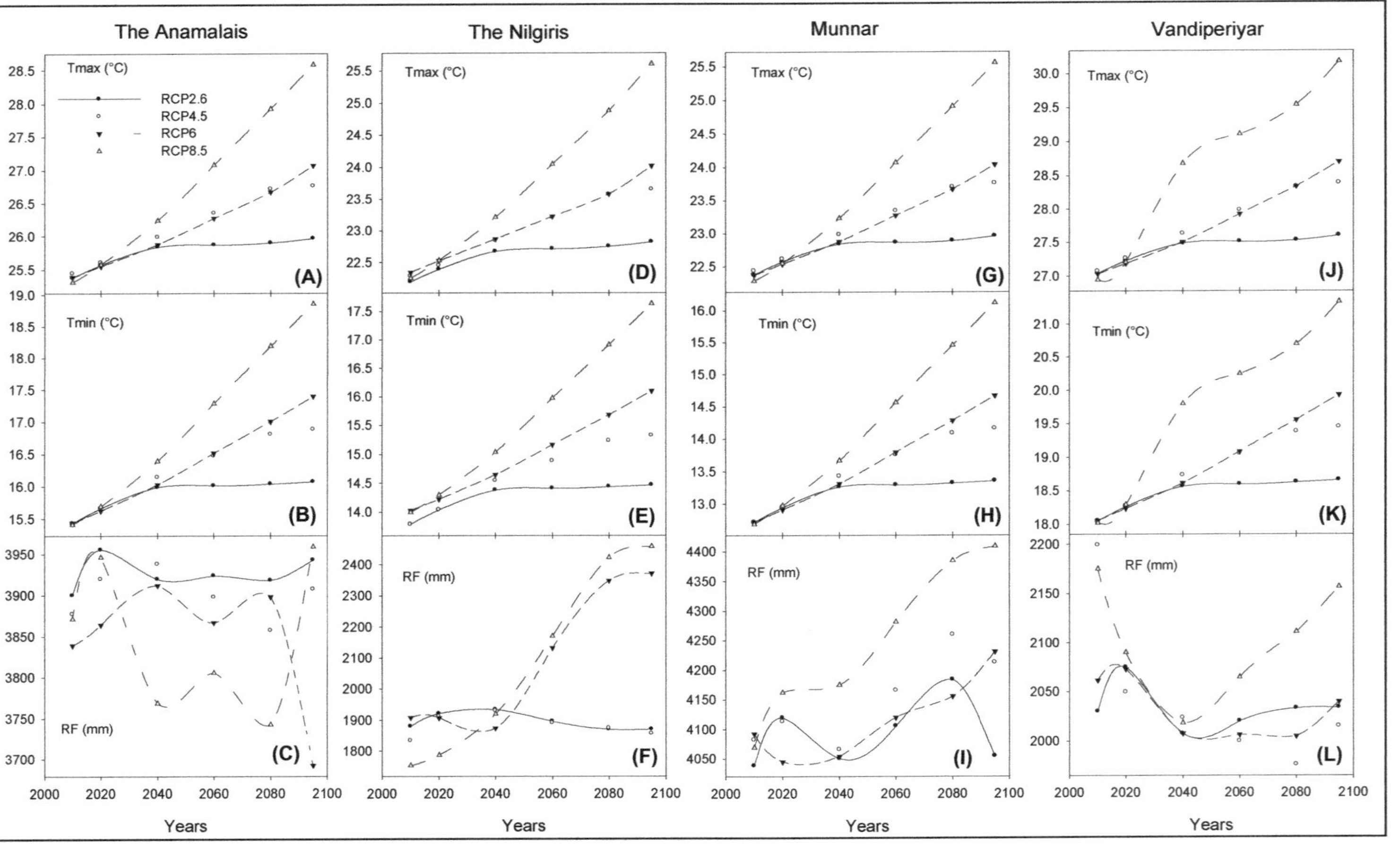

Figure 8.10: Projected Change in Temperature Maximum (a, d, g, j), Temperature Minimum (b, e, h, k) and Rainfall (c, f, I,l) in the Anamalais (a-c), Nilgiris (d-f), Munnar (g-l) and Vandiperiyar (j-L).

Table 8.2: Annual Yield and Rainfall Characteristics at the Study Site under Warm (W), Cold (C) and Neutral (N) ENSO Phases (1991-2010)

Particulars	*Yield ha⁻¹*				*Rainfall (mm)*			
	a	*b*	*c*	*D*	*a*	*b*	*c*	*D*
Annual								
C	2923	2306	2385	1908	3671	1706	4004	2194
N	2758	2387	2372	1948	4218	1904	4207	2075
W	2921	2305	2444	1921	3497	1966	3694	1782
C-N	165	–81	12	–41	–547	–198	–202	119
W-N	163	–82	71	–27	–721	62	–512	–294
χ^2 (df=2)	4.724 (0.094)	1.609 (0.447)	0.690 (0.708)	0.894 (0.640)	1.808 (0.405)	3.0900 (0.213)	2.644 (0.267)	2.858 (0.240)
SD								
C	184	224	223	359	679	122	200	405
N	134	258	177	220	1112	330	667	339
W	313	278	61	211	491	321	451	359

(a) The Anamalais; (b) The Nilgiris; (c) Munnar; and (d) Vandiperiyar of Idukki.

6.3. Impact of ENSO on Rainfall

Decrease in annual rainfall ranged from 17 to 721 mm during cold events as compared with the neutral years. The results indicated that warm ENSO phase increase rainfall at high elevation region and cold ENSO event increase rainfall of low elevation region. From the results, it was inferred that both the mid-elevation tea growing regions are on critical situations and both cold and warm phases of ENSO events affect negatively on the rainfall. Mean year-to-year annual rainfall variability (SD) was higher in the Anamalais (1112 mm) during the neutral years as compared with cold (679 mm) and warm (491 mm) years followed by Munnar. The Nilgiris and Vandiperiyar showed comparatively least difference in mean annual rainfall between ENSO and neutral years.

6.4. Impact of SST on Crop Yield and Rainfall

Correlation of sea surface temperature (SST) anomalies over the different sectors of the tropical Pacific Ocean with rainfall and crop production have been presented in Table 8.3. The positive and weak correlation between SST anomalies from June to August (JJA) was observed over the NINO3 sector and south Indian annual tea production anomalies of all the study locations. However, SST had negative relationship with rainfall except for the Nilgiris, which was significant at 1 per cent level. The correlation is positively stronger for the NINO1+2 regions (except few) than NINO3, NINO3.4 and NINO4. Our findings highlight that yield and production of south Indian tea production appeared to be less responsive to ENSO phase which may be due to improvements in production technology that mitigate problems associated with excess rainfall.

Table 8.3: Correlation of Normalized Rainfall, Annual Production (kg ha^{-1}) and Pacific SST Anomaly (JJA) over different Sectors (for NINO 3 region; correlation were worked for four seasons) for the period from 1991 to 2010; n=20a

Particulars	*NINO1+2*	*NINO3*				*NINO3.4*	*NINO4*
		DJF	*MAM*	*JJA*	*SON*		
Crop Yield/ha							
The Anamalais	0.475	0.310	0.182	0.323	0.000	–0.141	0.131
The Nilgiris	0.125	–0.364	–0.075	0.038	0.524	–0.049	0.019
Munnar	0.406	0.223	0.260	0.226	–0.341	–0.193	–0.026
Vandiperiyar	0.255	0.302	0.178	0.071	–0.241	–0.204	–0.073
Rainfall							
The Anamalais	–0.141	–0.156	–0.001	–0.010	0.205	0.331	0.204
The Nilgiris	0.350	0.040	–0.121	0.402	0.025	0.183	0.348
Munnar	–0.259	–0.162	0.007	–0.209	0.279	0.019	–0.060
Vandiperiyar	0.010	–0.141	0.086	–0.151	0.030	–0.085	–0.119

a: Correlation coefficient of 0.44 and 0.56 are significant at 5 per cent and 1 per cent respectively; DJF: December-February; MAM: March-May; JJA: June-August; SON: September-November.

7. Future Climate Projections using GCM Models

Keeping current climate as the baseline, we used historical climate data from www.worldclim.org database (Hijmans *et al.*, 2005) and calibrated with observed station data. The spatial resolution of the GCM results is inappropriate for estimating or quantifying the climate impacts on agriculture as in almost all cases the grid cells measure more than 100 km a side. It is also important to identify the climate models, which can at least simulate the present climate with a reasonable accuracy in mean, variability and change in critical climate variables. Downscaling is therefore, needed to provide high-resolution surfaces of expected future climates if the likely impacts of climate change on agriculture are to be more accurately forecasted. The method assumes that changes in climates are only relevant at coarse scales, and that relationships between variables are maintained towards the future (Ramirez and Jarvis, 2010). Downscaling method was applied on over 17 GCMs from the IPCC Fifth Assessment Report (2014) for the newest representative concentration pathways (RCPs) emission scenarios (IPCC, 2014).

7.1. Downscaling of Temperature

Figure 8.10 is based on the Coupled Model Intercomparison Project Phase 5 (CMIP5) ensemble (20) of 17 GCM models; changes calculated with respect to the 1991–2010 period. This comparison will help us to detect climate change that will occur in future and getting suitable management to decrease its effects. The results of the projected climate change over the south Indian tea growing regions for RCP using GCMs showed an increasing trend for maximum and minimum temperature in all regions. The increase in minimum temperatures is higher than that of maximum temperatures in all the locations under study. The changes in maximum and minimum temperature will be smallest for the RCP2.6 and largest in RCP8.5 for 2020, 2040, 2060, 2080 and 2095, respectively due to the radioactive forcing under each scenario. Means of minimum temperatures revealed that the increase in maximum temperature will be higher in Vandiperiyar (1.95 °C) during 2040-2060 and 2.98 °C during 2080-2100 at the Nilgiris. Similarly, the increase in minimum temperature was 2.0 °C during 2040-2060 in Vandiperiyar and was 2.98 °C during 2080-2100 in the Nilgiris. The increase of global mean surface temperature by the end of the 21st century (2081–2100) relative to 1986–2005 is likely to be 0.3 to 1.7 °C under RCP2.6, 1.1 to 2.6°C under RCP4.5, 1.4 to 3.1°C under RCP6.0 and 2.6 to 4.8 °C under RCP8.5.

7.2. Downscaling of Rainfall

The changes in precipitation are not uniform in all the locations. The rainfall pattern of Anamalais is less or has no change (SD ±150 mm). In Vandiperiyar, it is decreasing by 200 mm while in the Nilgiris and Munnar, it is increasing in the range between 600 and 400 mm respectively. The Nilgiris and Munnar are likely to experience an increase in annual mean rainfall under the RCP8.5 scenario. In mid and low altitude regions such as the Anamalais and Vandiperiyar, the mean precipitation is likely to decrease under the RCP8.5 scenario (Figure 8.10). Extreme precipitation events over most of the regions are likely to become more intense and frequent.

7.3. Land Suitability of the Tea Planting Regions under Future Climate Scenario

To explore the sustainability of the tea planting regions, it is necessary to first evaluate the land ecological suitability evaluation (LESE) based on geographic information system (GIS) (Brail and Klosterman, 2001; Chen *et al.*, 2010; Collins *et al.*, 2001; Moreno and Seigel, 1988). However, there have been limited studies which evaluate the land ecological suitability and rational spatial distribution of tea crops using GIS. The increasing needs of the tea industry and the shortage of resources are stimulating the demand for effective methods of LESE which can help decision makers and farmers preserve and explore more highly suitable regions and satisfy producers' demands for increasing profits (Geerts *et al.*, 2006; Kalogirou, 2002). Hijmans *et al.* (2012) have developed a mechanistic model based on the Ecocrop database to spatially predict crop suitability. The model essentially uses minimum, maximum, and mean monthly temperatures, and total monthly rainfall to determine a suitability index based on each parameter separately, to finally determine an overall suitability rating.

Figure 8.11 showed that marginal, very marginal, suitable, very suitable and excellent suitable area under current and future scenarios respectively. The results showed that there is a mild to strong shift in suitability and differed among regions, altitude, soil, climate and topography when compared with the baseline scenario. The total area of the study covered 1,08,901 ha including three districts and five regions. Highly suitable regions were identified as sites that are highly advantageous for the growth of tea crops under future scenario. These sites covered an area of

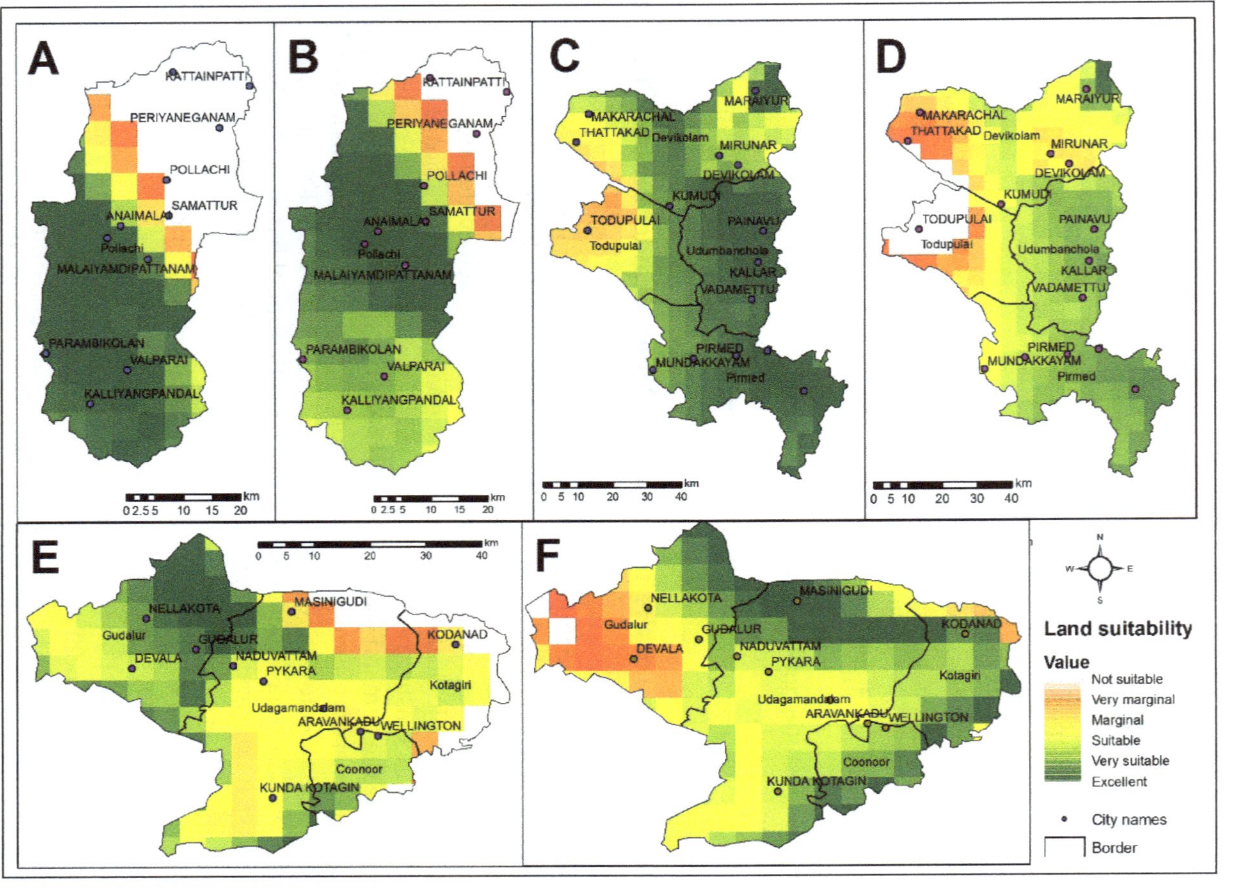

Figure 8.11: A Land Suitability Evaluation (LESE) Map of Tea Production Regions of South India. Land suitability under baseline (a, c, d); future scenario (b, d, f); a and b: the Anamalais; c and d Munnar and Vandiperiyar; e and F: the Nilgiris.

31,491 ha which represent 30.0 per cent total evaluation area and it is 6,027 ha less (4.4 per cent) than baseline. They were mainly located in the northern regions of the Anamalais and the Nilgiris along with the eastern regions of Vandiperiyar and Munnar. Regions deemed very suitable which covered 26,215 and 36,087 ha and accounted 59.4 per cent of the total evaluated area.

They were located in east border of Anamalais and Munnar and the central regions of Udhagamandalam and Gudalur. Very marginal to marginal areas were located in the southern and southwest regions of the Nilgiris *viz.*, Gudalur and Udhagamandalam. These areas have arguably the greatest need for new crop varieties that are tolerant to extreme climatic conditions, especially to drought. This result established that the reasonable land use management policies and land use planning measures should be developed to maintain sustainable development of tea cultivation in south Indian tea plantations.

8. Conclusions

Climate is changing at an unprecedented rate and the magnitude of change is highly variable from place to place, largely the regions that are already struggling from the impacts of irregular and extreme climate events. It is important to note that the extreme climatic conditions are increasing, which in turn instead of process pose process significant stress on crops. Areas suitable for cultivation of tea crops present day may not be suitable for cultivation in future and vice-versa. Hence it is important to identify such areas in high spatial resolution which will be of key input in developing future planning and sustain the productivity. Another chief need is to develop new crop varieties which can sustain the extreme region specific ecological conditions; however, varieties improved for cultivation in one region could be adopted for cultivation elsewhere, where they would face the same abiotic and biotic stresses. Therefore, the degree of fitness of developed cultivars/ varieties to environmental condition will vary among the regions and over the time. Blending the resistance traits of traditional varieties through modern plant breeding will ensure that new varieties are bred to suit specific requirements of local conditions and communities. Further analysis is needed to identify tolerant clones/varieties and areas to target for climate adaptation strategies, particularly for improved tolerant varieties against climate change. Adaptation of climate ready crops to changing climate conditions will also allow the cultivation to continue in current as well as new suitable areas.

Acknowledgements

The authors are thankful to Advisory Officers of all the Regional Centre for making available meteorological and yield data. This work was supported by the CSIR fourth paradigm institute (CSIR-4PI), formally CSIR Centre for Mathematical Modelling and Computer Simulation (C-MMACS), Banglore through SPARK programme.

References

Bannayan, M., Tojo Soler, C.M., Garcia y. Garcia, A., Guerra, L.C., and Hoogenboom, G. (2009). Interactive effects of elevated CO_2 and temperature on growth and development of a short-and long-season peanut cultivar. *Clim. Change*, 93(3-4), 389–406. doi:10.1007/s10584-008-9510-1.

Brail, R., and Klosterman, R. (2001). Planning support systems: Integrating geographic information systems, models, and visualization tools, *Journal of the American Planning Association*. ESRI Press, Redlands, CA.

Carr, M.K.V., and Stephens, W. (1992). Climate, weather and the yield of tea. In: Tea: *Cultivation to Consumption*. Springer Netherlands, Dordrecht, pp. 87–135. doi:10.1007/978-94-011-2326-64.

Chanda, N. (1957). Tea catechins. *Two and Bud*, 4(2), 5.

Chen, H., Liu, G., Yang, Y., Ye, X., and Shi, Z. (2010). Comprehensive evaluation of Tobacco ecological suitability of Henan province based on GIS. *Agric. Sci. China*, 9(4), 583–592. doi:10.1016/S1671-2927(09)60132-2.

Collins, M.G., Steiner, F.R., and Rushman, M.J. (2001). Land-use suitability analysis in the United States: Historical development and promising technological achievements. *Environ. Manage*, 28(5), 611–621. doi:10.1007/s002670010247.

De Costa, W.A.J.M., Mohotti, A.J., and Wijeratne, M.A. (2007). Ecophysiology of tea. *Brazilian J. Plant Physiol*, 19(4), 299–332. doi:10.1590/S1677-04202007000400005.

De Costa, W.A.J.M., Weerakoon, W.M.W., Herath, H.M.L.K., Amaratunga, K.S.P., and Abeywardena, R.M.I. (2006). Physiology of yield determination of rice under elevated carbon dioxide at high temperatures in a subhumid tropical climate. *F. Crop. Res*, 96(2-3), 336–347. doi:10.1016/j.fcr.2005.08.002.

Evans, L.T. (1997). Adapting and improving crops: the endless task. Philos. *Trans. R. Soc. B Biol. Sci*, 352(1356), 901–906. doi:10.1098/rstb.1997.0069.

Ewert, F., Rounsevell, M.D.A., Reginster, I., Metzger, M.J., and Leemans, R. (2005). Future scenarios of European agricultural land use. Agric. Ecosyst. *Environ*, 107(2-3), 101–116. doi:10.1016/j.agee.2004.12.003.

FAOSTAT, (2015). Food and Agriculture Organization of the United Nations. (Retrieved February 19, 2015) from http://faostat3.fao.org/faostat-gateway/go/to/download/Q/QC/E.

Geerts, S., Raes, D., Garcia, M., Del Castillo, C., and Buytaert, W. (2006). Agro-climatic suitability mapping for crop production in the Bolivian Altiplano: A case study for Quinoa. *Agric. For. Meteorol.* **139(3-4):** 399–412. doi:10.1016/j.agrformet.2006.08.018.

Hijmans, R.J., Cameron, S.E., Parra, J.L., Jones, P.G., and Jarvis, A. (2005). Very high resolution interpolated climate surfaces for global land areas. *Int. J. Climatol.* **25(15)**: 1965–1978. doi:10.1002/joc.1276.

Hijmans, R.J., Luigi, G., and Mathur, P. (2012). DIVA-GIS Version 7.5: Manual.

Horie, T. (1993). Predicting the effects of climatic variation and elevated CO_2 on rice yield in Japan. *J. Agric. Meteorol.* 48(5): 567–574. doi:10.2480/agrmet.48.567.

Horie, T., Nakagawa, H., Ohnishi, M., and Nakano, J. (1995). Rice production in Japan under current and future climates. In: Matthews, R.B., Kropff, M.J., Bachelet, D., van Laar, H. (Eds.), *Modeling the impact of climate change on rice production in Asia*. CAB International, Wallingford, UK, pp. 143–164.

IMAGE-team, (2001). The IMAGE 2.2 Implementation of the SRES scenarios: A comprehensive analysis of emissions, climate change and impacts in the 21st century. CD-ROM publication 481508018.

IPCC, (2014). Climate Change 2014: Synthesis Report. Contribution of Working Groups I, II and III to the Fifth Assessment Report of the Intergovernmental Panel on Climate Change. IPCC, Geneva, Switzerland.

IPCC, (2007). Climate Change 2007: Impacts, adaptation and vulnerability: contribution of Working Group II to the fourth assessment report of the Intergovernmental Panel, Genebra, Suíça. Cambridge University Press, Cambridge, New York, USA.

Kalogirou, S. (2002). Expert systems and GIS: an application of land suitability evaluation. Comput. *Environ. Urban Syst*, 26(2-3), 89–112. doi:10.1016/S0198-9715(01)00031-X.

Kiladis, G.N., and Diaz, H.F. (1989). Global Climatic anomalies associated with extremes in the southern oscillation. *J. Clim*, 2(9), 1069–1090. doi:10.1175/1520-0442(1989)002<1069: GCAAWE>2.0.CO_2.

Lobell, D.B., Field, C.B., Cahill, K.N., and Bonfils, C. (2006). Impacts of future climate change on California perennial crop yields: Model projections with climate and crop uncertainties. *Agric. For. Meteorol*, 141(2-4), 208–218. doi:10.1016/j.agrformet.2006.10.006.

Matthews, R., Horie, T., Kropff, M., Bachelet, D., Centenom, H., Shin, J., Mohandas, S., Singh, S., Zhu, D., and Lee, M. (1995). A regional evaluation of the effect of future climate change on rice production in Asia. In: Matthews, R.B., Kropff, M.J., Bachelet, D., van Laar, H. (Eds.), *Modelling the impact of climate change on rice production in Asia*. CAB international, Wallingford, UK, pp. 95–139.

Moreno, D., and Seigel, M. (1988). A GIS approach for corridor siting and environmental impact analysis. In: *Proceedings of GIS/LIS 1988 the Third Annual International Conference*. American Society for Photogrammetry and Remote Sensing, Falls Church, San Antonio, *Texas*, pp. 507–514.

NOAA, (2015a). Scripps CO_2 monitoring program. Scripps Inst. Oceanogr. Univ. Calif. - San Diego, USA. (Retrieved March 18, 2015) from ftp://aftp.cmdl.noaa.gov/products/trends/co2/co2_ annmean_mlo.txt.

NOAA, (2015b). NCEP/Climate Prediction Centre. El Niño South. Oscil. (Retrieved April 25, 2015) from http://www.cpc.ncep.noaa.gov/products/precip/CWlink/MJO/enso.shtml.

Owuor, P.O., Kamau, D.M., Kamunya, S.M., Msomba, S.W., Uwimana, M.A., Okal, A.W., and Kwach, B.O. (2011). Effects of genotype, environment and management on yields and quality of black tea. In: *Lichtfouse*, E. (Ed.), *Genetics, Biofuels and Local Farming Systems, Sustainable, Sustainable Agriculture Reviews.* Springer Netherlands, Dordrecht, pp. 277–307. doi:10.1007/978-94-007-1521-9_10

Raj Kumar, R., Edwin Raj, E., and Mohan Kumar, P. (2012). Changing pattern of southwest monsoon in the Anamalais. Planters' Chronicle, 108(1 and 2), 10–14.

Raj Kumar, R., and Mohan Kumar, P. (2010). Contributing factors of crop productivity in Tea: A case study in the Anamalais. Planters' Chronicle, 106(1), 15–23.

Raj Kumar, R., and Mohan Kumar, P. (2009). Climate change adaptation strategies in Agriculture: Influence of ecological variables on productivity in tea. In: Prasada Rao, G., Kesava Rao, A., Alexander, D. (Eds.), National Seminar on *"Climate Change Adaptation Strategies."* Centre for Climate Change Research, Kerala Agricultural University, Vellanikkara, Thrissur, Kerala, pp. 139–153.

Ramesh, K.V. and Goswami, P. (2014). Assessing reliability of regional climate projections: the case of Indian monsoon. *Sci. Rep.* 4, 4071. doi:10.1038/srep04071.

Ramesh, K. V., and Goswami, P. (2007). Reduction in temporal and spatial extent of the Indian summer monsoon. *Geophys. Res. Lett*, 34(23), 1-6. doi:10.1029/2007GL031613.

Ramirez, J., and Jarvis, A. (2010). Disaggregation of global circulation model outputs decision and policy analysis, Policy Analysis.

Reynolds, M.P., Rajaram, S., and Sayre, K.D. (1999). Physiological and genetic changes of irrigated Wheat in the post-green revolution period and approaches for meeting projected global demand. *Crop Sci,* 39(6), 1611. doi:10.2135/cropsci1999.3961611x

Rosenzweig, C., Singh, U., and Padilla, J.L. (1995). Simulating Rice response to climate change. In: Rosenzweig, C. (Ed.), *Climate Change and Agriculture: Analysis of Potential International Impacts. American Society of Agronomy,* pp. 99–121. doi:10.2134/asaspecpub59.c5

Saharia, U., and Bezbaruah, H. (1984). Effect of timing of fertilizer application on flowering and seed-setting of tea seed trees in northeast India. *Two and Bud,* 31(2), 12–14.

Wight, W. (1959). Nomenclature and classification of the tea plant. *Nature,* 183(4677), 1726–1728. doi:10.1038/1831726a0

2017, Impact of Climate Change on Plantation Crops Pages 145–155
Editors: K.B. Hebbar, S. Naresh Kumar & P. Chowdappa
Published by: ASTRAL INTERNATIONAL PVT. LTD., NEW DELHI

Chapter 9

Rubber

R. Krishnakumar, James Jacob, K.K. Jayasooryan and P.R. Satheesh

1. Introduction

Natural Rubber (*Hevea brasiliensis*) is a perennial rain fed tree crop that has been cultivated in India since the beginning of the 20th century (Sethuraj and Jacob, 2012). Traditionally this crop has been cultivated along the foothills of the Western Ghats up to an altitude of about 450-500 m above MSL in Kanyakumari district of Tamil Nadu and Kerala. In recent decades its cultivation has expanded to further North along the Western Ghats as well as in parts of North East India and pockets along the Eastern Ghats (Figure 9.1) (Krishan, 2013).

Several studies have shown that climate of the rubber growing regions have undergone major changes in recent years/decades (Jacob, 2013). The changes have been more spectacular in terms of rise in temperature (particularly daily maximum temperature) and alterations in the distribution of rainfall and to some extent a reduction in the amount of annual rainfall (Rajeevan *et al.*, 2008; Jacob, 2013.).

A number of studies have shown a clearly changing pattern of rainfall over different Indian terrains during the recent decades (Lal *et al.*, 2001; Rajeevan, 2001; Goswami *et al.*, 2006). The change in rainfall pattern of the spice and plantation regions of India is characterised by an increasing trend in south west monsoon (SWM) over northeast and south west regions and decreasing trend along other regions (Guhathakurta and Rajeevan, 2006). There was a significant reduction of 232.6 mm rainfall during SWM and an increase of 93.9 mm in northeast monsoon (NEM) during the last 135 years in Kerala. Percentage departure of the decadal mean rainfall from the normal indicated that the excess rainfall years during SWM were frequent in the recent decades when compared to the earlier decades from 1871 to 2005 (Krishnakumar *et al.*, 2009).

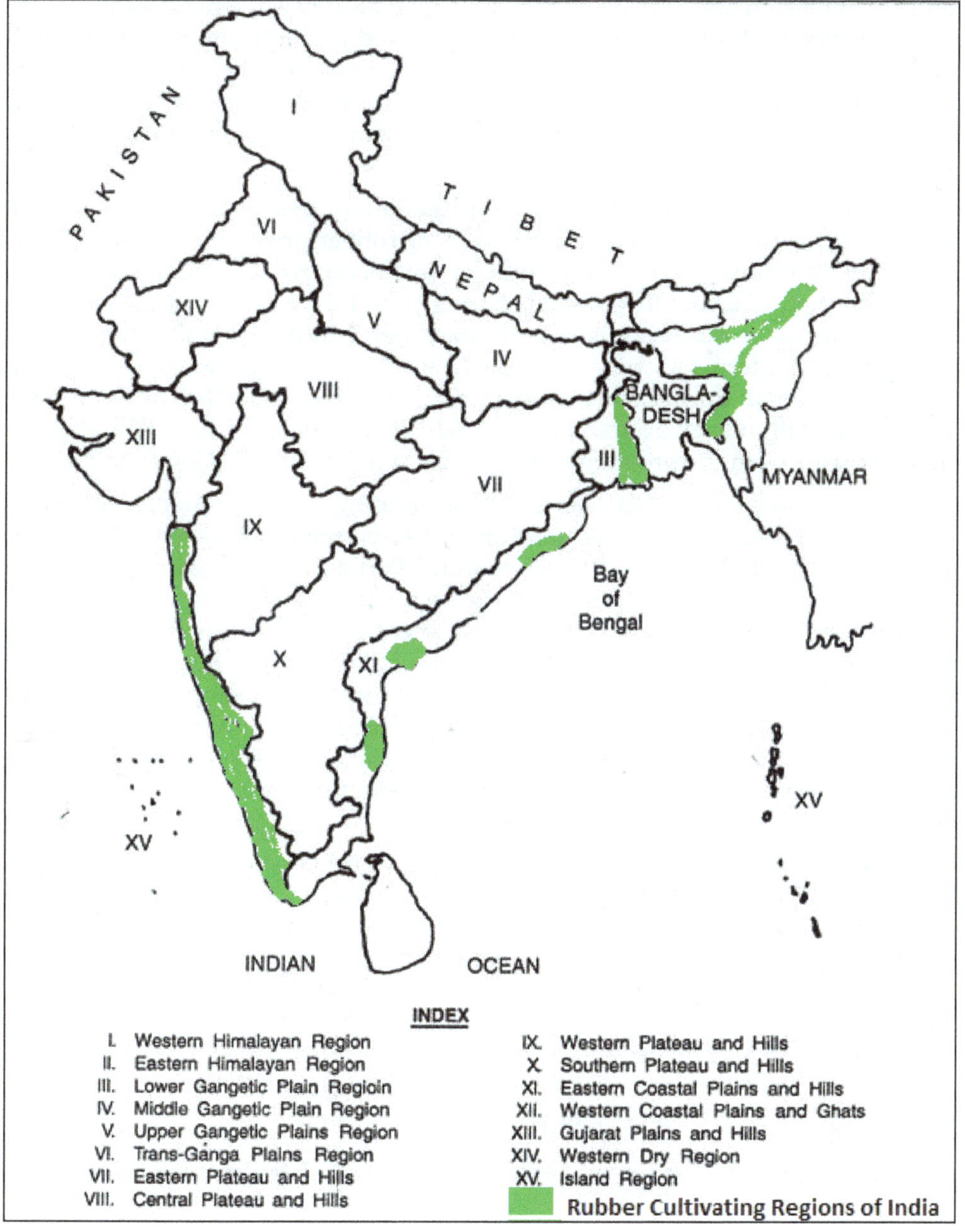

Figure 9.1: Agro Climatic Regions in India (categorised by the Planning Commission of India) showing where rubber is cultivated. Hatched areas only indicative and not drawn to scale.

Guhathakurta and Rajeevan (2006) showed that a significant decreasing trend in the monsoon rainfall for three sub-divisions, Jharkhand, Chhattisgarh and Kerala. Subdivisions like Gangetic West Bengal, Western Uttar Pradesh, Jammu and Kashmir, Konkan and Goa, Madhya Maharashtra, Rayalseema, coastal Andhra Pradesh and north interior Karnataka experienced a significant increasing trend. Goswami *et al.* (2006) showed an increasing trend in extreme rain events over central India during the monsoon season. At the same time, significant decreasing trend in the frequency of moderate events during the same period. But the annual mean rainfall showed no significant trend.

A change in precipitation pattern of Kerala requires special attention while discussing about climate change responses of rubber growing in the traditional area. Regarding the trends of precipitation in Kerala, Krishnakumar *et al.* (2008) have observed that monthly rainfall during June-July was decreasing while the same was increasing during august and September, indicating a shift in the monthly precipitation pattern. Pal and Al-Tabbaa, (2009) have reported an increasing trend in extreme rainfall during winter and autumn in Kerala while the spring seasonal extreme rainfall showed decreasing trend, which may lead to increasing probability of water scarcity in the pre-monsoon time and a delaying monsoon onset.

Studies have shown that even though a warming trend is evident across many regions of the country, it is not uniform and some regions show a cooling trend also (Kumar and Hingane, 1998; Kumar and Parikh, 1998; Singh and Sontakke. 2002; Gadgil and Dhorde, 2005). Kerala being the largest rubber cultivating region in the country, India Meteorological Department (IMD) has reported an increase of 0.8°C, 0.2°C and 0.5°C increases in maximum, minimum and mean temperatures respectively over Kerala. It is also reported that the maximum temperature of Kerala increased by 0.64°C while the minimum temperature rose 0.23°C during the period of 1956 to 2004 (Rao *et al.*, 2009). Jayasooryan *et al.* (2015) reported an increasing trend with respect to the indices of extreme temperature events in a rubber growing region in Kerala.

The long term climatic events in the traditional rubber growing regions in India showed a difference of approximately 2°C in maximum and minimum temperatures between two decades (Raj *et al.*, 2011; Satheesh, 2014). This is a good indicator of local climate variability in the traditional rubber growing regions. Decrease in the number of rainy days during the monsoon season with no change over the annual period in the traditional region suggests that the spread of rain fall is becoming more skewed in nature. Impacts of climate change are felt more strongly through changes in climate extremes. Temperatures are rising, rainfall is becoming more skewed and bright sunshine hour per day is decreasing.

2. Climate Change and Growth and Establishment of Rubber in the Field

Throughout the world, natural rubber is cultivated as a rain fed crop and irrigation is limited only to the nursery. However, in recent years many growers in Kerala had to provide life saving irrigation to young rubber plants during the first summer after field planting (RRII, 2015). Normally young plants are planted in the field in Kerala during the month of June-July when sufficient soil moisture is assured and sunshine intensity is usually low. In recent years there has been considerable rise in the day temperature, number of active sunshine hours and a reduction in the amount of rainfall received during June – July (Krishnakumar *et al.*, 2008). These changes can have an adverse effect on the initial survival and establishment of young rubber plants in the field. Due to these adverse changes in the recent years, an increasing proportion of growers are forced to give at least one irrigation during peak summer time in order to stop young plants from drying up.

Casualties posed by uncertain weather pattern on young plants and different mitigation measures adopted by farmers along the traditional rubber growing regions were studied by RRII. The study revealed that 16 per cent of farmers in Kerala are providing irrigation to young plants during the early stages of establishment. In Southern Kerala only 3 per cent of the farmers practising irrigation, while in Central Kerala 30 per cent farmers providing irrigation, which keeps the casualty below 1 per cent. Along the North-Central Kerala and Northern Kerala 19 and 20 per cent of farmers provide irrigation respectively. Visual drought symptoms due to soil moisture stress was manifested as yellowing in the early stages, which was seen in almost all the regions studied. Severe cases of yellowing, leaf tip dryness and leaf shedding were seen in North-Central Kerala, indicating low water availability during summer period (RRII, 2015).

3. Climate Change and Diseases of Rubber

There are a few important fungal diseases affecting natural rubber. Young rubber plants are affected by fungi such as *Colletotrichum, Phytophthora, Oidium, etc.* (Liyange and Jacob., 1992 Manju *et al.*, 2001 Gogoi and Dey., 2012). The most important disease affecting mature rubber tree is Abnormal Leaf Fall disease caused by *Phytophthora* spp. All these fungal diseases have a direct association with weather parameters such as rainfall, relative humidity and temperature. These important weather parameters have undergone significant changes in the rubber growing regions of India in the recent times.

Most parts of the traditional natural rubber growing regions of India, extending from Kanyakumari district in the South to Kasaragod district in the North received excess and prolonged rains during 2013. This led to severe incidence of Abnormal Leaf Fall (ALF) disease caused by the fungus, *Phytophthora* sp. A comparison of LAI during 2012 and 2013 (as studied using satellite images) showed that as the monsoon advanced, LAI decreased substantially in both years, but the reduction was much more substantial and prolonged in many districts during 2013 than 2012 reflecting increased leaf fall due to ALF disease in 2013 (Pradeep *et al.*, 2014). The decline was more pronounced in Central and Northern Kerala than in the South. Kanyakumari district is generally known to be free from ALF disease, but there was considerable leaf loss due to ALF in June 2012 and June and July 2013 even as the monsoon was unusually severe in 2013 (Pradeep *et al.*, 2014).

In the recent years, a new fungal disease become quite prevalent in parts of Karnataka namely *Corynespora*. Similarly *Colletotrichum* has become a major concern in parts of the most traditional rubber growing regions in Central Kerala (Manju *et al.*, 2014). It cannot be ascertained whether climate change has been responsible for the emergence of the new diseases, even as one would suspect that to be the case. There have been sporadic incidences of mealy bug infestation becoming a significant pest in rubber. These are thermophilic insects and it is likely that warming conditions will result in more incidences of this pest in coming years also.

4. Climate Warming and Rubber Yield

Among various parameters that can influence latex flow and thereby rubber yield, ambient temperature is perhaps the most important one. The cooler the ambient temperature, the longer the duration of latex flow and thus the higher the rubber yield. Thus yield is relatively poor during the summer months. It has been shown that summer season yield is only 60- 80 per cent of the winter yield (Rajagopal *et al.*, 2013; Reju *et al.*, 2014). According to Satheesh and Jacob, (2011), if both Tmax and Tmin rose by 1°C, NR productivity will reduce by 9-16 per cent in the agro-climatic conditions of Kerala and by 11 per cent in the hot and drought-prone North Konkan region. On the other hand, in the cold prone North Eastern India, there is hardly any reduction in NR productivity if both Tmax and Tmin went up by 1°C (Figure 9.2). The study also indicates that a persistent warming trend reported from Kerala could be reduce the NR productivity by 4-7 per cent and in North East India it may go up by as much as 11 per cent in the next decade.

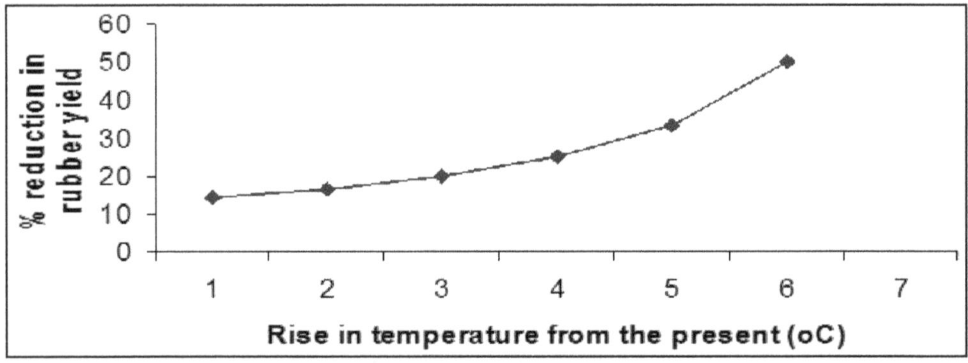

Figure 9.2: Estimated Per cent Reduction in Rubber Yield for Every Degree Rise in Temperature from the Present (Direct effect only) after Satheesh and Jacob, (2011).

Studies shown that as climate warms, rubber yield will decline in the traditional rubber growing regions because T max and T min had a negative impact on rubber yield (Jacob *et al.*, 2012). In Kerala and the Konkan regions, the prevailing temperatures are already at the higher side of the temperature threshold for rubber cultivation (Jacob *et al.*, 1999) and therefore any further warming can become harmful in these regions. However in the Northeast, climate warming can improve rubber yield, because T max had a positive impact on rubber yield. Therefore climate warming has different effects in different agro climatic regions (Satheesh and Jacob, 2011; Rey *et al.*, 2014). It is generally considered that severe winter condition is a limiting factor for growth and productivity of rubber in NE India (Jacob *et al.*, 1999) and therefore a warming trend in this region may improve the rubber yield.

Satheesh and Jacob (2011) made the attempt to assess the direct impact of climate warming on NR productivity. This study clearly indicate how Tmax and Tmin have been increasing in the past, how it has adversely affected the NR productivity in the past and what would be the raising temperatures might do to NR productivity in future in the different agro-climatic regions in India where the rubber is cultivated today.Climate change is obviously much more complex than daily variations in weather parameters such as daily maximum or minimum temperatures (Figure 9.3). Changes in cloud formation, wind, rainfall pattern, occurrence of extreme weather events, unexpected breaks in monsoon, spread of new pest and diseases *etc.* are important factors that can seriously influence rubber yield. Commercial productivity of natural rubber from growers' fields registered a steady increase over the years even as the local climate continued to warm, apparently contradicting the predictions. Increase in the area under the high yielding clone, RRII 105 that came into tapping over the years was responsible for masking the adverse impact of climate warming on productivity of rubber until recently (Satheesh and Jacob, 2015).

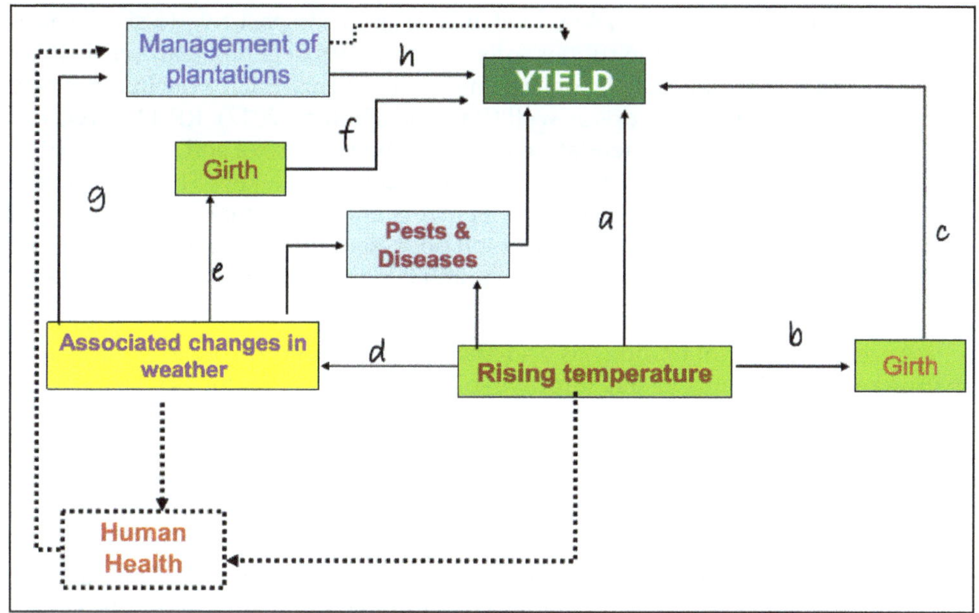

Figure 9.3: Schematic Chart Showing the Direct and Indirect Effects of Climate Warming on Growth and Yield of Rubber (after Satheesh and Jacob, 2011).

Using a different approach namely Ecological Niche Modelling, Ray *et al.* (2015) also found that in the coming years, if the present warming trend continues, more area will become stressful for rubber cultivation along the Western Ghats. More areas in North East India (where low temperature is presently a limiting factor for rubber cultivation) will become better suited for rubber cultivation if the warming trend continues (Ray, 2015; Satheesh, 2015).The two high altitude districts of Kerala namely Idukki and Wynad traditionally did not have much rubber cultivation in the past. In recent years, these two districts witnessed considerable warming (Rao *et al.*, 2009) which may be one reason why rubber cultivation is expanding to these districts now.

Extreme weather events have been on the rise in recent years (Guhathakurta *et al.*, 2011; Jain and Kumar, 2012). Tmax mean, hot day frequency, temperature of hottest day *etc.* shows a steady increase during the last 40 years. The magnitude of increase in maximum temperature is greater than that of minimum temperature (Jayasooryan *et al.*, 2015). It is almost impossible to categorically state if any single extreme weather event is the result of climate change, either global or regional. However, there is a clear rising trend in the number of occurrence of such events in recent years, particularly number of hot days and number of days with extremely high intensity rainfall.

5. Climate Change, Soil Organic Matter Content and Floral Diversity

Rising temperature can cause increased loss of organic matter from the soil due to oxidation (Yang *et al.*, 2005; Li *et al.*, 2013). In the past five to six decades, organic carbon content of the rubber soils of Kerala has declined (Saha *et al.*, 2009) and warming conditions may be one main reason for this. Such a decline has been generally noticed in other crops as well (Ray and Thomas, 2012). RRII has recently come out with a few recommendations that will help improve soil organic carbon levels. Minimum disturbing of the soil (zero tillage) is the key. There is no need to make large pits to plant a young rubber. It is enough to make a small pit just sufficiently large enough to contain the soil bole (zero pitting). Leguminous cover crops or commercial intercrops may be planted in young rubber holdings following principles of zero tillage.

Another important finding has been that allowing natural flora to grow inside a mature rubber holding will help improve soil organic carbon status as well as reduce soil acidity (Nath and Chaudhuri, 2010). This will improve the floral biodiversity status of rubber holdings which can have positive impacts on the populations of butterflies, honey bees *etc.* which are generally on the decline of several reasons including climate change. Several studies have demonstrated the significant ecological services provided by natural rubber plantations which include restoration of denuded ecosystems high rates of CO_2 sequestration (Annamalainathan *et al*,. 2011; Ambily *et al.*, 2012), improvement to the physical, chemical and biological properties of soil *etc.* (Wauters *et al.*, 2008).

6. Carbon Sequestration by Rubber Plantations

Carbon dioxide accumulation in the atmosphere has contributed the maximum to global warming compared to the other greenhouse gases. A mature rubber plantation can sequester as much as 35 to 45 ton CO_2/ha/yr (Annamalainathan *et al.*, 2011). It is estimated that entire rubber plantation in India can fix as much as 4-5 per cent of the CO_2 emitted by the automobiles in the country every year. Fast growing tree species like natural rubber are efficient in sequestering atmospheric CO_2. However, trees alone will not be able to off set the current rate of accumulation of this gas in the atmosphere even as the world emits increasing quantities of greenhouse gases into the atmosphere year after year.

7. Natural Rubber and Synthetic Rubbers

Adopting a low carbon growth trajectory is crucial to reducing emissions without compromising economic growth. Rubber is a key industrial raw material essentially needed to drive the economy. While natural rubber is a renewable raw material that is produced from rubber plantations in an eco-friendly manner, synthetic rubbers (SR) are produced from petroleum stocks. Therefore, when SRs are manufactured, huge quantities of CO_2 and other greenhouse gases are emitted into the atmosphere unlike natural rubber production (George *et al.*, 2006). Thus natural rubber is the obvious choice over SRs when it comes to low carbon growth. A fast growing economy like India requires increasing quantities of rubber – either natural or synthetic or both - and climate change puts a major obstacle in increasing domestic supply of natural rubber. Therefore, developing climate resilient smart clones of natural rubber is a top priority of research.

References

Ambily, K.K., Meenakumari, T., Jessy, M.D., Ulaganathan, A. and Nair, N.U. (2012). Carbon sequestration potential of RRII 400 series clones of *Hevea brasiliensis*. *Natural Rubber Research*, 25(2), 233-240.

Annamalainathan, K., Satheesh, P.R., and Jacob, J. (2011). Ecosystem flux measurements in rubber plantations. *Natural Rubber Research*, 24(1), 8-37.

Gadgil, A., and Dhorde, A. (2005). Temperature trends in twentieth century at Pune, India. *Atmospheric Environment*, 35, 6550-6556.

George, B., Varckey, J.K., Thomas, K.T., Jacob, J., and Mathew, N.M. (2006). Carbon dioxide emission reduction though substituting synthetic elastomers with natural rubber. In: Kyoto protocol and the rubber industry. (pp. 133-146). Rubber Research Institute of India, Kottayam, India.

Gogoi, N.K., and Dey, S.K. (2012). Prevalence of diseases and pests for rubber (Hevea brasiliensis) in Tripura. *Rubber Science*, 25, 189-199.

Goswami, B. N., Venugopal, V. D., Sengupta, D., Madhusudan, M.S., and Xavier, P. K. (2006). Increasing trend of extreme rain events over India in a warming environment. *Science*, 314, 1442-1445.

Guhathakurta, P., and Rajeevan, M. (2006, May). Trends in the rainfall pattern over India. NCC Research Report published by National Climate Centre, Pune.

Guhathakurta, P., Sreejith, O. P., and Menon, P. A. (2011). Impact of climate change on extreme rainfall events and flood risk in India. *Journal of Earth System Science*, 120, 359–373.

Jacob, J. (2013). Chlorophyll fluorescence –based non-invasive techniques to detect and quantify climate stress on plants. *Journal of Plantation Crops*, 41, 1-7.

Jacob, J., Annamalainathan, K., Alam, B., Sathik, M.B.M., Thapliyal, A.P. and Devakumar, A.S. (1999). Physiological constraints for cultivation of Hevea brasiliensis in certain unfavourable agroclimatic regions of India. *Journal of Natural Rubber Research*, 12, 1-16.

Jacob, J., Satheesh, P.R., and Ray, D. (2012). Climate warming and natural rubber productivity. Paper presented at the Kerala Environment Congress (KEC-2012), Thiruvananthapuram, India.

Jain, S.K. and Kumar, V. (2012). Trend analysis of rainfall and temperature data for India. *Current Science*, 102, 37-49.

Jayasooryan, K.K., Satheesh, P.R., Krishnakumar, R., and Jacob, J. (2015). Occurrence of extreme weather events- A probable risk on natural rubber cultivation. *Journal of Plantation Crops*, 43, 218-224.

Krishan, B. (2013). Growth and yearly yield potential of a few RRII 300 series, IRCA and other clones of Hevea brasiliensis under the dry-humid climate of Odisha, eastern India. *Rubber Science*, 26, 250- 259.

Krishnakumar, K.N., Prasada Rao, G.S.L.H.V., and Gopakumar, C.S. (2009). Rainfall trends in twentieth century over Kerala, India. *Atmospheric Environment*, 43, 1940-1944.

Kumar, K. S., and Parikh, J. (1998). Climate Change Impacts on Indian Agriculture: The Ricardian Approach. In Measuring the Impact of Climate Change on Indian Agriculture, World Bank. *Technical Paper No*. 402.

Kumar, R.K., and Hingane, L.S. (1988). Long-term variations of surface air temperature at major industrial cities of India. *Climatic Change*, 13, 287-307.

Lal, M., Nozawa, T., Emori, S., Harasawa, H., Takahashi, K., Kimoto, M., AbeOuchi, A., Nakajima, T., Takemura, T., and Numaguti, A. (2001). Future climate change: implications for Indian summer monsoon and its variability. *Current Science*, 81, 1196-1207.

Li, Y., Deng, X., Cao, M., Lei, Y., and Xia, Y. (2013). Soil restoration potential with corridor replanting engineering in the monoculture rubber plantations of Southwest China. *Ecological Engineering*, 51, 169–177.

Liyanage, A.D.S., and Jacob, K. (1992). Diseases of economic importance in rubber. In: *Natural rubber: biology, cultivation and technology.* (Developments in Crop Science No. 23, pp. 324-359). Elsevier, The Netherlands.

Manju, M.J., Benagi, V.I., Shankarappa, T.H., Jacob, K.C., and Vinod, K.K. (2014). Dynamics of *Corynespora* leaf fall and *Colletotrichum* leaf spot diseases of rubber plants (*Hevea brasiliensis*). *Journal of Mycology and Plant Pathology*, 44(1), 108-112.

Manju, M.J., Idicula, P.I., Jacob, C.K., Vinod, K.K., Prem, E. E., Suryakumar M. and Kothandaraman R. (2001). Incidence and severity of *Corynespora* leaf fall (CLF) disease of rubber in coastal Karnataka and north Malabar region of Kerala. *Indian Journal of Natural Rubber Research*, 14(2), 137 – 141.

Nath, S., and Chaudhuri, P. S. (2010). Human-induced Biological Invasions in Rubber (*Hevea brasiliensis*) Plantations of Tripura (India) - *Pontoscolex corethrurus* as a Case study, *Asian Journal of Experimental Biology*, 1, 360–369.

Pal, I., and Al-Tabbaa, A. (2009). Trends in seasonal precipitation extremes - An indicator of 'climate change' in Kerala, India. *Journal of Hydrology*, 367, 62–69.

Pradeep, B., Meti, S., and Jacob, J. (2014). Satellite-based remote sensing technique as a tool for real time monitoring of leaf retention in natural rubber plantations affected by abnormal leaf fall disease. *Rubber Science*, 27(1), 1-7

Rajagopal, R., Thomas K.U. and Karunaichamy, K. (2013). Effect of panel changing on long term yield response of Hevea brasiliensis (Clone RRII 105) under different frequencies of tapping and stimulation. *Rubber Science*, 26, 259-273.

Rajeevan, M. (2001). Prediction of Indian summer monsoon: status, problems and prospects. *Current Science*, 81, 1451-1458.

Rajeevan, M., Bhate, J. and Jaswal, A. K. (2008). Analysis of variability and trends of extreme rainfall events over India using 104 years of gridded daily rainfall data. *Geophysical Research Letters*, 35, L18707.

Rao, G.S.L.H.V., Kesava Rao, A.V.R., Khishnakumar K.N., and Gupakumar, C.S. (2009). Impact of climate change on food and plantation crops in the humid tropics of India. *In Impact of Climate Change on Agriculture*, ISPRS Archives, Ahmedabad, India.

Ray, D. (2015). Assessing distribution potential of *Hevea brasiliensis* using ecological niche modelling approach. Ph. D Thesis submitted to Indian Institute of Technology, Kharagpur, India.

Ray, D., Behera, M.D., and Jacob, J. (2014). Indian Brahmaputra valley offers significant potential for cultivation of rubber trees under future climate. *Current Science*, 107(3), 461-469.

Ray, D., Behera, M.D., and Jacob, J. (2015). Predicting the distribution of rubber trees (*Hevea brasiliensis*) through ecological niche modelling with climate, soil, topography and socioeconomic factors. *Ecological Research* (Online).

Ray, J. G., and Thomas, B. T. (2012). Fertility characteristics of oxic dystrustepts under natural forest, rubber, and teak plantations in different seasons, Kerala, South India. *Communications in Soil Science and Plant Analysis*, 43, 2247–2261.

Reju, M.J., Mydin, K.K., and Abraham, T. (2014). utilization of wild *Hevea* germplasm in India : long term yield and growth of certain WXA hybrids. *Rubber Science*, 27(2), 215-224.

RRII (2015). Annual Report 2013-2014. Rubber Research Institute of India, Rubber Board, Kottayam, Kerala.

Saha, S. K., Nair, P. K. R., Nair, V. D., and Kumar, B. M. (2009). Soil carbon stock in relation to plant diversity of home gardens in Kerala, India. *Agroforestry Systems*, 76, 53–65.

Satheesh, P.R. (2014) Impact of climate change on Indian plantation sector with special reference to natural rubber. Ph. D Thesis submitted to Mahatma Gandhi University, Kottayam, India.

Satheesh, P.R., and Jacob, J. (2011). Impact of climate warming on natural rubber productivity in different agro-climatic regions of India. *Natural Rubber Research,* 24, 1-9.

Satheesh, P.R. and Jacob, J. (2015). Climate warming causes marked decline in natural rubber production in traditional rubber growing regions of India. *Rubber Science,* 28(1), 1-7.

Sethuraj, M.R and Jacob, J. (2012). Thrust areas of future research in natural rubber cultivation. *Rubber Science.* 25, 123-138.

Singh, N., and Sontakke, N. A. (2002). On climate fluctuations and environmental changes of the Indo Gangetic plains in India. *Climate Change,* 52, 287-313.

Singh, P. (2006). Final report of the working group on agro-climatic zonal planning including agriculture development in north-eastern India for xi five year plan (2007-12). Planning Commission of India.

Wauters, J.B., Coudert, S., Grallien, E., Jonard, M., and Ponette, Q. (2008). Carbon stock in rubber tree plantations in Western Ghana and Mato Grosso (Brazil). *Forest Ecology and Management.* 255, 2347-2365.

Yang, J.C., Huang, J.H., Tang, J.W., Pan Q.M., and Han. X.G. (2005). Carbon sequestration in rubber tree plantations established on former arable lands in Xishuangbanna, SW China. *Acta Phytoecologica Sinica,* 29, 296-303.

2017, Impact of Climate Change on Plantation Crops *Pages* **157–168**
Editors: **K.B. Hebbar, S. Naresh Kumar & P. Chowdappa**
Published by: **ASTRAL INTERNATIONAL PVT. LTD., NEW DELHI**

Chapter 10

Carbon Sequestration in Plantation Crops

K.B. Hebbar, D. Balasimha and S. Naresh Kumar

1. Introduction

Carbon (C) sequestration in the agriculture sector refers to the capacity of agriculture lands to remove carbon dioxide from the atmosphere. Carbon dioxide is absorbed by trees, plants and crops through photosynthesis and stored as carbon in biomass in tree trunks, branches, foliage and roots and soils (EPA, 2008b). Forests and stable grasslands are referred to as carbon sinks because they can store large amounts of carbon in their vegetation and root systems for long periods of time. Compared to normal tropical agricultural crops, tree crop plantations have a significantly larger sequestration potential and are able to sequester C for longer periods (typically for more than 30 years) with smaller annual fluctuations. Many annual crops such as maize can fix more C than forestry systems in comparable time, but their biomass usually decomposes rapidly, and the rate and return of assimilated C to the atmosphere are very fast (Liguori *et al.*, 2009). Soils are the largest terrestrial sink for carbon on the planet. The ability of agriculture lands to store or sequester carbon depends on several factors, including climate, soil type, type of crop or vegetation cover and management practices.

Most of the plantation crops are trees and have long life span and hence store large amount of carbon in their vegetation. This knowledge is important as tree crop plantations could be a more feasible mitigation solution in many parts of developing countries compared to pure afforestation and reforestation projects, since tree crop plantations also provide work, income and food, especially when established in smallholder systems where local people have control over production. As such, plantations may also be an important element in increasing adaptive capacity in the sense of adapting to climate change and other pressures that are being faced by local communities in developing countries. Use of tree crop plantation to sequester C and and increase sustainable development would link climate change mitigation and adaptation, and an enhancement of this link has been called for by several authors (Ayers and Huq 2009; Halsnæs and Verhagen 2007; Klein *et al.*, 2007; Klein *et al.*, 2005; Tol 2005; Verchot *et al.*, 2007) Information on C sequestration in tree-crop plantation monoculture systems is limited. Some studies have been conducted, mainly in India on coconut (Naresh Kumar 2009; Kasturi Bai 2009), arecanut (Balasimha 2009, Balasimha and Naresh Kumar, 2013), rubber (James 2009), tea (Mohan and Rajkumar 2009), coffee (Devakumar 2009) and oil palm (Suresh and Kochbabu 2009) and from Southeast and East Asia, on the C content of oil palm (*e.g.* Chase and Henson 2010; Foong-Kheong *et al.*, 2010; Germer and Sauerborn 2008) and rubber plantations (*e.g.* Cheng *et al.*, 2007; Song and Zhang 2010). With regard to Africa, only a very few studies have been conducted on tree-crop plantation monoculture systems, for example, Wauters *et al.* (2008) on rubber and Duguma *et al.* (2001) on cocoa in an agroforestry system. Farming system, which integrates diverse crops – perennials, semi-perennials and annuals – animals, fishery and forestry enhances the productivity per unit area, time and inputs.

2. Carbon Sequestration of Plantation Crops

The plantation crops play a pivotal role in the Indian economy. The statistics on major plantation crops are given in Table 10.1.

Table 10.1: Area, Production and Productivity of Plantation Crops in India

Sl.No.	*Crop*	*Area('000 ha)*	*Production ('000 tonnes)*	*Productivity (kg/ha)*
1.	Arecanut[1]	445	729	1,164.0
2.	Cashew[2]	1027.0	725.0	706.0
3.	Cocoa[3]	78.0	16	475.0
4.	Coconuts[4]	2141.0	21665.0 (number)	10119.(number/ha)
5.	Coffee[5]	343.0	348.0	1014
6.	Tea[6]	567.0	1200	190.3
7.	Rubber[7]	758	919.0	1351.0
8	Oil Palm[8]	95.0	102.0	30-35(t/ha ffb)

**Source*: 1 DASD; 2,3 DCCD; 4 CDB; 5 FAO; 6 Tea Board; 7 CMIE; 8 Oil palm World Annual 2009

2.1. Arecanut and Cocoa

The areca-cocoa mixed crop not only gives a sustainable production, but also serves as a good system for biomass production and carbon accumulation. Arecanut is grown either as mono-plantation or intercropped with other plantations like cocoa, banana, etc, whereas, cocoa is grown only as intercrop of either coconut or arecanut. Presently, cocoa is grown in an area of 78000 hectares with about 16,000 tonnes production. There is an ambitious programme to bring more areas under cocoa with a projected production potential of about 30,000 tonnes during next five years. On the other hand arecanut is being consumed at local and national level. However, the income levels of farmers are highly fluctuating owing to market dynamics. These suggest the importance of stable income from arecanut-cocoa plantation systems. The net primary productivity in the areca-cocoa system and areca high density crop systems in terms of biomass production, carbon stocks and carbon sequestration indicate a very good potential of carbon sequestration.

Table 10.2: Average Annual Biomass and Carbon Stock in Plantation Crops

Crop	*Biomass (t/ha/yr)*	*Carbon Stock (t/ha/yr)*
Arecanut+Cocoa	6.92	2.88
Coconut	34.00	13.6
Cashew	4.27	1.71
Coffee	11.69	5.85
Oil palm	21.00	7.35
Rubber	–	7.62
Tea	20.12	8.05

Table 10.3: Total Estimated CO_2 Sequestration by Different Plantation Crops in India (Source; Naresh Kumar, 2015)

Crop	*Total Area under the Crops in India (million ha)*	*Total Estimated CO_2 Sequestration potential in India (% of total potential from plantations)*
Arecanut	0.445	8.5
Cocoa	0.078	2.2
Coconut	2.14	44.1
Cashew	1.02	17.9
Oil Palm	0.095	0.4
Coffee	0.343	6.0
Rubber	0.758	13.8
Tea	0.67	7.2
Total	5.446	100.01

Areca - cocoa system had a standing biomass of 23.15, 54.09 and 87.10 t ha^{-1} in 5th, 8th and 15th years of growth, respectively at CPCRI Regional Station Vittal (75° E longitude, 12°N latitude, 90m altitude) in lateritic soil with 5.4 pH (Table 10.1). Cocoa plants were spaced at 2.7 x 5.4m with in areca plantation spaced at 2.7 x 2.7m. A simple parametric regression model was developed for estimation of arecanut and cocoa standing biomass. Data on pod and bean yield (in case of cocoa) and nut yield (in case of arecanut) were collected from the experimental plot. All data collected as mentioned above were summed to get total above ground biomass production. Harvest index was calculated as ratio of bean dry weight (in case of cocoa) and dehusked nut dry weight (in case of arecanut) to annual total dry matter increment.

Carbon in different plant part samples like tissues of stem, leaf, twigs, pods, husk, nuts and beans was estimated by combustion method (Balasimha 2009, Balasimha and Naresh Kumar 2010, Balasimha and Naresh Kumar 2013).

Annual increments in biomass or net primary productivity ranged from 1.38-2.66t ha^{-1} in cocoa and 3.34-7.11 t ha^{-1} in areca (Balasimha and Naresh Kumar 2013). (Table 10.2). The study has thus revealed that the biomass and primary productivity is considerable with areca-cocoa mixed crop and comparable to any agroforestry systems involving cocoa. These methods on above ground biomass and carbon estimations provide useful basic information on these aspects and can be used as models in future. Other areca based models have also been worked out. The model involving multispecies cropping systems studied at Vittal also contribute considerably to carbon stocks and CO_2 sequestrations (Tables 10.3).

Similar studies on slightly different models at North East of India at Kahikuchi showed considerable carbon sequestrations (Tables 10.4 and 10.5).

Table 10.4: Standing Biomass and Carbon of Areca and Cocoa System

Years	*Cocoa*		*Areca*		*Total*	
	Biomass t/ha	*Carbon t/ha*	*Biomass t/ha*	*Carbon t/ha*	*Biomass t/ha*	*Carbon t/ha*
3	3.02	1.21	2.93	1.23	6.01	2.44
5	11.01	4.4	15.56	6.56	23.15	10.96
8	17.19	6.88	36.90	15.5	54.09	23.38
15	26.82	10.73	60.28	25.32	87.1	36.05

Table 10.5: Annual Net Primary Productivity, Carbon Stock Increments and CO_2 Sequestration in Areca-Cocoa System

Parameter	*Annual Increment (t ha^{-1}) in Areca*			*Annual Increment (t ha^{-1}) in Cocoa*		
	3-5 years	*6-8 years*	*9-15 years*	*3-5 years*	*6-8 years*	*9-15 years*
Biomass	4.21	7.11	3.34	2.66	2.06	1.38
Carbon stock	1.78	2.98	1.40	1.06	0.83	0.55
Net CO_2 sequestration*	6.53	10.94	5.14	3.89	3.05	2.02

* 1 CO_2 t.ha^{-1} = C (t.ha^{-1}) x 3.67

2.2. Coconut

Carbon sequestration potential of coconut was found to be quite high compared to most of the tree crops. Naresh Kumar (2007; 2008a and b, ; 2009; 2015) analyzed whole plant (above ground) carbon stocks and sequestrations and worked-out carbon sequestration for coconut plantations under different agro-climatic zones. Apart from these, variations due to cultivar, agro-climatic zone and management were also analyzed. The results indicate that annual C sequestration in coconut above ground biomass varied from 15 to 35 tCO_2/ha/year depending on cultivar, agro-climatic zone, soil type and management. Annually sequestered carbon stocked in to stem in the range of 0. 3 to 2.3 tCO_2/ha/year. Standing carbon stocks in 16 year old coconut cultivars in different agro-climatic zones varied from 15 to 60 tCO_2/ha/year. Annual carbon sequestration by coconut plantation is higher in red sandy loam soils and lowest in littoral sandy soils (Naresh Kumar, 2007 and 2008a and b, 2009; 2015).

Table 10.6: Annual Biomass, Carbon Stock and CO_2 Sequestration in Areca high Density System at Vittal

Parameter	*Biomass (t/ha/year)*	*Carbon Stock (t/ha/year)*	*Net CO_2 Sequestration* (t/ha/year)*
Areca (1300)*	4.86	1.994	7.134
Pepper (1300)	1.027	0.410	1.507
Banana (210)	2.683	1.073	3.938
Cocoa (210)	0.683	0.273	1.003
Clove (180)	0.195	0.078	0.286
Total	9.448	3.779	13.869

*Figures in parentheses indicate number of plants/ha.

Table 10.7: Annual Biomass, Carbon Stock and CO_2 Sequestration in Areca High Density System at Kahikuchi-Model I

Parameter	*Biomass (t/ha/year)*	*Carbon Stock (t/ha/year)*	*Net CO_2 Sequestration* (t/ha/year)*
Areca (1372)*	6.28	2.51	9.21
Pepper (1372)	0.93	0.372	1.37
Clove (228)	0.58	0.232	0.851
Citrus (438)	0.71	0.284	1.042
Total	8.5	3.39	12.47

* Figures in parentheses indicate number of plants/ha.

Table 10.8: Annual Biomass, Carbon Stock and CO_2 Sequestration in Areca High Sensity System at Kahikuchi-Model II

Parameter	*Biomass (t/ha/year)*	*Carbon Stock (t/ha/year)*	*Net CO_2 Sequestration* (t/ha/year)*
Areca (1372)*	6.13	2.45	8.99
Pepper (1372)	0.90	0.36	1.32
Nutmeg (228)	0.78	0.312	1.15
Citrus(438)	0.69	0.376	1.02
Total	8.5	3.49	12.48

* Figures in parentheses indicate number of plants/ha

Productivity per unit area of land can be increased by adopting high density multi species cropping system (HDMSCS). The principle of multistoried cropping is to use the basic production inputs such as light, water and nutrients to the maximum extent with minimum soil deterioration. This is attained by growing suitable crops having different canopy patterns and root systems. Crops with deep and shallow penetrating root systems will utilize soil nutrients and moisture available in the zones of their penetration. High and low canopied crops will have minimum mutual competition for solar energy. Hence maximum utilization of solar radiation can be realized by appropriate combination of crop. Multi-storied cropping thus is an effective method for increasing agricultural production per unit area. A variety of crops are grown in the interspaces of coconut ranging from short season annuals to perennial tree species. Biomass production of various crops in a cropping system gives a rough indication of their compatibility. The beneficial changes in rhizosphere of the main crop and intercrops not only improve the soil micronutrients, but also result in increasing yields considerably and also help in carbon sequestration. Utilization of land, airspace and inputs with maximum sustainable returns are the advantages of HDMSCS. The observations clearly revealed that the system not only gave an additional income from the increased coconut yield (Bavappa, *et al.*,

Table 10.9: Biomass and Carbon Stock of Coconut Monocrop and Component Crops

Crop.	*Biomass t/ha (Dry wt. basis)*	*Carbon (t/ha)*
Coconut (175)		
Root	18	7.20
Leaf	10.5	4.20
Stem	16.9	6.76
Bunches, nuts and other bio-materials	23.6	9.44
Total biomass	69 t/ha	27.60
Tapioca (930)		
Shoot	7.34	2.93
Tuber	1.07	0.42
Total	8.41	1.675
E. Foot Yam (930)		
Shoot	0.38	0.152
Yam	1.45	0.580
Total	18.3	0.732
Banana (300)		
Pseudostem (Including leaf)	0.71	0.284
Fruit/bunch	0.81	0.324
Total	1.52	0.608
Pineapple (6500)	14.31	5.73

Value in bracket indicates population.

1986) as well as from the yield of inter crops, but also the system will function as a strong candidate for cabon sinks for the overall benefit of the farmer. Estimation of carbon stock in an 18 year old coconut plantation with tapioca, elephant foot yam, banana and pineapple as intercrops revealed that the total biomass (including the economic produce) was about 86.3 t/ha Taking an average of 13.14 t CO_2/ha/yea, it is estimated that coconut mono-plantations can sequester about 27.6 million tonnes of CO_2/year in India (Naresh Kumar, 2015), which is equivalent to 0.013 per cent of India's total GHG emission during 2010 (India emitted 2,136.8 million tonnes of CO_2 eq in 2010, BUR, 2015).

2.3. Rubber

A mature rubber plantation can sequester as much as 35 to 45 ton CO_2/ha/yr (Annamalainathan *et al.*, 2011). Fast growing tree species like natural rubber are efficient in sequestering atmospheric CO_2. Carbon sequestration starts in natural rubber plantation from the first year of planting onwards and the end of 21 years, the plantation fixed a total of 164.18 t carbon per ha, equivalent to 601.99 t CO_2 per ha making the annual rate of CO_2 sequestration 28.7 t CO_2/ha/year (Jacob 2009).

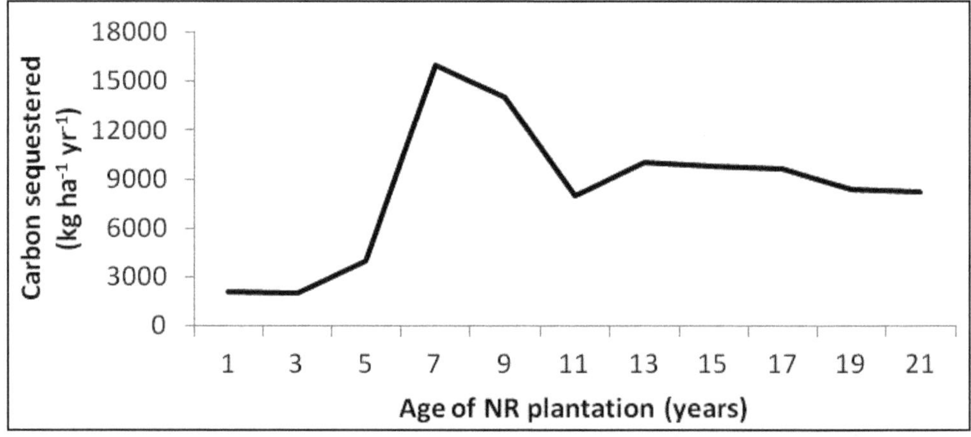

Figure 10.1: C-sequestration of Natural Rubber at different Age Groups.

Total C Sequestered = 164.18 T C/ha (in 21 Years)

Annualized Mean = 164.8/21 =7.818 T C/ha

That is equal to 28.7 T CO_{2e}/ha/year

The rate of CO_2 sequestration by natural rubber holdings varied from 28.8 to 40.1 t CO_2/ha/year in the different countries, possibly due to the differences in the agro-climatic conditions prevailing in these countries. Nevertheless, natural rubber plantations fixed much more carbon than the mature natural forests (Dixon *et al.*, science, 1994, 263, 184-190). It has been reported that the mature Amazonian forests fix just 3X t C/ha/year. Taking a modest mean rate of 30 t CO_2/ha/year, the world's over 10.4 million hectare of rubber plantations fix as much as 312 million t CO_2 every year.

2.4. Tea

It has been reported that tea plants are highly heterogenic in nature due to their free natural hybridization. They also exhibit variation in their metabolic activities, principally the photosynthetic carbon dioxide (CO_2) assimilation. Values presented on the basis of Pn rate varied from carbon dioxide sequestrated by biomass. However, estimated per cent net carbon dioxide sequestrated by tea sector (CO_2 sequestrated – less emission) was > 87 per cent (Mohan Kumar and Rajkumar 2009).Though the tea plants exhibited variations in their metabolic activities, on an average, they could produce 12.5 t biomass/ha/year. In addition, shade trees contribute a biomass of 7.6 t/ha/yr. Computing the overall biomass produced per unit area, it accounts 20.12 t/ha/yr. It may be noted that the inter crops like pepper/citrus species cultivated are not included in this computation. As per the biomass produced per unit area, the carbon stock was derived which accounts for 8.05 t C/ha/yr or 29.54 t of CO_2 equivalents. As per the land cover in Indian tea sector, 15.36 million t CO_2/ha/yr sequestrated which is a significant CDM sink besides providing livelihood to over a million people in the country directly and preserving the ecological balance. Total world's tea production is 3.2 million kg with a mean yield of 1145 kg/ha from an area of 27,94,759 ha. C absorption @ 18.9 t/ha or 69.36 t/ha as CO_2 (African conditions) is equivalent to CO_2 sequestration of 194.2 million t CO_2 or 54.96 as C worldwide.

2.5. Coffee

Coffee based agroforestry system is one of the most potential land use systems in India. Unlike the most other countries, both the species of coffee cultivation in India is done under the shade of tree. This is the practice followed from the time of inception of coffee cultivation in India. From the results it was found that, highest carbon accumulation in the above ground biomass of shade trees was found in native tree species (141.81, 131 t/ha) compared to silver oak (78.47, 102 t/ha) trees under evergreen and moist deciduous vegetation, respectively. The carbon content of litter biomass under these situations were 1.71, 0.98 and 1.02, 0.92 t/ha, respectively. The soil organic carbon content was 273, 260 and 282, 286 t/ha, respectively. The total carbon sequestration found in coffee plantation ranged from a minimum of 304.27 t/ha to 366.41 t/ha which is quite high compared to even evergreen vegetation of natural forest (Devakumar 2009).

The field study on partitioning of biomass and harvest index (HI) in twelve to fourteen years old arabica cultivars by collar pruning plants at the base level indicated that the most of the synthesized photosynthates translocated towards the development of the main stem to an extent of 32 per cent followed by branches/primaries (28 per cent), fruit biomass (Sink) (22 per cent) and leaf biomass (18 per cent) in arabica cultivar. Similarly the mean of 3 years observations on harvest index and leaf area index was 0.22 and 2.29 respectively. Also, the root shoot ratio in 20 to 40 years old uprooted field established arabica cultivars was presumed 30.09 to 39.28 per cent (Mean of 34.69 per cent).

Based on the above results, the biomass production is extrapolated as follows:

National productivity: 839.6 kg/ha

Total fruit biomass productivity: 1679.2 kg/ha/yr (22 per cent translocation of photosynthates)

Total biomass of coffee: 7.633 tons/ha/yr (Total translocated photosynthates)

Total biomass including root: 11.69 tons/ha/yr

Total biomass of coffee in India: 40,08,756 tons/yr

2.6. Cashew

The amount of carbon sequestered by cashew plantations under normal density planting (156 trees/ha) and high density planting (625 trees/ha) is presented. Carbon in the above ground biomass as well as root biomass of cashew trees is included in this analysis. The amount of carbon sequestered in cashew was estimated by a carbon content of 40 per cent of biomass. Based on research undertaken at Directorate of Cashew Research, Puttur (Karnataka), it was found that, cashew genotype (VTH-174) trees of 7 years old sequestered about 2.2 fold higher carbon under high density planting system as compared to normal density planting system. Carbon storage by cashew has been estimated as 32.25 and 59.22 t CO_2/ha at 5th and 7th years of growth, respectively under high density planting. The extent of C sequestered will depend on the amounts of C in standing biomass, age of the crop, tree density, variety etc.

2.7. Oil Palm

Oil palm is an introduced crop under Indian conditions and is being grown to an extent of 0.08 million ha. The potential areas identified for the cultivation of oil palm in different states of India is approximately 1.04 million ha. The standing above ground biomass in the different oil palm hybrids (mature plantations) grown under Indian conditions ranged from 55.08 to 91.58 t ha^{-1}. The amount of carbon sequestered by the hybrids ranged between 17.98 and 35.44 t $C.ha^{-1}$. It has been reported that mature oil palm hybrids accumulated about 20-22 $t_{dm}.ha^{-1}.yr^{-1}$ under Indian conditions, which can sequester 7.35 $tC.ha^{-1}.yr^{-1}$, which roughly comes to 26.98 $tCO_2.ha^{-1}.yr^{-1}$. In real terms, if we take the current area of oil palm in India in to consideration, the estimated carbon storage by the oil palm would be approximately 2.158 million t $CO_2.yr^{-1}$ from 0.095 m ha. The estimated storage of carbon by oil palm under Indian conditions could be to the tune of 28.05 million t $CO_2.yr^{-1}$ from the potential 1.04 m ha identified. However the above estimates may vary depending upon the age of plantations, source of planting material, soil type, management condition, agro climatic region, (Suresh and Kochu Babu, 2009).

Table 10.10: Carbon Sequestration Potential of Oil Palm at different Age Groups

Age (years)	*Biomass ($t\ ha^{-1}$)*	*Carbon Sequestered ($t\ C\ ha^{-1}$)*
1-3	14.5	2.527
4-8	40.3	17.113
9-13	70.8	16.455
14-18	93.4	21.327
19-24	113.2	21.104
>25	102.5	20.753

3. Conclusions

Tree crop plantations can be a potential path for coupling climate change mitigation and economic development by providing C sequestration and supplying wood and non-wood products to meet domestic and international market requirements. At the same time they provide work, income and food. The magnitude of mitigation depends on natural, social barriers, economic factors and time frame. While most of the Clean Development Mechanism (CDM) projects are awarded to industries who can justify the reduction of CO_2 emission and get Certified Emission Reduction unit or CER and get it traded, those who have farms, small plantations, fruit orchards, plantations and field crops do not get any support. This is a question of emitters getting the advantage, while the absorbers of CO_2 are left off. Therefore, the mitigate it is essential to highlight at appropriate forum about need for monetary compensation to the farmers.

References

Annamalainathan, K., Satheesh, P.R. and Jacob, J. (2011). Ecosystem flux measurements in rubber plantations. *Natural Rubber Research*, 24(1), 8-37.

Ayers, J.M. and Huq, S. (2009). The value of linking mitigation and adaptation: a case study of Bangladesh. *Environ Manage* 43,753–764.

Balasimha, D. (2009). Arecanut and Cocoa. In proceedings of the panel meeting on carbon sequestration in plantation crops and trading under CDM. Indian Society for Plantation Crops and Central Plantation Crops Research Institute Kasargod, August 4, 2009.

Balasimha, D., and Naresh Kumar, S. (2010). Net primary productivity, carbon sequestration and carbon stocks in areca- cocoa mixed cropping system. *Proc 16th International Cocoa Research Conference, Bali, Indonesia*, pp. 215-226.

Balasimha, D. and Naresh Kumar, S. (2013). Net primary productivity, carbon sequestration and carbon stocks in areca- cocoa mixed cropping system. *J. Plantation Crops*, 41, 8-13.

Bavappa, K.V.A., Kailasam, C., Abdul Khader, K.B, Biddappa, C.C., Khan, H.H., Kasturi Bai, K.V., Ramadasan, A., Sundararaju, P., Bopaiah, B.M., Thomas, G.V., Misra, L.P., Balasimha, D., Bhat, N.T. and Shama Bhat, K. (1986). Coconut and arecanut based high density multispecies cropping systems. *J. Plantation Crops*, 14: 74-87.

BUR (2015). India: First Biennial Update Report to the United Nations Framework Convention on Climate Change: India, Ministry of Environment, Forest and Climate Change, Government of India. p 180.

Chase, L.D.C., and Henson, *I.E.* (2010). A detailed greenhouse gas budget for palm oil production. *Int J Agr Sustain*, 8,199–214.

Cheng, C., Wang, R. and Jiang, J. (2007). Variation of soil fertility and carbon sequestration by planting Hevea brasiliensis in Hainan Island, China. *J Environ Sci (China)*, 19,348–352.

Corley, R.H.V.(1983). Potential productivity of tropical perennial crops. *Expl. Agric*,19, 217-237.

Devakumar, A.S. (2009). Coffee. In proceedings of the panel meeting on carbon sequestration in plantation crops and trading under CDM. Indian Society for Plantation Crops and Central Plantation Crops Research Institute Kgd, August 4, 2009.

Dixon, R.K., Solomon, A.M., Brown, S., Houghton, R.A., Trexier, M.C.and Wisniewski, J. (1994). Carbon pools and flux of global forest ecosystems. *Science*, 14,263(5144),185-90.

Duguma, B., Gockowski, J. and Bakala, J. (2001). Smallholder Cacao (Theobroma cacao Linn.) cultivation in agroforestry systems of West and Central Africa: challenges and opportunities. *Agroforestry Systems*, 51,177–188.

EPA. EPA's Report on the Environment (ROE) (2008 Final Report). U.S. Environmental Protection Agency, Washington, D.C., EPA/600/R-07/045F (NTIS PB2008-112484). 2008.

Foong-Kheong, Y., Sundram, K. and Basiron, Y. (2010). Mitigating climate change through oil palm cultivation. *Int J Global Warm*, 2,118–127.

Germer, J. and Sauerborn, J. (2008). Estimation of the impact of oil palm plantation establishment on greenhouse gas balance. *Environ Dev Sustain*, 10,697–716.

Halsnæs, K. and Verhagen, J. (2007). Development based climate change adaptation and mitigation conceptual issues and lessons learned in studies in developing countries. *Mitig Adapt Strat Glob Change*, 12, 665–684.

James Jacob, (2009). Rubber. In proceedings of the panel meeting on carbon sequestration in plantation crops and trading under CDM. Indian Society for Plantation Crops and Central Plantation Crops Research Institute Kasargod, August 4, 2009.

Kasturi Bai, (2009). Coconut. In proceedings of the panel meeting on carbon sequestration in plantation crops and trading under CDM. Indian Society for Plantation Crops and Central Plantation Crops Research Institute Kasargod, August 4, 2009.

Klein, R.J.T., Huq, S., Denton, F., Downing, T.E., Richels, R.G., Robenson, J.B. and Toth, F.L. (2007). Inter-relationships between adaptation and mitigation. Climate Change 2007: Impacts, Adaptation and Vulnerability. Contribution of Working Group II to the Fourth Assessment Report of the Intergovernmental Panel on Climate Change. Cambridge University Press, Cambridge, UK.

Klein, R.J.T., Schipper, E.L. and Dessai, S. (2005). Integrating mitigation and adaptation into climate and development policy: three research questions. *Environ Sci Pol*, 8,579–588.

Liguori, G., Gugliuzza, G., Inglese, P., 2009. Evaluating carbon fluxes in orange orchards in relation to planting density. *J Agric Sci*, 147,637–645.

Manoj Kumar, P. and Rajkumar, R. (2009). Tea. In proceedings of the panel meeting on carbon sequestration in plantation crops and trading under CDM. Indian Society for Plantation Crops and Central Plantation Crops Research Institute Kasargod, August 4, 2009.

Naresh Kumar, S. (2007). Carbon sequestration potential of coconut plantations. ICAR NEWS in Spectrum section. July-Sept. 13 (3): 15-16.

Naresh Kumar, (2008a). Modeling impacts and adaptations in coconut to climate change in different agro-climatic zones and field response of coconut to increased temperatures and CO_2. Project Annual Report.

Naresh Kumar, S. and Aggarwal, P.K. (2008b). Impact of climate change on coconut plantations. In: Climate change and Indian Agriculture (P.K. Aggarwal ed.), ICAR Publication pp. 24-27.

Naresh Kumar, S. (2009). Carbon sequestration in coconut plantations. In Global Climate Change and Indian Agriculture-case studies from ICAR Network Project (PK Aggarwal ed.), ICAR, New Delhi Pub., pp.129-134.

Naresh Kumar, S. (2015). Carbon sequestration and Beyond by Plantations: Exploring scope for Added Income. In: Agroforestry: Present Status and Way Forward (Dhayani, SK., Newaj, R., Alam B., Dev, I eds). Biotech books Pub., pp. 193-205

Song, Q. and Zhang, Y. (2010). Biomass, carbon sequestration and its potential of rubber plantations in Xishuangbanna, Southwest China. Shengtaixue Zazhi, 29,1887–1891.

Suresh, K. and Kochu Babu, K. (2009). Oil Palm. In proceedings of the panel meeting on carbon sequestration in plantation crops and trading under CDM. Indian Society for Plantation Crops and Central Plantation Crops Research Institute Kasargod, August 4, 2009.

Tol, R.S.J. (2005). Adaptation and mitigation: trade-offs in substance and methods. *Environ Sci Pol,* 8,572–578.

Verchot, L., van Noordwijk, M. and Kandji, S. (2007). Climate change: linking adaptation and mitigation through agroforestry. *Mitig Adapt Strat Glob Change,* 12,901–918.

Wauters, J., Coudert, S. and Grallien, E. (2008). Carbon stock in rubber tree plantations in Western Ghana and Mato Grosso (Brazil). *For Ecol Manage,* 255,2347–2361.

2017, Impact of Climate Change on Plantation Crops Pages 169–188
Editors: K.B. Hebbar, S. Naresh Kumar & P. Chowdappa
Published by: ASTRAL INTERNATIONAL PVT. LTD., NEW DELHI

Chapter 11

Phenotyping Tools to Understand Effects of Climate Change

V.S. John Sunoj, K.B. Hebbar
and P.V. Vara Prasad

1. Plantation Crops in India

The major plantation crops in India are coconut, arecanut, cocoa, rubber, tea, coffee, cashew, oil palm and rubber. The organized development of some of the plantation crops like tea, coffee and rubber started at colonial era. At the same time, coconut and arecanut were already established before colonial era. According to the statistics from Food and Agriculture Organization (FAO) in 2014, India stands one among the top countries with highest production of coconut (11,078,873Mt), coffee (318,200Mt), tea (1,208,780 Mt), cashew (753,000Mt) and rubber (900,000 Mt). Number of industries are in India based on plantation crops. For an instance, India is one of the major producer of coconut. Coconut mainly consumed by the domestic market and it's also used for the edible oil and coir production. Coconut plantations in India provides livelihood for 10 million people including farmers and small and large scale industrialist. Coconut is grown in over 200 districts of different states in India, the major producing area is 20 districts, which add almost 70 per cent of national production (Naresh Kumar and Aggarwal, 2013).At the same time, India is one of the leading exporter of coffee, tea, cashew and rubber. Plantation crops have a major role in Indian economy because of the large and small scale farmers and industries in the growing areas based on the production of planation crops.

Global agriculture production facing threat due to the climate variability. Global climate change models predict daily mean temperature to increase by about 1.0 to 3.7°C by end of the 21st century with increase in the variability in rain fall and increased frequency of extreme weather events like heat waves, cold waves, drought and floods (IPCC, 2013).Coastal and hilly area considered as the major vulnerable areas to climate change, where the majority of the plantation crops are grown in India. The change in climate can impact on plantation crops more than the seasonal crop because of the perennial nature and it can adversely affect the economy of particular region. Additionally, farmer are investing more precious resources like land, inputs and time on plantation crops than other annual crops (Hebbar *et al.*, 2013; Naresh Kumar and Aggarwal, 2013).Coconut is one of the major plantation crops in India and it is projected that productivity in high yield areas in India mainly Andhra Pradesh, Odissa, south and central parts of West Bengal, Gujarat, plains of Karnataka, and eastern and southeastern parts of Tamil Nadu is going to decrease under future climate change conditions due to drought and high temperature stress (Naresh Kumar and Aggarwal, 2013). At the same time, it is predicted that coffeeand rubber plantationsfacing ill effects of climate variability and going to receive the negative impact of drought due to the decrease in rain fall and high temperature in future (Majumder *et al*, 2014; Ovalle-Rivera *et al.*, 2015). The tea production in north east part of India is predicted to decrease in year 2050 due to increase in average temperature (Dutta, 2014). Selection of genotypes which are appropriate for the future climatic condition are really important for plantation crop because of the perennial nature. It is really challenging to manage an established plantation which are not suitable for the changed climatic condition and a financial burden for the farmers, so planting scientifically proven tolerant genotypes can help the farmers to save time and money. In this chapter we are explaining different tools with physiological, morphological and biochemical background for screening high temperature and drought tolerance in plantation crops. We acknowledge that threshold of physiological and biochemical changes due to the individual effect of each stress factors or combination of both can vary in different crops. Hence, the screening tools are explained in the generalized fashion to fit for individual effect of each stress factors or combination of both and diverse response of different crops.

2. Screening Tools for Temperature and Moisture Deficit Tolerance

2.1. Vegetative Stage

2.1.1. Photosynthesis and Leaf Night Respiration

The physiological response among crops arediverse under drought and high temperature stress conditions. Mainly gas exchange fluctuates under stress conditions and in turn results in changes in photosynthesis (Prasad *et al.*, 2006). Increased demand for photosynthates for the maintenance due to the damage caused by stress can decrease the growth respiration and increase the maintenance respiration. Hence, respiration effect the plant growth, development and yield. Photosynthesis and night respiration are not directly correlated, but both are important for plant growth and yield. Photosynthesis and night respiration are two important traits for screening the plants under stress conditions (Prasad *et al.*, 2011; Djanaguriraman *et al.*, 2013). Portable photosynthesis systems (Infra-red gas analyzers; IRGA; *e.g.* Li-6400 xt, LiCor. Lincoln, Nebraska, USA) are widely used to measure the photosynthesis and respiration.

Under drought condition photosynthesis decreases due to stomatal and non stomatal limitation (Zhou *et al.*, 2007). Stomatal limitation involves the partial or complete closing of stomata and that decreases the flow of CO_2 to the mesophyll cells. The initial decrease in the photosynthesis in drought stress is mainly due to the closure of stomata which is associated with content of abscisic acid (ABA). The plant hormone ABA, transfer signal of stress to the all parts of plants and that leads to the stomatal closure and related changes (Stoddard *et al.*, 2006). Non stomatal limitations will appear in later stages as the severity of drought increases causing changes in the biochemical processes essential for the photosynthesis. The main biochemical changes includes regeneration of ribulosebisphosphate (RuBP) and ribulose1,5-bisphosphate carboxylase/oxygenase (Rubisco) protein content (Bota *et al.*, 2004), decreased Rubisco activity (Parry *et al.*, 2002), impairment of ATP synthesis, and photophosphorylation or decreased inorganic phosphorus (Prasad *et al.*, 2008). There are some contrasting results showing the occurrence of non stomal limitation in initial stage leading to a temporary increase in intercellular CO_2 causing stomatal closure (Briggs *et al.*, 1986). The high temperature is also an important environmental stress that negatively affect the photosynthesis. Each crop have different threshold for the optimum temperature and temperature increase beyond that level can decrease photosynthesis. Under high temperature conditions, photosynthesis is related to the Rubisco and its dependence to subtract CO_2 and oxygen and due to the decreased solubility of oxygen to a lesser extent than CO_2 inducing photorespiration and decreased photosynthesis. Rubisco and Rubisco activase are inhibited under high temperature condition and that lead to the decreases in photosynthesis (Lea and Leegood, 1999; Crafts-Brandner and Salvucci, 2000; Prasad *et al.*, 2004).

Plant respiration is equally important as photosynthesis while considering plant growth and developments under high temperature and drought stress conditions. A combination of day time photosynthesic rate and leaf night respiration rate can give an output of carbon gain and carbon loss under stress conditions (Sunoj *et al.*, 2016). The photosynthesis occurs only in day time and mainly occurs in leaf, but the respiration is a continuous process (day and night) and occurs in all the organs. About 30 to 70 per cent of carbon gained by photosynthesis is lost through respiratory metabolism (Mohammed and Tarpley, 2011). Glucose, fructose and maltose are the reducing sugars and sucrose is the major transported non-reducing sugar in plants and respiration oxidizes these carbon rich carbohydrates, to generate energy for the growth and maintenance. In addition, leaf night respirationcomplete dependence on degradation of starch and sucrose is documented (Atkin *et al.*, 2000). Respiration is a group of two portions that includesmaintenance respiration and growth respiration. Maintenance respiration is the most responsive to the changes in growing conditions. The main functions of respiration is turn over protein and lipids and maintenance of ion transport across the membranes. Under stress conditions, maintenance respiration increases to manage these important functions. The high leaf night respiration due to the damage caused by stress uses carbohydrates for the maintenance and that decreases allocation of carbohydrates towards biomass accumulation (Halford *et al.*, 2010). Increased leaf night respiration under high temperature conditions reported in several crops (Loka and Oosterhuis 2010; Djanaguriraman *et al.*, 2013).Contrasting results are observed in leaf night respiration under drought condition. Researchers reported no change, decrease or increase in leaf night respiration in different crops at different temperatures and it proves that changes in night respiration depends on the leaf temperature (Atkin and Macherel, 2009; Gauthier *et al.*, 2014).

2.1.2. Chlorophyll Fluorescence

Chlorophyll florescence is used to evaluate the plant health status and photochemical efficiency of photosystem II (PS II; Fv/Fm) is routinely used as an indicator of the degree of stress (Laxman *et al.*, 2013; Djanaguriraman *et al.*, 2013). PS II plays an important role in photosynthesis under stress conditions.Chlorophyll fluorescence is widely used for quantifying the impact of drought and temperature stress. High temperature and drought can damage the thylakoid membrane and that causes malfunction of PSII light harvesting complex (Djanaguriraman *et al.*, 2011). PSII is more sensitive to high temperature condition as compared to drought. There are several types of chlorophyll fluorescence techniques used in broader fields related to biophysics, biochemistry and physiology.The most commonly used technique in the field of abiotic stress research are pulse amplitude modulation (PAM) fluorometryand fast chlorophyll fluorescence technique (Chlorophyll Fluorometers; *e.g.* OS 30+, OptiScience, Hudson, USA).

Photosynthetic antenna molecules present in the thylakoid membranes absorb light energy through light harvesting pigments chlorophyll and carotenoids (Horton and Ruban, 2004). Light energy absorbed by the chlorophyll can be used for three processes such as to drive photosynthesis (photochemistry), dissipated as heat or emitted as red fluorescence. These three process occur in competitions and decrease in one process can increase other two. Therefore measuring chlorophyll fluorescence can provide details on photochemical efficiency and heat dissipation. Using above fluorescence techniques can provide the informations like minimal fluorescence (F0), maximum fluorescence (Fm), variable fluorescence (Fv = Fm-F0), photochemical efficiency of PSII (Fv/Fm; which represents maximum quantum yield of PSII), (F0/Fm; which represents damage to thylakoid membranes), OJIP fluorescence curve, yield, photochemical quenching (PQ; quenching in the form of photochemistry), non-photochemical quenching (NPQ; quenching in the form of heat dissipation and fluorescence) and linear electron transport in PSII (ETR).The important informations used in abiotic stress screening are Fv/Fmratiothat indicates stress and F0/Fm ratio indicates damage to thylakoid membranes due to stress (Prasad *et al.*, 2008). The ratio of Fv/Fm in a healthy plant ranging from 0.78- 0.84 is considered as the maximum quantum yield of PSII photochemistry. The plants give values less than this range considered as under stress at high temperature and drought conditions. Plants severally affected with stress can give values below the mentioned range because of the damage to PSII (Djanaguriraman *et al.*, 2013; Laxman *et al.*, 2013).

2.1.3. Chlorophyll Index

Foliar chlorophyll content is a good indicator of plant stress and plant health because of its effects on photosynthesis and growth (Datt, 1999). Environmental (drought and high temperatures) and nutrient (particularly N) stresses commonly cause loss of leaf chlorophyll content leading to poor photosynthesis, growth, biomass, and economic yield. Drought and high temperature can decrease leaf chlorophyll index and intensity of decrease varies with different crops (Asharef and Harris, 2013). Degradation of chlorophyll can affect the photosynthesis, because of importance of chlorophyll as a light harvesting pigment. Slower degradation of chlorophyll content identified in some crops are termed as stay green types and the stay green character and chlorophyll retention in leaves considered as the tolerance of particular genotype to high temperature and drought (Fokar *et al.*, 1998; Araus *et al.*, 2012).Stay green genotypes are mostly expressed under drought condition and the genotypes with this particular trait can maintain the photosynthes is longer than the senescent type. Yield benefits are high in stay green genotypes as compared with the senescent type (Borrell *et al.*, 2001; Jordan *et al.*, 2003; Hebbar *et al.*, 2013). Stay green like trait is not identified in any of the plantation crops and investigations are required in this particular aspect while screening genotypes for drought and high temperature stress. Decrease in chlorophyll index under high temperature condition is mainly due to the reduced chlorophyll biosynthesis or increased chlorophyll degradation or effect of both. Numerous enzymes involved in the chlorophyll biosynthesis get deactivated under high temperature and that leads to the cessation of chlorophyll biosynthesis. For an instance, the activity of 5-aminolevulinate dehydratase (ALAD), the first enzyme of pyrrole biosynthetic pathway, decreased under high-temperature conditions (Mohanty *et al.*, 2006; Asharef and Harris, 2013). Under drought condition decrease in chlorophyll index is reported in several crops. However, studies on chlorophyllase and peroxidase shown that the decrease in chlorophyll is due to the elevated break down of chlorophyll as compared to the slow synthesis (Harpaz-Saad *et al.*, 2007; Kaewsuksaeng, 2011). The chlorophyll content can be quantified by destructive and non-destructive methods. Destructive methods includes the wet lab analysis using spectrophotometers (Lichtenthaler and Buschmann, 2001). Using chlorophyll meters (eg. SPAD 502, Konica Minolta Inc., Japan) is time saving, nondestructive and reliable for large number of plant in breeding population in a limited time frame. In coconut a portable non-destructive meter (atLeaf chlorophyll meter) was validated using experimental data from plant type (tall vs. dwarf and green foliage vs. colored foliage), cropping systems, fertigation method, soil type to establish the relationship between at Leaf chlorophyll meter measurements and chlorophyll and N concentration were quantified. The atLeaf chlorophyll meter provided good estimates for chlorophyll concentrations in coconut, with coefficients of determination greater than 0.81 (Hebbar *et al.*, 2016).

2.1.4. Leaf Water Relation

Water is a major component for majority of the crops and it is no exception for plantation crops.The water availability for the plantation crop getting decreased due to the irregular pattern and intensity of precipitation and increased demand for fresh water for domestic consumption. Changing climate will make situation more serious. In future, availability and quality of water going to be a serious issue around the globe (IPCC 2008). Under a situation like this creates the need for utilizing water resource in efficient ways. Hence, screening for diverse genotypes in plantation crops or suitable irrigation practices has to be implemented. While screening genotypes for traits related to efficient use of available water under drought and high temperature condition, traits like water use efficiency (WUE) become more important. WUE is the amount of water used to produce the dry matter in plants through evapotranspiration and it involves a series of above and below ground morphological, physiological and biochemical modification in plants. Gravimetric method is commonly used for calculating the water use efficiency in different crops and it is laborious and time consuming in most of the plantation crops because of the canopy spread and size. Intrinsic water use efficiency can be calculated using the gas exchange measurements. Instantaneous water use efficacy (P_N/E) is the ratio of photosynthesis rate (P_N) and transpiration (E). On the other hand, intrinsic water use efficiency (P_N/gs) also be calculated from the ratio of photosynthesis rate (P_N) and stomatal conductance (*gs*). The intrinsic and instantaneous water use efficiency provides information regarding the carbon fixed by photosynthesis in relation to the water lost by stomatal conductance and transpiration or capacity of plants to preserve water and maximizing carbon fixing (Rymbai *et al.*, 2014).All over again, calculating water use efficiency from gas exchange measurements are time consuming for large breeding populations.

Hence, scientists have depended on the surrogate methods of assaying the WUE. Measurement of carbon-isotope discrimination (CID) is an indicator of instantaneous water use efficiency (P_N/E) and it can be used as screening tool for the high temperature and drought (Richards *et al.*, 2010; Djanaguriraman *et al.*, 2011).It is already been used as a tool for screening large population in breeding programs to identify the genetic variation in transpiration efficacy and use for further breed to improve water use efficiency and yield (Condone *et al.*, 2004). The main advantage of this technique is collection of samples at the end of growing season and it can provide the details of instantaneous water use efficacy of entire growth season. There are two isotopic form of carbon (^{12}C and ^{13}C) in atmosphere. ^{12}C is more preferred by plants for photosynthesis as compared to ^{13}C. Carbon isotope discrimination ($\Delta^{13}C$) is ratio of $^{13}C/^{12}C$ in plant material relative to the value of the same ratio in the air which used by plants for photosynthesis. The ratio of $^{13}C/^{12}C$ in plants tissue is less than the ratio of $^{13}C/^{12}C$ in the atmosphere, which indicates the discrimination against heavy isotope of carbon (^{13}C) present in the atmospheric CO_2 during photosynthesis. The differences in the discrimination is due to the stomatal limitation and enzymatic activities (Condone *et al.*, 2004). Rubisco prefers $^{12}CO_2$ as compared with $^{13}CO_2$ because of the isotope effect.This isotopic discrimination is remain in the plant dry matter, which can be used to calculate discrimination. Increased photosynthesis and decreased stomatal conductance causes decrease in intercellular CO_2 resulting in decrease in discrimination against $^{13}CO_2$ (Farquhar *et al.*, 1982) or it can be explained as the increase in intercellular CO_2 implies greater carbon isotope discrimination for the heavier ^{13}C isotope.The CID is not a good tool for the studies where the ambient $^{13}C/^{12}C$ ratio is unknown (eg. Growth chambers and greenhouses without CO_2 controls).

Relative water content and leaf water potential are other methods to calculate the leaf water status in plant and it is related to several leaf physiological traits such as leaf turgor, growth, stomatal conductance, transpiration, photosynthesis and respiration. Leaf water content expresses the relative amount of water present on the plant tissues. Alternatively, water potential measures the energetic status of water inside the leaf cells (Yamasaki and Dillenburg, 1999).

2.1.5. Osmotic Adjustments

Osmotic adjustments are really important physiological mechanism to combat with stress conditions, mainly under water deficit condition (Tangpremsri *et al.*, 1995; Morgan, 2000). Plants can resist dehydration due to stress by reducing the cellular osmotic potential by accumulating certain organic compounds of low molecular mass, generally referred to as compatible osmolytes (O'Neill, 1983; Wahid *et al.*, 2007). Different crops responds in different ways to accumulate variety of osmolytes such as sugars and sugar alcohols (polyols), amino acids, tertiary and quaternary ammonium compounds, and tertiary sulphonium compounds (Sairam and Tyagi, 2004). Plants can enhance the stress tolerance by the accumulation of such solutes and changing osmotic potential in leaves. Examples for some of the osmolytes which identified in wide range of crops are sugars, glycinebetaine (amphoteric quaternary amine), proline (immino acid) and 4-aminobutyric acid (GABA; non-protein amino acid). The accumulation of cellular solutes have a key role in turgor maintenance and cell volume during stress conditions and consequently withstand the water loss and it can help to increase the biomass and yield under water stress condition (Blum, 2005). At the same time, including osmotic potential as a tool for drought stress is questioned by Serraj and Sinclair (2002), pointing towards the survival and yield of crops. In annual crops, occurrence of osmotic adjustments happens at the late stage and it helps the plant to survive, but not helps to maintain the yield. On the other hand, plantation crops are perennial in nature and with a greater recovery rate from the stress condition can make the osmotic adjustments important trait for screening stress tolerance. Leaf osmotic potential measurements can provide information regarding the capacity of particular genotypes which can withstand under stress conditions. Osmometer (*e.g.*, Wescor VAPRO Model 5600, Wescor, Inc., Logan, USA) is widely used to measure the osmotic potential (Laxman *et al.*, 2013).

2.1.6. Cell Membrane Thermo-stability

Cellular membrane modification is a result of various environmental stresses includes high temperature and drought, which result in the partial or total dysfunction of cellular membrane. Persistent functioning of cellular membrane under stress condition is essential for the proper photosynthesis and respiration (Prasad *et al.*, 2011; Djanaguriraman *et al.*, 2013). Membrane damage is associated with increased fluidity of membrane lipids, lipid peroxidation, and protein degradation in various metabolic processes under high temperature condition (Savchenko *et al.*, 2002). At the same time, exact cell membrane structural and function modifications caused by stress is not well understood.The reactive oxygen species (ROS; singlet oxygen [$^1O_2$], superoxide radicals [O^{2-}], hydrogen peroxide [H_2O_2] and hydroxyl radical [OH^-]) induced peroxidation of membrane lipid (in terms of malondialdehyde

[MDA] content) is considered as the manifestation of stress induced damage at the cellular level. The damage due to the lipid peroxidation of membrane can cause increased membrane fluidity (Jain *et al.*, 2001). Cell membrane thermo stability is considered as the surrogate method to assume the proper functioning of enzymatic (superoxide dismutase [SOD], peroxidase [POX] and catalase [CAT]) and non-enzymatic(proline, phenols and flavonoids) ROS quenching activity under high temperature and drought conditions (Sunoj *et al.*, 2014). The cellular dysfunction under high temperature and drought stress conditions will result in increased leaf permeability and leakage of ions, which canbe calculated by measuring the changes in electric conductivitydue to the leaked ionsfrom the leaves (Tripathy *et al.*, 2000;Bajji *et al.*, 2001; Lu etal., 2003). Measuring the changes in the lipid composition of membrane lipids are another method to assess the drought and high temperature stress (Repellin *et al.*, 1997; Narayanan *et al.*, 2015).

2.1.7. High Throughput Screening for High Temperature and Drought Stress

2.1.7.1. Infra-Red Thermography for Monitoring Canopy Temperature

Canopy temperature is an important pointer towards the plant water status and it can be used as a screening tool for genotypic selection for the drought and high temperature tolerance. The widely used methods for the infra-Red thermography for monitoring canopy temperatureare infra-Red thermometers, field mounted Infra -Red sensors (eg. Smart crop sensors; Smart field, Inc., USA) and infra-Red cameras (hand held, field mounted or Unmanned Arial Vehicle [UAV] mounted). Basic principle behind the infrared thermal imaging is all the objects above 0 degrees Kelvin (-273 °C), emit infrared energy. The Infra-Red energy emitted from the measured object is converted into an electrical signal by the imaging sensor in the camera and displayed on a monitor as a color or monochrome thermal images and in thermometers it is shown as digits.

Canopy temperature is influenced with different microclimatic and environmental factors such as solar radiation, cloud, wind, atmospheric temperature, relative humidity and soil moisture. At the same time, canopy temperature is determined by several physiological, biochemical and morphological traits as well, such as epicuticular wax, leaf angle, transpiration, stomatal conductance, root growth (root biomass and length), capability of plants to move water through the vascular system, and source and sink metabolism (Mason and Singh, 2014). Under high temperature and drought conditions, which is linked with vapor pressure deficit that increases the loss of water from leaves through transpiration. Increased transpiration helps to keep the cooler canopy temperature and it is possible when plants are grown under fully irrigated conditions or when plant have a better root growth. Growth and development of root systems are really important for absorbing water in these conditions to keep the canopy cooler for longer duration of time (Prasad *et al.*, 2008). Improved root system under water limited condition (moderate drought) and high temperature can help tolerant genotypes to absorb water from the deeper soil. Under these stressed conditions, measuring canopy temperature and stomatal conductance can be considered as the surrogate measurement of root growth. Lower canopy temperature and increased stomatal conductance indicates the increased root growth and that can help plant absorb water from deep soil to keep the canopy temperature cooler, which help plants to minimize damage due to drought stress.

2.1.7.2. Spectral Reflectance Signature

The fraction of solar radiation enter to the atmosphere of earth is transmitted, absorbed or reflected. Different vegetation or any other materials (eg. soil, turbid water and clear water) have capacity to absorb and reflect at different wave length and that makes unique spectral reflectance signature for that particular vegetation or material. Theoretically, spectral reflectance signature can be used to identify the vegetation or material, if the sensing system has sufficient spectral resolution to distinguish particular spectrum of the targeted vegetation or material. This is the basic principle for multispectral remote sensing.High temperature and drought stress changes canopy spectral reflectance signature of different vegetation and these change were recorded using spectroradiometer (Dobrowski *et al.*, 2005; Genc *et al.*, 2013). The data of canopy spectral reflectance can be used to calculate the different vegetative indices (Vina *et al.*, 2011) such as Simple Ratio (SR), normalized difference vegetation index (NDVI), enhanced vegetation index (EVI), green atmospherically resistant vegetation index (GARI), wide dynamic range vegetation index (WDRVI), green and red-edge chlorophyll indicesand MERIS terrestrial chlorophyll index, MTCI. Among the vegetative indices, NDVI is commonly used by researchers to calculate the effect of high temperature and drought stress conditions on plants.

2.1.7.3. Normalized difference Vegetation Index (NDVI)

Normalized difference vegetation index (NDVI) is most popular mathematical algorithms used to calculate vegetation indices (VIs). The NDVI is used to calculate the crop nutrient deficiency, vegetative greenness, indicator of biotic and abiotic stress and canopy photosynthetic rate (Raun *et al.*, 2001; Rodriguez *et al.*, 2004). It is also useful for precision agriculture, weed detection for herbicide spraying, and rate and timing of nitrogenous fertilizer application (Pietragalla and Vega, 2011). The popularity and advantage of NDVI is availability of data from different sources such as hand held sensors (eg. Green seeker), data from spectroradiometer, camera (handheld, UAV mounted and field mounted; *e.g.* Canon Power shot SX280HS) and satellite. In abiotic stress research, NDVI is a good method to calculate intensity of drought and high temperature stress (Peter *et al.*, 2002; Hazratkulova *et al.*, 2012). NDVI is calculated by measuring light reflectance in the red (600–700 nm) and near infra-Red (700-900 nm) regions of spectrum. The healthy plant will absorb most of the red light because of the high photosyntheticactivity (chlorophyll absorbs red and blue light for the photosynthesis) and reflects most of the NIR light. The formula used to calculate the NDVI is given below.

$$NDVI = (R_{NIR} - R_{Red}) / (R_{NIR} + R_{Red})$$

Where, R_{NIR} is the reflectance of NIR radiation and R_{Red} is the reflectance of visible red radiation in the spectrum.

2.2. Reproductive Stage

2.2.1. Pollen Viability, Germination and Stigma Receptivity

The reproductive success is controlled by the particular reproductive stages such as developmental stage of micro (pollen development) and megasporogensis (stigma development), anthesis, pollen tube growth, fertilization, and early embryo development. Incongruous circumstances (biotic or abiotic) at the time of any above these stages can directly affect the yield. These stages are highly susceptible to drought and high temperature stresses. Changes in theses stages can decrease fertilization or increase early embryo abortion, leading to lower crop yield (Boyer and Westgate, 2004; Prasad *et al.*, 2006). Keeping this in mind, pollen viability, pollen germination and stigma receptivity can be used as a tool for assessing the genotypes to make selections for the tolerant genotype under stresses at reproductive stage. Even though, complete physiological and biochemical mechanisms which interferes pollen viability, germination and stigma receptivity is not clear and these three processes are important for the reproductive success of any crop. When discussing the issues of pollen and pollination, there involves three major steps, 1) pollen adhesion, 2) pollen hydration and 3) pollen polarization and germination. The complete mechanism for each step is not clear. The nature of contact at pollen adhesion, transport of water, nutrients and other small molecules to the pollen grain from stigma at pollen hydration, perceiving signals and transduction to select a single point for pollen tube emergence at pollen polarization and germination and pollen tube invasion to stigma remains unclear (Edlund *et al.*, 2004). There are some hypotheses projected as accountable for the dysfunction of pollen under drought and high temperature stress, some of them are 1) premature dislocation of microspores (Saini *et al.*, 1984); 2) abnormal vacuolization leads to dysfunction of tapetal cells (Lalonde *et al.*, 1997); 3) lack of endothecial development and damage of tapetal cells at early stages (Ahmed *et al.*, 1992) 4) changed carbohydrate metabolism and accumulation (Saini, 1997; Jain *et al.*, 2007); 5) loss of gametophyte viability due to oxygen starvation in the developing microspores. At the same time some researches are showing the important role of hormones such as abscisic acid (ABA) and cytokinin. The plant hormone ABA is important under stress condition to prevent the water loss, improving the root hydraulic conductivity and maintain root growth, which delays the dehydration. ABA plays an important role in the signal transduction of stress and that leads to sterility or abortion. The cytokinin under stress condition disrupt the cellular and nuclear integrity of cells in the periphery of the endosperm (Jones and Setter, 2000; Prasad *et al.*, 2008). At high temperature, probabilities of changes in pollen function and anatomy is high. Pollen shape, pollen viability, pollen germination, pollen tube growth, thickness of tapetum and exain layer altered under high temperature stress. Further, pollen tapetal cells were vacuolated and showed autolysis, generative cell acquired a fusiform shape and the chromatin showed condensation (Prasad *et al.*, 2000; Prasad *et al.*, 2006; Djanaguriraman *et al.*, 2013; Djanaguriraman *et al.*, 2013a).

The stigma is the first reproductive structure that comes in contact with pollen grain on the way to fertilization. Stigma receptivity is defined as the ability of stigma to facilitate pollen germination and that is also an important part for reproductive success in various crops. The effective pollination period is closely linked with the duration (longevity) of stigmatic receptivity and it can vary among different crops (Sanzol and Herrero, 2001). High temperature and drought stress can play a major role in longevity of stigma receptivity. High temperature reduces stigma receptivity by lack of support to pollen for adhesion, elongation and penetration (Hedhly *et al.*, 2003).

2.2.2. Crop Yield and Yield Indices

High or sustainable yield are the ultimate aim for the success of the screening of genotypes for stress tolerance. In plantations crops, final output from the screening criterion will take more time as compare to other seasonal crops and that adds more importance while executing above mentioned screening tools. Final yield and related traits can also be considered for the screening while conducting long term experiments and it is necessary for the successful conclusion of other screening methods. In other crops like vegetables, legumes and cereals, fruit (expressed as grain, seed or nut) set per cent, size and number are considered as important traits for the selection of tolerant genotypes and high temperature and drought stress reported to decrease the yield in vegetables, legumes and cereals (Sato *et al.*, 2000; Prasad *et al.*, 2008). These yield related traits are generally get affected when the stress occurs at the time of reproductive developments (micro and megasporogensis) (Prasad *et al.*, 2006; Valliyodan and Nguyen, 2006).

The relative yield performance of genotypes in drought stressed and more favorable environments seems to be a common starting point in the identification of traits related to drought tolerance and the selection of genotypes for use in breeding for dry environments (Clarke *et al.*, 1992 ; Hebbar et al., 2016)). To differentiate drought resistance genotypes, several selection indices have been suggested on the basis of a mathematical relationship between favorable and stress conditions (Clarke *et al.*, 1982; Huang, 2000). Tolerance (TOL) (McCaig and Clarke, 1982; Clarke *et al.*, 1992), mean productivity (MP) (McCaig and Clarke, 1982), stress susceptibility index (SSI) (Fischer and Maurer, 1978), geometric mean productivity (GMP) and stress tolerance index (STI) (Fernandez, 1992) have all been employed under various conditions.

The drought intensity index (*DII*) was calculated using the following formula of Ramirez-Vallejo and Kelly, 1998.

$$DII = (1 - Xs \div \text{Xi})$$

Where, *Xs* is the mean experiment yield of all genotypes grown under stress, and *Xi* is the meanexperiment yield of all genotypes grown under non-stress conditions. Values exceeding 0.7 would indicate severe drought.

Ramirez-Vallejo and Kelly (1998) also concluded that the most effective approach to breed beans for resistance to drought would be based first on selection for high geometric mean seed yields followed by selection for low Fischer Maurer drought susceptibility index values. The Fischer and Maurer drought susceptibility index (*DSI*) is calculated as follows:

$$DSI = (1 - Ys \div Yi) \div DII$$ (Fischer and Maurer, 1978)

Schneider *et al.* (1997) showed the geometric mean (*GM*) of seed yield to be the best predictor of genotype performance in stress and non-stress environments. They recommended a breeding strategy that involved genotypic selection based first on *GM*, followed by selection based on seed yield in the stress environment. The *GM* is calculated as follows:

$$GM = \sqrt{Ys \times Yi}$$

Where, Ys is the mean seed yield of a line under drought stress and Yi is the mean seed yield of the line grown under non-stress. The square root of the product ($Ys \times Yi$) from two treatments is used to calculate the *GM* for an individual genotype.

Drought resistance can only be estimated by comparing the performance of breeding lines under stress and non-stress (irrigated) conditions. Using data from the two water treatments (rainfed and irrigated) at Arsikere district of Karnataka state, India, drought intensity index, different susceptibility indices and means to assist in selection of drought resistant genotypes of coconut was calculated (Hebbar, unpublished data). Arasikere was under severe drought for the years 2011, 2012 and 2013. Yield data was collected from rainfed and irrigated coconut orchards of experimental farm at Arasikere for different genotypes as shown in the Table 11.1.

Table 11.1: The Relative Yield Performance of Coconut Genotypes in Rainfed and Irrigated Conditions of Arasikere, Karnataka and the Calculation of Tolerant Indices

Cultivar	*Water Treatments*		*Per cent Reduction*	*DSI*	*GM*
	Irrigated	*Rainfed*			
WCT	52	4	92	1.14	14
LCT	63	19	70	0.86	35
ADOT	54	5	91	1.12	16
Sanramon	74	2	97	1.20	12
WCTXGBGD	180	25	86	1.06	67
BS1	44	21	52	0.65	30
PHOT	73	0	100	1.23	0
WCTXCOD	85	5	94	1.16	21
CODXWCT	72	3	96	1.18	15
TPT	62	25	60	0.74	39
Zangiber	130	36	72	0.89	68
Java	47	20	57	0.71	31
Mean	78	14	81	1.00	29

In this experiment, *GM* was high for cv.Zangiber and WCT x GBGD which also had high yield under stress and non-stress conditions.Drought susceptibility index (*DSI*) was low for BS1 and Java, but they were low in nut production. Caution in using this index is advised as certain genotypes with the lowest *DSI* rankings had the lowest overall yield potential (White and Singh, 1991). Small yield differences between the stress and non-stress treatments produce low *DSI* values even though the potential yield of the line is low. Therefore in coconut *GM* is the best indicator of drought tolerance and can be used in breeding programs across different environments.

2. Conclusions

Maintaining growth and yield under high temperature and drought is the major challenges to plantation crops. Climate variability is going to enhance in future and it can impact more on the plantation crops because the major growing areas in India are vulnerable to climate change. Plantation crops are experiencing the ill effects of adverse climate than annual crops. Selection of tolerant genotypes to high temperature and drought using scientific tools are really important to maintain the growth and yield in future climatic conditions. Farmers are investing their valuable time, land and inputs for plantation crops and outputs from that is crucial for the financial stability of famers and related industries. Integration of physiological, biochemical, molecular, breeding and modeling tools can play a major role in the selection of tolerant genotypes. Additionally, better understanding of physiological, biochemical, molecular and genetic basis of the mechanisms promoting the tolerance to high temperature, drought and combined effect of both will improve the capacity to enhance the yield or maintenance of yield under hostile environmental conditions.

References

Ahmed, F.E., Hall, A.E., and DeMason, D.A. (1992). Heat injury during floral development in cowpea (*Vignaunguiculata*, Fabaceae). *American Journal of Botany*, 79,784–791.

Araus, J. L., Serret, M. D., and Edmeades, G.O. (2012). Phenotyping maize for adaptation to drought. *Frontiers in Physiology*, 3(305), 1-20.

Ashraf, M., and Harris, P. J. C. (2013). Photosynthesis under stressful environments: An overview. *Photosynthetica*, 51 (2), 163-190.

Atkin, O.K., and Macherel, D. (2009). The crucial role of plant mitochondria in orchestrating drought tolerance. *Annals of Botany*, 103, 581–597.

Atkin, O. K., Millar, A. H., Gardestrom, P., and Day, D. A. (2000). Photosynthesis, carbohydrate metabolism and respiration in leaves of higher plants. In: *Photosynthesis: Physiology and Metabolism*, (pp. 153–175). Academic Publishers. Netherlands.

Bajji, M., Kinet, J. M., and Lutts, S. (2001). The use of the electrolyte leakage method for assessing cell membrane stability as a water stress tolerance test in durum wheat. *Plant Growth Regulation*, 00, 1–10.

Blum, A. (2005). Drought resistance, water use efficiency, and yield potential–are they compatible, dissonant, or mutually exclusive? *Australian Journal of Agriculture Research*, 56,1159–1168.

Borrell, A. K., Hammer, G., and van Oosterom, E. (2001). Stay-green: A consequence of the balance between supply and demand for nitrogen during grain filling. *Annals of Applied Biology*, 138,91–95.

Bota, J., Medrano,H., and J. Flexas. (2004). Is photosynthesis limited by decreased Rubiscoacivity and RuBP content under progressive water stress? *New Phytologist*, 162,671–681.

Boyer J. S. and Westgate, M. E. (2004). Grain yields with limited water. *Journal of Experimental Botany*, 55(407), 2385- 2394.

Briggs, G. M., Jurik, T. W., and Gates, D. M. (1986). Non-stomatal limitation of carbon dioxide assimilation in three species during natural drought conditions. *PhysiologiaPlantarum*, 66,521–526.

Clarke, J.M., De Pauw, R.M., and Townley-Smith, T.M. (1992). Evaluation of methods for quantification of drought tolerance in wheat. *Crop Science.* 32, 728–732.

Condon, A. G., Richards, R.A., Rebetzke, G. J., and Farquhar, G.D. (2004). Breeding for high water-use efficiency. *Journal of Experimental Botany*, 55,2447–2460.

Crafts-Brandner, S. J., and Salvucci, M. E. (2000). Rubiscoactivase constraints the photosynthetic potential of leaves at high temperatures. *Proceedings of the National Academy of Sciences*, 97,13430–13435.

Djanaguiraman, M., Prasad, P. V. V., Boyle, D. L., and Schapaugh, W. T. (2011). High temperature stress and soybean leaves: Leaf anatomy and photosynthesis. *Crop Science*, 51, 2125-2131.

Djanaguiraman, M., Prasad, P. V. V., and Schapaugh, W. T. (2013). High day- or nighttime temperature alters leaf assimilation, reproductive success, and phosphatidic acid of pollen grain in soybean [*Glycine max* (L.) Merr.]. *Crop science*, 51, 2125–2131.

Djanaguiraman, M., Prasad, P. V. V., Boyle, D. L. and Schapaugh, W. T. (2013a). Soybean pollen anatomy, viability and pod set under high temperature stress. *Journal of Agronomy and Crop Science*, 199, 171–177.

Dobrowski, S. Z., Pushnik, J. C., Zarco-Tejada, P. J., and Ustin, S.L. (2005). Simple reflectance indices track heat and water stress-induced changes in steady-state chlorophyll fluorescence at the canopy scale. *Remote Sensing of Environment*, 97,403 – 414.

Datt B (1999). A new reflectance index for remote sensing of chlorophyll content in higher plants: tests using *Eucalyptus* leaves. *Journal of Plant Physiology.* 154, 30–36.

Dutta, R. (2014). Climate change and its impact on tea in Northeast India. *Journal of water and climate change*, 5 (4), 625-632.

Edlund, A. F., Swanson, R., and Preuss, D. (2004). Pollen and Stigma Structure and Function: The Role of Diversity in Pollination. *The Plant Cell*, 16,84-97.

FAO. (2014). http://faostat3.fao.org/home/E

Farquhar, G. D., O'Leary, M. H., and Berry J. A. (1982). On the relationship between carbon isotope discrimination and intercellular carbon dioxide concentration in the leaves. *Australian Journal of Plant Physiology*. 9,121-137.

Fernandez, G.C.J. (1992). Effective selection criteria for assessing stress tolerance. In: *Proceedings of the International Symposium on Adaptation of Vegetables and Other Food Crops in Temperature and Water Stress*, (pp.257-270). AVRDC Publication, Tainan, Taiwan.

Fischer, R.A., and Maurer, R. (1978). Drought resistance in spring wheat cultivars. Part 1: grain yield response. *Australian Journal of Agricultural Research*. 29, 897–912.

Fokar, M., Nguyen, H.T., and Blum, A. (1998). Heat tolerance in spring wheat. I. Estimating cellular thermo tolerance and its heritability. *Euphytica*, 104, 1-8.

Gauthier, P. P. G., Crous, K. Y.,Ayub, G.,Duan, H.,Weerasinghe,L. K., David, S. E.,Mark, G., Tjoelker, M. G., Evans,J. R., Tissue,D. T., and Atkin,O. K. (2014). Drought increases heat tolerance of leaf respiration in *Eucalyptus globulus*saplings grown under both ambient and elevated atmospheric [CO_2] and temperature. *Journal of Experimental Botany*, 65(22), 6471–6485.

Genc, L., Inalpulat, M., Kizil, U., Mirik, M., Smith, S. E., and Mendes, M. (2013). Determination of water stress with spectral reflectance on sweet corn (*Zea mays* L.) using classification tree (CT) analysis. Zemdirbyste-Agriculture, 100(1), 81–90.

Halford, N. G., Curtis, T. Y., Muttucumaru, N., Postles, J., and Mottram, D.S. (2011). Sugars in crop plants. *Annals of Applied Biology*, 158, 1–25.

Harpaz-Saad, S., Azoulay, T., and Arazi, T. (2007).Chlorophyllase is a rate-limiting enzyme in chlorophyll catabolism and is posttranslationally regulated. *Plant Cell*, 19, 1007-1022.

Hazratkulova, S., Sharma, R. C., Alikulov, S., Islomov, S., Yuldashev, T., Ziyaev, Z., Khalikulov, Z., Ziyadullaev, Z., and Turok, J. (2012). Analysis of genotypic variation for normalized difference vegetation index and its relationship with grain yield in winter wheat under terminal heat stress. *Plant Breeding*, 131, 716—721.

Hebbar, K.B., Balasimha, D., and Thomas, G.V. (2013). Plantation Crops Response to Climate Change: Coconut Perspective. In: *Climate resilient Horticulture: Adaptation and mitigation strategies*, (pp. 177-187). Spinger, New Delhi.

Hebbar, K. B., Rane, J., Ramana, S., Panwar, N. R., Ajay, S., Subba Rao, A., and Prasad, P. V. V. (2014). Natural variation in the regulation of leaf senescence and relation to N and root traits in wheat. *Plant and soil*. 378,99-112.

Hebbar, K.B., Subramanian, P., Sheena, T.L., Shwetha,K., Sugatha, P.,Arivalagan, M., and Prasad, P.V.V. (2016). Chlorophyll and nitrogen determination for coconut (*Cocos nucifera*) using a non-destructive meter. *Journal of Plant Nutrition* 39: 1610-1619.

Hedhly, A., Hormaza, J. I., and Herrero, M. (2003). The effect of temperature on stigmatic receptivity in sweet cherry (*Prunusavium* L.). *Plant, Cell and Environment*, 26, 1673–1680.

Horton, P., and Ruban, A. (2004). Molecular design of the photosystem II light-harvesting antenna: photosynthesis and photoprotection. *Journal of Experimental Botany*, 56(411), 365-376.

Huang, B. (2000). Role of root morphological and physiological characteristics in drought resistance of plants. In: *Plant Environment Interactions*, (pp. 39-64). Marcel Dekker Inc., New York, USA.

IPCC, (2008). Climate change and water. In: *IPCC technical paper VI of the secretariat*, (pp. 1–210). IPCC, Geneva.

IPCC, (2013). Summary for Policymakers. In: *Climate Change 2013: The Physical Science Basis. Contribution of Working Group I to the Fifth Assessment Report of the Intergovernmental Panel on Climate Change*, (pp.1–27).Cambridge University Press, Cambridge, United Kingdom and New York, USA.

Jain, M., Mathur, G., Konl, S., and Sarin, N. B. (2001). Ameliorative effects of proline on salt stress lipid peroxidation in cell lines of groundnut (*Arachis hypogea* L.). *Plant Cell Reports*, 20:463–468.

Jones, R.J., and Setter. T.J. (2000). Hormonal regulation of early kernel development. In: Physiology and modeling kernel set in maize, (pp. 25–42). ASA, Madison, WI.

Jain, M., Prasad, P. V. V., Boote, K. J., Allen Jr, L. H., and Chourey, P. S. (2007). Effects of season-long high temperature growth conditions on sugar-to-starch metabolism in developing microspores of grain sorghum (*Sorghum bicolor* L. Moench). *Planta*, 227, 67–79.

Jordan, D. R., Tao, Y., Godwin, I. D., Henzell, R. G., Cooper, M., and McIntyre, C. L. (2003). Prediction of hybrid performance in grain sorghum using RFLP markers. *Theoretical and Applied Genetics*, 106, 559–567.

Kaewsuksaeng, S. (2011). Chlorophyll degradation in horticultural crops. Walailak *Journal of Science and Technology*, 8, 9-19.

Lalonde, S., Beebe, D., and Saini, H.S. (1997). Early signs of disruption of wheat anther development associated with the induction of male sterility by meiotic-stage water deficit. *Plant Reproduction*, 10, 40–48.

Laxman, R. H., Srinivasa, R. N. K., Bhatt, R. M., Sadashiva, A. T., Sunoj, J. V. S., Geeta, B., Pavithra, C. B., Manasa, K. M., and Dhanyalakshmi, K. H. (2013). Response of tomato (*Lycopersicon esculentum* Mill.) genotypes to elevated temperature. *Journal of Agrometerology*, 15, 38–44.

Lea, P.J., and Leegood, R.C. (1999). Plant Biochemistry and Molecular Biology. John Wiley, Chichester.

Lichtenthaler, H. K., and Buschmann, C. (2001). Chlorophylls and carotenoids: measurement and characterization by UV–VIS spectroscopy. In: *Current Protocols in Food Analytical Chemistry*, (pp. F4.3.1–F4.3.8).John Wiley and Sons, Inc., NewYork.

Loka, D.A., and Oosterhuis, D.M. (2010). Effect of high night temperatures on cotton respiration, ATP levels and carbohydrate content. *Environmental and Experimental Botany*, 68, 258–263.

Lu, S., Guo, Z., and Peng, X. (2003). Effects of ABA and S-3307 on drought resistance and antioxidative enzyme activity of turf grass. *Journal of Horticulture Science and Biotechnology*, 78, 663–666.

Majumder, A., Datta, S., Choudhary, B. K., and Majumdar, K. (2014). Do Extensive Rubber Plantation Influences Local Environment? A Case Study from Tripura, Northeast India. *Current World Environment*, 9(3), 768-779.

Mason, R. E., and Singh, R. P. (2014). Considerations when deploying canopy temperature to select high yielding wheat breeding lines under drought and heat stress. *Agronomy*, 4, 191-201.

McCaig, T.N., and Clarke, J.M.(1982). Seasonal changes in nonstructural carbohydrate levels of wheat and oats grown in semiarid environment. *Crop Science*. 22, 963–970.

Mohanty, S., Baishna, B. G., and Tripathy, C. (2006). Light and dark modulation of chlorophyll biosynthetic genes in response to temperature. *Planta*, 224, 692-699.

Mohammed, A.K., and Tarpley, L. (2011). Effects of high night temperature on crop physiology and productivity. In: *Plant growth regulators provide a management option, global warming impacts - Case studies on the economy, human health, and on urban and natural environments*, ISBN, 978-953-307-785-7.

Morgan, J.M. (2000). Increase in grain yield of wheat by breeding for an osmoregulation gene: Relationship to water supply and evaporative demand. *Australian Journal of Agriculture Research*, 51, 971–978.

Narayanan, S.,Prasad, P.V. V., and Welti, R. (2015). Wheat leaf lipids during heat stress: II. Lipids experiencing coordinated metabolism are detected by analysis of lipid co-occurrence. *Plant, Cell and Environment*, doi:10.1111/pce.12648.

Naresh Kumar, S., and Aggarwal, P.K. (2013). Climate change and coconut plantations in India: Impacts and potential adaptation gains. *Agricultural Systems*, 117,45–54.

O'Neill, S. D. (1983). Role of Osmotic Potential Gradients during Water Stress and Leaf Senescence in Fragariavirginiana. *Plant Physiology*, 72, 931-937.

Ovalle-Rivera, O., Laderach, P., Bunn, C., Obersteiner, M., and Schroth, G. (2015). Projected shifts in *Coffeaarabica* suitability among major global producing regions due to climate change. PLOS ONE, | DOI,10.1371/journal.pone.0124155.

Parry, M. A. J., Androlojc, P. J., Khan, S., Lea, P. J., and Keys, A. J. (2002). Rubisco activity: Effects of drought stress. *Annals of Botany* (London), 89,833–839.

Peters, A.J., Elizabeth, A., Waltershea, L.J., Andres, V., Mlchael, H., and Mark, D. S. (2002).Drought Monitoring with NDVI-Based Standardized Vegetation Index. *Photogrammetric Engineering and Remote Sensing*, 68,71-75.

Pietragalla. J., and Vega, A.M. (2012). Normalized difference vegetation index. In: *Physiological Breeding II: A Field Guide to Wheat Phenotyping*, (pp. 37-40). D.F. CIMMYT. Mexico.

Prasad, P.V. V., Boote, K.J., and Allen Jr, L. H. (2006). Adverse high temperature effects on pollen viability, seed-set, seed yield and harvest index of grain-sorghum [Sorghum bicolor (L.) Moench] are more severe at elevated carbon dioxide due to higher tissue temperatures. *Agricultural and Forest Meteorology*, 139, 237–251.

Prasad, P. V. V., Boote, K. J., Vu, J. C. V., and Allen Jr, L.H. (2004). The carbohydrate metabolism enzymes sucrose-P synthase and ADG-pyrophosphorylase in phaseolus bean leaves are up-regulated at elevated growth carbon dioxide and temperature. *Plant Science*, 166, 1565–1573.

Prasad, P. V. V., Craufurd, P. Q., Summerfield, R. J., and Wheeler, T. R. (2000). Effects of short episodes of heat stress on flower production and fruit-set of groundnut (*Arachishypogaea* L.). *Journal of Experimental Botany*, 51, 777–784.

Prasad, P. V. V., and Djanaguiraman, M. (2011). High night temperature decreases leaf photosynthesis and pollen function in grain sorghum. *Functional Plant Biology*, 38, 993–1003.

Prasad, P. V. V., Staggenborg, S. A., and Ristic, Z. (2008). Impacts of drought and/ or heat stress on physiological, developmental, growth, and yield processes of crop plants. In: *Response of crops to limited water: Understanding and modeling water stress effects on plant growth processes, Advances in Agricultural Systems Modeling Series* 1, (pp. 301-355). ASA, CSSA, SSSA, Madison, WI.

Ramirez-Vallejo, P. and Kelly, J.D. (1998). Traits related to drought resistance in common bean. *Euphytica*, 99,127-136.

Raun, W. R., Solie, J. B., Johnson, G. V., Stone, M. L., Lukina, E. V., Thomason, W. E. and Schepers, J. S. (2001). In-season prediction of potential grain yield in winter wheat using canopy reflectance. *Agronomy Journal*, 93(1), 131–138.

Repellin,A., Thi,A.T.P., Tashakorie,A., Sahsah,Y., Daniel,C., Zuily-Fodi,Y.l. (1997). Leaf membrane lipids and drought tolerance in young coconut palms (*Cocosnucifera* L.). *European Journal of Agronomy*, 6(1-2), 25-33.

Richards, R.A., Rebetzke, G.J., Watt, M., and Condon, A.G. (2010). Breeding for improved water productivity in temperate cereals: phenotyping, quantitative trait loci, markers and the selection environment. *Functional Plant Biology*, 37, 85–97.

Rodriguez, M. G., Reynolds, M. P., Escalante-Estrada, J. A. and Rodriguez-Gonzalez, M. T. (2004). Association between canopy reflectance indices and yield and physiological traits in bread wheat under drought and well-irrigated conditions. *Australian Journal of Agricultural Research*, 55(11),1139–1147.

Rymbai, H., Laxman, R. H., Dinesh, M. R., Sunoj, J. V. S., Ravishankar, K.V., and Jha A.K. (2014). Diversity in leaf morphology and physiological characteristics among mango (*Mangiferaindica*) cultivars popular in different agro-climatic regions of India. *ScientiaHorticulturae*, 176, 189–193.

Saini, H.S. (1997). Effects of water stress on male gametophyte development in plants. *Plant Reproduction*, 10: 67–73.

Saini, H.S., Sedgley, M., and Aspinall, D. (1984). Developmental anatomy in wheat of male sterility induced by heat stress, water deficit or abscisic acid. *Australian Journal of Plant Physiology*, 11, 243–253

Sairam, R. K., and Tyagi, A. (2004). Physiology and molecular biology of salinity stress tolerance in plants. *Current Science*, 86, 407–421.

Sanzol, J., and Herrero, M. (2001). The "effective pollination period" in fruit trees. *ScientiaHorticulturae*, 90, 1–17.

Sato, S., Peet, M. M. and Thomas, J. F. (2000). Physiological factors limit fruit set of tomato (*Lycopersiconesculentum*Mill.) under chronic, mild heat stress. *Plant, Cell and Environment*, 23, 719–726.

Savchenko, G., Klyuchareva, E., Abramchik, L. and Serdyuchenko, E. (2002) Effect of periodic heat shock on the inner membrane system of etioplasts. *Russian Journal of Plant physiology*, 49, 349–359.

Schneider, K.A., Rosales-Serna, R., Ibarra-Perez, F., Cazares-Enriquez, B., Acosta-Gallegos, J.A., Ramirez-Vallejo, P., Wassimi, N. and Kelly, J.D. (1997). Improving common bean performance under drought stress. *Crop Science*, 37,43-50.

Stoddard, F. L., Balko, C., Erskine, W., Khan, H. R., Link, W. and Sarker, A. (2006). Screening techniques and sources of resistance to abiotic stresses in cool-season food legumes.*Euphytica*, 147, 167–186.

Serraj, R. and Sinclair, T. R. (2002). Osmolyte accumulation: Can it really help increase crop yield under drought conditions? *Plant, Cell and Environment*, 25, 333–341.

Sunoj, V. S. J., Naresh Kumar, S. and Muralikrishna, K. S. (2014). Effect of elevated CO2 and temperature on oxidative stress and antioxidant enzymes activity in coconut (*Cocosnucifera* L.) seedlings. *Indian Journal of Plant Physiology*, 19(4),382-387.

Sunoj, V.S.J, Shroyer, K.J., Jagadish, S.V.K. and Prasad, P.V.V. (2016). Diurnal temperature amplitude alters physiological and growth response of maize (*Zea mays* L.) during the vegetative stage. *Environmental and Experimental Botany*.130,113-121.

Tangpremsri, T., Fukai S. and Fischer, K.S. (1995). Growth and yield of sorghum lines extracted from a population for differences in osmotic adjustment. *Australian Journal of Agriculture Research*, 46,61–74.

Tripathy, J.N., Zhang, J., Robin, S., Nguyen, T.T. and Nguyen, H.T. (2000). QTLs for cell-membrane stability mapped in rice (*Oryza sativa* L.) under drought stress. *Theoretical and Applied Genetics*, 100,1197-1202.

Valliyodon, B. and Nguyen, H.T. (2006). Understanding regulatory networks and engineering for enhanced drought tolerance in plants. *Current Opinion in Plant Biology*. 9,189–195.

Vina, A., Gitelson, A. A., Anthony, L., Nguy-Robertson, and Peng, Y. (2011). Comparison of different vegetation indices for the remote assessment of green leaf area index of crops. *Remote Sensing of Environment,* 115,3468–3478.

Wahid, A., Gelani, S., Ashraf, M. and Foolad, M.R. (2007). Heat tolerance in plants: An overview. *Environmental and Experimental Botany,* 61,199–223.

White, J.W. and Singh, S.P. (1991). Breeding for adaptation to drought. In: *Common beans: Research for crop improvement,* (pp.501-551). C.A. B. International. Wallingford, U.K. and CIAT, Cali, Colombia.

Yamasaki, S. and Dillenburg, L. R. (1999). Measurements of leaf relative water content in *Araucaria angustifolia*.RevistaBrasileira de Fisiologia Vegetal, 11(2), 69-75.

Zhou, Y., Lam, H. M. and Zhang, J. (2007). Inhibition of photosynthesis and energy dissipation induced by water ad high light stresses in rice. *Journal of Experimental Botany*. 58, 1207–1217.

2017, Impact of Climate Change on Plantation Crops *Pages* **189–207**
Editors: **K.B. Hebbar, S. Naresh Kumar & P. Chowdappa**
Published by: **ASTRAL INTERNATIONAL PVT. LTD., NEW DELHI**

Chapter 12

Biotechnological Tools to Mitigate Climate Change

K.N. Nataraja, R.S. Sajeevan and K.H. Dhanyalakshmi

1. Climate Change and its Impact on the Productivity of Perennial Crops

Anthropogenic activities have contributed for the present changes in environmental conditions. Now the issues related to the climate change represents the major research challenge faced by conservation biologists, and crop plant biologists. According to the Intergovernmental Panel on Climate Change (IPCC), elevated green-house gases like CO_2, and associated shifts in temperature and rainfall are expected to alter plant eco-physiology, distribution and their interactions with other organisms (IPCC 2014). The IPCC concluded that anthropogenic climate change will continue to have a strong effect on life cycle of plants and species interactions (IPCC 2014; Settele *et al.*, 2015). There are several research papers and reviews on effect of climate change on plant growth, reproduction and phenology (Parmesan and Hanley, 2015). Climate change alters seed germination (Milbau *et al.*, 2009; De Frenne *et al.*, 2013), leaf emergence (Slayback *et al.*, 2003; Jeong *et al.*, 2011), flowering and fruiting (Fitter and Fitter, 2002; Xia and Wan, 2013). Fu *et al.* (2015) noticed increase in 'heat requirement' for leaf flushing over time in 13 temperate trees and some trees may even extend the growing season. The magnitude of phenological responses is known to differ among trophic levels and global meta-analysis data suggest asynchrony between interacting trophic levels (predator–prey and insect–host, Parmesan and Hanley, 2015). These observations suggest multiple ill effects of climate change on growth behavior of perennial trees and plantation crops. There can be poor seed set and/or altered reproductive growth due to changed environmental conditions in plantation crops. A few studies examined to test how floral rewards are likely to be affected by climate warming or drought generated interesting data sets and interactions with pollinators could well be profound (Mu *et al.*, 2015; Scaven and Rafferty, 2013). From this context, there is an urgent need to examine the cellular and system level responses to climate change using modern biotechnological tools and techniques. Some of the developments and advancement in various fields of molecular biology and biotechnology, that can adopted for crop improvement of plantation crops are discussed.

2. Biotechnological Tools for Trait Specific Gene Discovery and Functional Validation

To meet the goal of evolving plantation crops suitable for cultivation in the current scenario of climate change, applications of different biotechnological tools in understanding diverse aspects of plant physiology are numerous. There are tools and techniques that help in identifying the genetic factors controlling crop adaptation (via genomics) and elucidating the function of important genes (via transcriptomics), gene products (via proteomics) as well as whole pathways (via transcriptomics, proteomics and metabolomics). Speedy advances are being noticed in the techniques used in these fields and novel tools are being developed. Some of the recent advances in genomics have led to Next Generation Sequencing (NGS) platforms, which have been used for the generation of valuable genomic resources and large amount of datasets are deposited in public databases (Table 12.1). Effective use of this information would aid in targeted crop improvement in a timely manner.

Structural genomics, a branch of genomics that deals with the genome sequencing has contributed for cloning and characterization of a large set of genes in many crop plants. Genome information of many of the plantation crops are yet to be generated, although there are attempts in this direction. Genome sequence information of the African oil palm (*Elaeis guineensis*), the predominant source of oil, has been generated and over 34,000 genes have predicted (Singh *et al.*, 2013). Similarly, the draft genome sequence of Hevea brasiliensis, a natural rubber yielding tree has been generated and the gene prediction models showed 68,955 gene, of which 12.7 per cent are unique to Hevea (Rahman *et al.*, 2013). The draft genome of *Theobroma cocoa* is now available, analysis of which might yield information on novel genes (Argout *et al.*, 2011). A database called COCOA Genome database is available at www.cacaogenomedb.org to accelerate the research and development programs by providing genomics, genetics and breeding resources. In coconut, the complete chloroplast genome information is available for a dwarf plant (Huang *et al.*, 2013). This type of genome information would be highly useful for developing genomic resources required for targeted crop improvement in plantation crops.

In the absence of genome sequences, several alternative approaches are followed to gather information on genes and their relevance. Information on expressed sequence tags (ESTs), cDNA libraries, microarrays and serial analysis of gene expression (SAGE) datasets that are available in different databases, can now be analyzed and be used to identify genes. Analysis of the genome-wide sequence variation by re-sequencing improved the availability of information that can be used

Table 12.1: Information on Some of the Important Plant Genome Databases Available which would be Useful for Generating Genomic Information Required for the Targeted Manipulation of Traits

Crop	*Scientific Name*	*Database*	*Web Portal*
Maize Maize	*Zea mays* L.	Genome Database	http://www.maizegdb.org/
Soybean	*Glycine max* (*L.*) Merrill.	*Soybean database*	*http://www.proteome.dc.af frc.go.jp/Soybean/*
Rice	*Oryza sativa* L.	–TIGR Rice Genome Annotation Project-Rice Genome Research Project – cultivated rice	http://blast.jcvi.org/eukblast/index.cgi?project=osa1http://rgp.dna.affrc.go.jp/
Western poplar	*Populus trichocarpa* v3.0	JGI – Phytozome 11	https://phytozome.jgi.doe. gov/pz/portal.html#!info? alias=Org_Ptrichocarpa
Potato	*Solanum tuberosum* L.	Spud DB – Potato Genomic Resources	http://solanaceae.plantbio logy.msu.edu/pgsc_download.shtml
Multiple genomes	–	PlantGDB – Resources for Plant Comparative Genomics	http://www.plantgdb.org/
Multiple genomes-	*Arabidopsis thaliana, Hordeum vulgare, Triticum aestivum, Zea mays,* Physcomitrella patens, and *Oryza sativa Japonica,*	*Gramene* -A comparative resource for plants	http://www.gramene.org/
EucalyptusEucalyptus	Eucalyptus camaldulensis L.	camaldulensis GB	http://www.kazusa.or.jp/e ucaly/
Cacao GB	*Theobroma cacao* L.	*Cacao GB*	*www.cacaogenomedb.org*

to develop DNA markers, required for molecular breeding. Molecular markers such as simple sequence repeat (SSR) and Single Nucleotide Polymorphism, also known as SNPs, identified from whole genome programs are valuable recourses for genetic analyses in many crops (Perera *et al.*, 1998; Argout *et al.*, 2011; Preethi *et al.*, 2015; Suwastika *et al.*, 2015; Perez-de-Castro *et al.*, 2012; Dhanapal and Govindaraj, 2015).

2.1. Targeted Gene Expression Studies to Identify Candidate Genes

The gene discovery for crop improvement programs was tedious in early days, due to the lack of efficient biotechnological tools (Parvathi and Nataraja, 2016). Knowing when and where a gene product is expressed can provide important clues to its biological function. The gene expression analysis allows the identification of genes and the study of their relationship with cellular processes. The first and foremost step is to identify the trait/s associated with the phenotype and subsequently, candidate gene/s linked to the trait. Gene expression analysis under different developmental stages as well as under growth conditions at various time points is being used to identify novel genes in agricultural crops. Such efforts have also been extended for woody species and perennial plants. For example, in Poplars (Populus species), drought responsive genes have been identified by correlating the physiological responses to gene expression (Arshad *et al.*, 2011). Analysis of gene expression has been attempted in cocoa, in order to understand the biological events under flooding stress (Bertolde *et al.*, 2014). In oil palm, the relevance of regulatory genes under various growth and developmental stages as well as stress conditions have been studied (Omidvar *et al.*, 2013).

Gene expression analysis approaches such as suppression subtractive hybridization (SSH) (Diatchenko *et al.*, 1999; Kong *et al.*, 2005; Li *et al.*, 2010; Yang *et al.*, 2011), differential display (DD) (Alves *et al.*, 1998; Liu and Baird, 2003; Maqbool *et al.*, 2008; Venkatachalam *et al.*, 2009) and microarrays analysis (Lorenz *et al.*, 2011, Atkinson *et al.*, 2013; Zhang *et al.*, 2014, Sham *et al.*, 2015) have been used in crop plants and plantation crops with varied degrees of success. The cultivation of tea, one of the most popular non-alcoholic drink worldwide, has been severely affected by water deficit, and SSH libraries have been used to identify genes that are relevant for establishing drought tolerance during field drought conditions (Gupta *et al.*, 2012). A SSH library constructed from the tender roots under drought conditions revealed 13.04 per cent of differentially regulated genes associated with various stresses. In a total of 123 putative droughtresponsive genes identified, ubiquitin-proteasome, glutathione metabolism and sugar metabolism pathways and several transcription factors were included (Das *et al.*, 2012). In coconut, Liang *et al.* (2014) suggested that the SSH based identification offers platforms for understanding the molecular basis of pulp development and hence a chance for improving its attributes under changing climate scenario. In cocoa, the differentially expressed genes from SSH library had been used as a strategy to understand the mechanisms of witches broom disease (Leal *et al.*, 2007).

2.2. Microarrays to Analyze Expression Pattern of known Genes

The early history of DNA arrays was started in 1975's, when Grunstein and Hogness created the colony hybridization method. The membrane based arrays and protocols were used in a variety of applications including: cloning genes of specific interest, identifying SNP's (Miller and Barnes, 1986), cloning genes that are differentially expressed between two samples (Crampton *et al.*, 1980) and physical mapping (Craig *et al.*, 1990). The core principle behind microarrays is hybridization between two DNA strands. With the advent of technology, the tedious manual processes of microarrays were automated using robotics to simplify the development of array clones from micro-titer plates onto membrane (Craig *et al.*, 1990; Lennon and Lehrach, 1991). The microarray technology has been used as a useful tool for the analysis of genome-scale gene expression analysis (Schena *et al.*, 1995; Eisen and Brown, 1999). This technology using EST's was first demonstrated by analysis of 48 Arabidopsis genes for differential expression in roots and shoots (Schena *et al.*, 1995).

By the late 1990's and early 2000's, cDNA's became widely available and complete genome sequences of some organisms and along with the advancement in robotics allowed to develop cDNA or ORF arrays that represent the vast majority of genes in a genome. In the early and mid2000's, tremendous amount of progress happened in the field of DNA array technology, both in the form of production as well as detection methods. In oligo-array, a modified form of microarray, a relatively shorter 25 to 80 bp in length were used compared to long DNA's 500 to 5000 bp and fluorescent based detection were developed. The low- or high-intensity fluorescence indicates the intensity of gene expression in contrast genotypes subjected to a specific condition, such as stress or pathogen attack (Kuo *et al.*, 2004). Oligo nucleotide arrays have been utilized in rubber for identification of genes associated with tapping panel dryness (Qin *et al.*, 2012). This method allows the simultaneous analysis of thousands of genes of interest, and the identification of both their presence and differential expression, the latter allowing inferences to be made regarding the possible function of specific genes (Wullschleger and Difazio, 2003; Al-Taweel and Fernando, 2011).

2.3. Global Transcriptome Analysis to Discover Novel Genes

The development of DNA sequencing by Sanger *et al.* (1977) revolutionized the biological sciences. The first generation 'Sanger DNA sequencing' was expensive and low throughput. The introduction of 'Next Generation Sequencing' (NGS) technology in 2005, made a lot of changes in the sequencing world. The NGS technology can determine millions of sequences in a single run and can significantly reduce the running cost, make the process simple in an automated pipeline. Global transcriptome analysis approaches through the use of various NGS technologies are the most recent approaches for gene and pathway discovery. In recent years, whole transcriptome or RNA sequencing (RNA-Seq) approaches are extensively used to discover novel genes and also examine their relevance (Thumma *et al.*, 2012; Liu *et al.*, 2015; Li *et al.*, 2015; Garg *et al.*, 2016). Sufficient background information on the technical aspects and advances in the area of high throughput RNA sequencing has been reported. The first successful NGS approach was 454 sequencing developed by Margulies *et al.* (2005). This method is based on a massively-parallel pyro-sequencing technology that can generate more than a million reads from a single run (http://www.454.com). In the meantime, 454 sequencing was superseded by new NGS technologies from Illimina (http://www.illumina.com), Life technologies (http://

www.lifetechnologies.com), HeliScope (http://www.helicosbio.com), PacBio RSII (http://www.pacificbiosciences.com) and Starlight (http://www.lifetechnologies.com). All these NGS technologies helped to sequence the entire transcriptome of a species or a tissue at particular condition or a single cell transcript profiling using a very small amount of RNA. The transcripts shortlisted through the transcriptome analysis coupled with real time PCR (RT-PCR), is one of the widely and most effective strategies to discover new genes for crop improvement (Howald *et al.*, 2012; Dhanyalakshmi *et al.*, 2016). Some of these techniques and technologies have been employed in plantation crops to understand the mechanisms associated with environmental responses. For example, in date palm (Phoenix dactylifera L.), a plantation crop cultivated mostly in arid climates and deserts, efforts have been made in the direction of generating genomic and transcriptomic information using pyro-sequencing platform (Roche/454GS FLX Titanium). Fourteen cDNA libraries from different tissues have been generated with 67,651 non-redundant contigs and 3,01,978 singletons, when pooled and assembled. Over 52,725 contigs have been annotated based on the plant databases and 45 contigs based on functional domains referencing to the Pfam database (Zhang *et al.*, 2012). Similarly, in coconut (Cocos nucifera) using Illumina sequencing platform, about 57,304 unigene have been sequenced from mixed tissues where 68.2 per cent of the unigene were successfully annotated based on Genbank non-redundant (Nr) protein database, where 99.9 per cent of unigene were novel compared to the released coconut EST sequences (Fan *et al.*, 2013). Yet another study in coconut through de novo transcriptome sequence assembly data from the leaf tissues by Huang *et al.* (2014) generated large volume of genomic resources in coconut. In rubber, transcriptomic studies have been attempted to study the molecular responses under pathogen infection (Paez *et al.*, 2015) and tapping panel dryness (Liu *et al.*, 2015). Transcriptomic studies in cocoa (Theobroma cacao L) developed from 56 cDNA libraries constructed from different organs, different genotypes and different environmental conditions generated 1,49,650 valid EST sequences corresponding to 48,594 unigene, 12,692 contigs and 35,902 singletons (Argout *et al.*, 2008). One of the major areas focused in cocoa through transcriptomics approach is the witches broom disease (da Hora Junior *et al.*, 2012). List of some of the important genomic resources generated in the plantation crops are presented in the Table 12.2. All these would be useful resources for future functional genomics studies and improvement programs required to generate climate resilient plantation crop.

2.4. Functional Validation of Genes Linked to Specific Traits

Targeted trait improvement requires precise understanding of the functions of the genes identified by different approaches. In recent years, function genomics, has become a major field of molecular biology. The functional genomics aims at making use of data produced by genomic and transcriptomic analysis to describe gene functions and interactions. The functional genomics also focuses on the dynamic aspects such as regulation of gene expression and protein–protein interactions. These studies also examine a range of processes including epigenetic regulation, in an attempt to answer relevant biological questions. Using different types of bioresources and targeted gene manipulation (both by up-regulation and down regulation), attempts are being made to assign the functions to genes discovered, although information available in plantation crops are limited.

Table 12.2: Transcriptomic Resources Available in Plantation Crops

Crop	*Tissue Used for Generating Resource*	*Reference*
Coconut	Spear leaves, young leaves, fruit flesh	Fan *et al.*, 2013
	Leaves	Huangh *et al.*, 2013
	Seed Leaves	Huangh *et al.*, 2014
	Embryonic calli derived from plumule	Rajesh *et al.*, 2016
Rubber	Latex	Chow *et al.*, 2007
	Bark	Li *et al.*, 2012
	Bark	Mantello *et al.*, 2013
	Bark	Lui *et al.*, 2015
	Latex	Chao *et al.*, 2015
	Bark	Li *et al.*, 2016
		Peaz *et al.*, 2015
		Lui *et al.*, 2015
Oil Palm	Mesocarp and Leaves	Bourgish *et al.*, 2011
	Fruit	Tranbagar *et al.*, 2011
	Young and mature leaves, root, fruit, flowers and offshoots	Zhang *et al.*, 2012
	Seed and Fruit	Dussert *et al.*, 2013
	Flower and Fruit	Shearman *et al.*, 2013
	Leaves	Lei *et al.*, 2014
	Spear leaf saples	Xiao *et al.*, 2014
Cocoa	Seed	Argout *et al.*, 2008
		Da Hora Junior *et al.*, 2012
Tea	Tender shoots, young leaves, mature leaves, stems, young roots, flower buds and immature seeds	Shi *et al.*, 2011
	Flowers	Tan *et al.*, 2013
	Leaves	Paul *et al.*/2014
	Tender shoots, young leaves, flower buds and flowers	Zhang *et al.*, 2015
Coffee	Leaves and fruits	Yuyama *et al.*, 2016

2.4.1. T-DNA Insertional Mutagenesis to Down Regulate the Gene Expression

Insertional mutagenesis strategies based on transposon systems (Parinov *et al.*, 1999) and TDNA tagging (Krysan *et al.*, 1999) have been found to be successful in Arabidopsis. Traditionally, T-DNA (Krysan *et al.*, 1999) and transposon-based insertional mutant populations (Parinov *et al.*, 1999; Tissier *et al.*, 1999) have provided useful resources for the analysis of phenotypes and discovery of gene/s inked to the phenotype. A large numbers of mutant have been generated by different groups, institutes and resources that are available for researchers. For example, Ds insertions (http://genetrap.cshl.org), the SALK T-DNA insertions (http://signal.salk.edu/tabout.html), the Arabidopsis knockout facility Madison (http://nasc.nott.ac.uk) and the institute national de la recherché agronomique lines (http://www.Arabidopsis.org/abrc/inra.html), provide useful information on mutant populations. Even though insertion mutants currently provide the best method of obtaining a null mutant for a given gene, this approach has limitations. Plants have a high degree of gene redundancy and about 6000 of the genes have highly significant sequence similarity to another gene, and thus functionally redundant. Even other genes with lower sequence similarity may also fall into functionally redundant group (The Arabidopsis Genome Initiative, 2000). Many mutations are lethal, often at the embryo state, and thus it is not possible to observe the phenotype of the adult mutant plant and some mutations gives weak phenotype. Hence to determine the functions of essential genes, weak or inducible mutations are more likely to be useful (Helliwell *et al.*, 2002).

2.4.2. RNA Interference (RNAi) to Down-Regulate Gene Function

RNA silencing (post transcriptional gene silencing, PTGS) is well suited to the systematic analysis of some of the plant genes and their function. Here, a gene fragment is introduced into a cell as dsRNA or as DNA that will generate dsRNA. The dsRNA activates the DICER/RISC process so that the properties of the affected cell reflect a loss of function in corresponding gene (Thakur, 2003, Nethra *et al.*, 2006). This process can be induced experimentally with high efficiency and targeted to a single specific gene or a multigene family.

In plants an effective method of silencing is through introduction of an endogenous gene construct encoding a hairpin RNA (hpRNA). It consists of an inverted repeat of a fragment of the gene sequence separated by a spacer. The construct may be generated in standard binary plant transformation vectors in which the hairpin encoding region is generated de novo for each gene (Helliwel and Waterhouse, 2003). Based on the high frequency and consistent heritability, Ma and Mitra (2002) developed RNAi vectors (pAM696 series) with intrinsic direct repeat, which may act as primary determinant of PTGS and these are referred to as direct repeat-induced PTGS (driPTGS). These vectors were effectively used to silence genes in tobacco. Several other groups also developed efficient RNAi vectors and tested in different plants (Zentella *et al.*, 2002). Inducible gene silencing using the RNAi constructs under alc (ethanol-inducible silencing) system minimized the problems associated with constitutive RNA silencing and enabled to dissect primary and secondary effects of PTGS at temporal and spatial resolution (Chen *et al.*, 2004). Further, wide range of Gateway ready RNAi vectors with many different combinations suitable for different systems are available from plant systems biology (PSB) and Flanders Inter university Institute for Biotechnology (VIB) and these resources can be accessed at www.plantgenetics.rug.ac.be/gateway, www.psb.ugent.be/gateway and www.vib.be. For gene silencing studies with abiotic stress genes the RNAi vectors with stress-inducible expression (using promoters like Rd29A, ABRE, *etc.*) would be useful.

Large scale silencing projects using transformation with constructs that contain hairpin RNA (hpRNA) and high throughput vectors are available. There is an extensive collection of RNA silencing-based transformants, PCR primers, and vectors for Arabidopsis chromatin-associated genes (www.chromdb.org) (Robertson, 2004). The dsRNA-mediated suppression of genes by vectors that produce sense and antisense transcripts has been successfully employed (Waterhouse *et al.*, 1998; Chuang and Meyerowitz, 2000). RNAi technology has been attempted in tea to produce low caffeine tea by silencing the caffeine synthase gene (Mohanpuria *et al.*, 2011).

2.5. Targeted Manipulation of Specific Traits

2.5.1. Transgenic Approach to Manipulate or Pyramid Traits

The conventional plant breeding has contributed significantly by developing several high yielding varieties in plantation crop plants. The long generation cycles, high heterozygosity and inbreeding depression limits the introgression of traits from wild species and relatives. Genetic engineering is the most potent biotechnological approach dealing with transfer of specially constructed gene assemblies through various transformation techniques. Tools of recombinant DNA technology facilitated development of transgenic plants. Transgenic crops were first commercialized in 1994 in developed countries and the technology helped the framers effectively manage pests and increase crop productivity with significant reduction in the cost of cultivation (Prado *et al.*, 2014). The development of transgenic crops is complex multi-step process (Figure 12.1), which is based on sound knowledge on traits and genes. The process of transgenic crop development is well coordinated to create a transgenic plant with desired beneficial phenotypes.

As genetic improvement efforts for specific traits need to be precise and directional, there is a need to combine various disciplines such molecular biology, plant physiology and tissue culture techniques for rapid success in plantation crops. In fact, transgenisis has emerged as a novel approach for carrying out "single gene" or "gene pyramiding" for the improvement of crop plants (Prado *et al.*, 2014), which can be extended to plantation crops to incorporate adaptive traits required under changing climate scenario. Identification and precise characterization of trait specific genes is the prerequisite for such an approach. There are scopes and opportunities for targeted modifications in perennial plantation crops as significant developments in plant genetic modification have been achieved in the last 15 years. For example, through targeted engineering of stress response pathways, salt tolerance has been addressed in tree species like poplars. Over expression of genes like AtSTO1 [encoding an enzyme in the abscisic acid biosynthesis pathway (9-cis-epoxycarotenoid dioxygenase3)], TaMnSOD (ROS scavenger), AtNHX1 (vacuolar Na/H antiporter) are some of the classical examples towards improving salt tolerance in poplars (Lawson and Michler, 2014; Jiang *et al.*, 2012, Zang *et al.*, 2014). Similarly, transgenic approach has been adopted in rubber to improve drought tolerance by over expressing a superoxide dismutase gene (MnSOD) (Jayashree *et al.*, 2003). These findings suggest that manipulation of cellular tolerance through

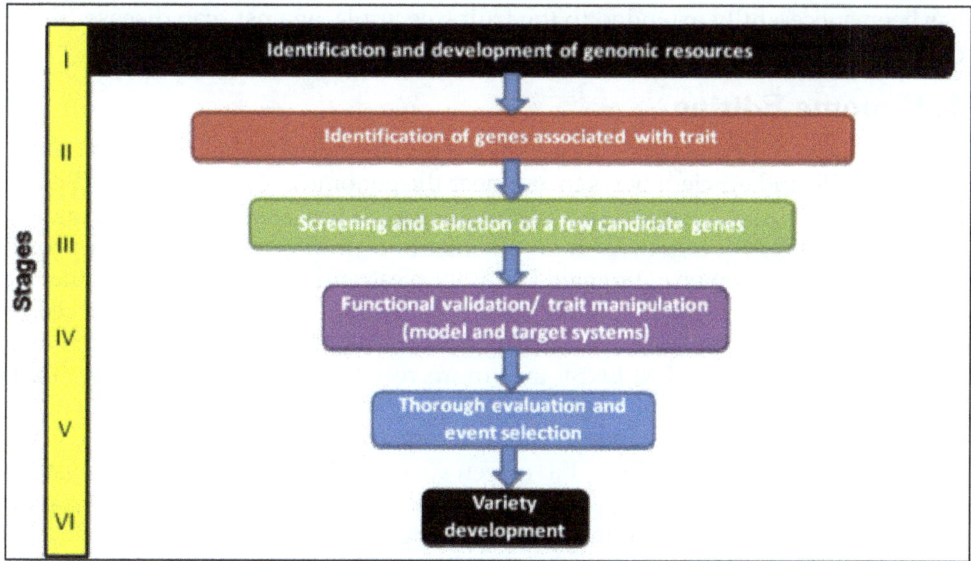

Figure 12.1: General Approaches followed for Gene Discovery and Targeted Manipulation of Traits in Field Crops, which can be Adopted for Plantation Crops to Improve Traits Required for Climate Resilience (Stages I-VI).

Stage I represents the identification of the relevant traits to be addressed in the improvement program and the development of the adequate genomic resources through diverse approaches. Stage II identifies a large set of genes associated with the selected traits of interest. Stage III represents the identification of a few candidate genes linked to the traits of interest from stage II. At stage IV, functional validation of the selected genes is done using over-expression or down regulation approaches in model systems as well as target systems. Stage V involves evaluation of the transgenic plants through targeted molecular and physiological approaches and identification of the promising events. Stage VI concentrates on the leads to develop variety, passing through field trial and addressing regulatory issues.

transgenic approach can immensely benefit plant growth and productivity under stress in perennial plantation crops.

3. Intragenesis and Cisgenesis

The limitations in the utilization of transgenic technologies after meeting with the regulatory issues following the safety issues have led to the identification of alternative technologies for transgenics. This prompted the development of intra-genesis and cis-genesis, which implies on the use of genetic material derived from the same species itself or closely related species for transformation (Jacobsen and Nataraja, 2008). This also focuses on the avoidance of foreign sequences such as selection genes and vector-backbone sequences. In intra-genesis, new gene combinations are allowed by the in vitro rearrangements of functional genetic elements (Holme *et al.*, 2013). The cis-genesis concept has been successfully tested in barley using its own phytase gene (Holme *et al.*, 2012). Recently, this technology has been successfully extended to the fruit tree apple, against scab disease (Krens *et al.*, 2015).

4. Genome Editing

The day to day advance in genome engineering tools have brought about more reliable and efficient ways to engineer the genomes in a more targeted ways. Genome editing with sitespecific nucleases is useful in genome engineering and targeted transgene integration in an efficient and precise manner. The approach involves the introduction of targeted DNA doublestrand breaks using an engineered nuclease, and stimulating DNA repair mechanisms. This has been evidenced by the emergence of tools like zinc finger nucleases (ZFN), transcriptionactivator like effector nucleases (TALENS), and more recently by the clustered regularly interspaced short palindromic repeat (CRISPR)/CRISPR-associated protein 9 (Cas9) system (CRISPR/Cas9 systems) (Townsend *et al.*, 2009; Belhaj *et al.*, 2013, Gaj *et al.*, 2013; Xing *et al.*, 2014). These have been successfully applied in model systems as well as crops like Arabidopsis, tobacco, maize, rice, wheat, potato, and tomato (Xiong *et al.*, 2015). The CRISPR/Cas9 technology has many advantages over other genome editing methods and hence is gaining popularity.

5. Conclusions

In developing countries, climate change would add to the burden of addressing issues related to the food and nutrition security. In addition, the predicted change might have serious problems on production and productivity of the major plantation crops like coconut, arecanut, oil palm, cashew, tea, coffee and rubber and also the minor plantation crop, cocoa. Increased incidence of abiotic stresses, an accumulative result of climate change would certainly limit production of the major plantation crops in India. Stresses like drought and high temperature affect vegetative growth and vigor, and also triggers floral abnormalities and poor fruit set. It also indirectly increases the incidence of pests and diseases in the plants. In plantation crops where there is a limitation of bioresources for crop improvement, modern biotechnological approaches are needed to evolve climate resilient plant types. From this context, 'Omics' based approaches would be useful to understand the climate responsive biological pathways and gens linked to the trait. The dissection of molecular mechanisms related to signal transduction and transcriptional regulation under diverse environmental conditions might help in engineering plantation crops.

References

Al-Taweel, K., and Fernando, W. G. D. (2011). Differential gene expression is a promising tool for understanding host– pathogen interactions. *The Americas Journal of Plant Science and Biotechnology*, 5, 1-10.

Alves, J. D., Tara, T., Toai, V., and Kaya, N. (1998). Differential display: a novel PCR-based method for gene isolation and cloning. Revista Brasiliera de Fisiologia Vegetal, 10(2). 161-164.

The Arabidopsis Genome Initiative (2000). Analysis of the genome sequence of the flowering plant Arabidopsis thaliana. *Nature*, 408(6814), 796-815.

Argout, X., Salse, J., Aury, J., Guiltinan, M. J., Droc, G., *et al.* (2011). The genome of cocoa. *Nature Genetics*, 43, 101-108.

Argout, X., Fouet, O., Wincker, P., Gramacho, K., Legavre, T., Sabau, X., Risterucci, A. M. *et al.* (2008). Towards the understanding of the cocoa transcriptome: production and analysis of an exhaustive dataset of ESTs of Theobroma cacao generated from various tissues and under various conditions. *BMC Genomics*, 9, 512.

Arshad, M., Biswas, K., Mattsson, J., Bisgrove, S., and Plant, A. (2011). Identification of genes that contribute to drought stress tolerance in Populus. *BMC Proceedings*, 5(Suppl 7), P79.

Atkinson, N. J., Lilley, C. J., and Urwin, P. E. (2013). Identification of genes involved in the response of Arabidopsis to simultaneous biotic and abiotic stresses, *Plant Physiology*, 162, 2028-2041.

Belhaj, K., Chaparro-Garcia, A., Kamoun, S., and Nekrasov, V. (2013). Plant genome editing made easy: targeted mutagenesis in model and crop plants using the CRISPR/Cas system. *Plant Methods*, 20139, 39.

Bertolde, F. Z., Almeida A. A. F., and Pirovani, C. P. (2014). Analysis of gene expression and proteomic profiles of clonal genotypes from Theobroma cacao subjected to soil flooding. *Plos One*, 9(10), e108705.

Chen, J. C., Jiang, C. Z., Gookin, T. E., Hunter, D. A., Clark, D. G., and Reid, M. S. (2004). Chalcone synthase as a reporter in virus-induced gene silencing studies of flower senescence. *Plant Molecular Biology*, 55, 521-530.

Chuang, C. F., and Meyerowitz, E. M. (2000). Specific and heritable genetic interference by double-stranded RNA in Arabidopsis thaliana, *Proceedings of National Academy of Sciences*, USA, 97(9), 4985-4990.

Craig, A. G., Nizetic, D., Hoheisel, J. D., Zehetner, G., and Lehrach, H. (1990). Ordering of cosmid clones covering the herpes simplex virus type I (HSV-I) genome: a test case for fingerprinting by hybridization. *Nucleic Acids Research*, 18, 2653-2660.

Crampton, J., Humphries, S., Woods, D., and Williamson, R. (1980). The isolation of cloned cDNA sequences which are differentially expressed in human lymphocytes and fibroblasts. *Nucleic Acids Research*, 8, 6007-6017.

da Hora Junior B. T., Poloni Jde, F., Lopes, M. A., Dias, C. V., Gramacho, K. P., Schuster, I., Sabau, X., Cascardo, J. C., Mauro, S. M., Gesteira Ada, S., Bonatto, D., and Micheli, F. (2012). Transcriptomics and systems biology analysis in identification of specific pathways involved in cacao resistance and susceptibility to witches' broom disease. *Molecular Biosystems*, 8(5), 1507-19.

Das, A., Das, S., and Mondal, T. K. (2012). Identification of differentially expressed gene profiles in young roots of tea [Camellia sinensis (L.) O. Kuntze] subjected to drought stress using suppression subtractive hybridization. *Plant Molecular Biology Reports*, 30, 1088-1101.

De Frennea, P., Rodríguez-Sánchezb, F., Coomesb, D. A., Baetena, L., Verstraetena, G., *et al.* (2013). Microclimate moderates plant responses to macroclimate warming. *Proceedings of National Academy of Sciences*, USA, 110(46), 18561-18565.

Dhanapal, A. P., and Govindaraj, M. (2015). Unlimited thirst for genome sequencing, data interpretation, and database usage in genomic era: The road towards fast-track crop plant improvement. *Genetics Research International*, 2015, 684321.

Dhanyalakshmi, K. H., Naika, M. B. N., Sajeevan, R. S., Mathew, O. K., Shafi, K. M., Sowdhamini, R., and Nataraja, K. N. (2016). An approach to function annotation for Proteins of Unknown Function (PUFs) in the transcriptome of Indian mulberry. *PLoS One*, 11(3), e0151323.

Diatchenko, L., Lukyanov, S., Lau, Y. F., and Siebert, P. D. (1999). Suppression subtractive hybridization: a versatile method for identifying differentially expressed genes. *Methods in Enzymology*, 303, 349-80.

Eisen, M. B., and Brown, P. O. (1999). DNA arrays for analysis of gene expression. *Methods in Enzymology*, 303, 179-205.

Fan, H., Xiao, Y., Yang, Y., Xia, W., Mason, A. S., *et al.* (2013). RNA-Seq analysis of Cocos nucifera: transcriptome sequencing and de novo assembly for subsequent functional genomics approaches. *Plos One*, 8, e59997.

Fitter, A. H., and Fitter R. S. R. (2002). Rapid changes in flowering time in British plants. *Science*, 296, 1689-1691.

Fu, Y. H., Piao, S., Vitasse, Y., Zhao, H., De Boeck, H., Liu, Q., *et al.* (2015). Increased heat requirement for leaf flushing in temperate woody species over 19802012: effects of chilling, precipitation and insolation. *Global Changes in Biology*, 21, 2687-2697.

Gaj. T., Gersbach, C. A., and Barbas, C. F., 3rd. (2013). ZFN, TALEN, and CRISPR/Casbased methods for genome engineering. *Trends in Biotechnology*, 31, 397-405.

Garg, R., Shankar, R., Thakkar, B., Kudapa, H., Krishnamurthy, L., Mantri, N., Varshney, R. K., Bhatia, S., and Jain, M. (2016). Transcriptome analyses reveal genotype and developmental stage-specific molecular responses to drought and salinity stresses in chickpea. *Scientific Reports*, 6. 19228.

Gupta, S., Bharalee, R., Bhorali, P., Bandyopadhyay, T., Gohain, B., Agarwal, N., Ahmed, P., Saikia, H., Borchetia, H., Kalitha, M. C., Handique, A. K., and Das, S. (2012). Identification of drought tolerant progenies in tea by gene expression analysis. *Functional and Integrated Genomics*, 3. 543-563.

Helliwell, C. A., and Waterhouse, (2003). Constructs and methods for high-throughput gene silencing in plants. *Methods*, 30(4), 289-295.

Helliwell, C. A., Wesley, S. V., Wielopolska, A. J., and Waterhouse, P. M. (2002). High throughput vectors for efficient gene silencing in plants. *Functional Plant Biology*, 29, 1217-1225.

Holme, I. B., Wendt, T., and Holm, P. B. (2013). Intragenesis and cisgenesis as alternatives to transgenic crop development. *Plant Biotechnology Journal*, 11, 395-407.

Holme, I. B., Dionisio, G., Brinch-Pedersen, H., Wendt, T., Madsen, C. K., Vincze, E., and Holm, P. B. (2012). Cisgenic barley with improved phytase activity. *Plant Biotechnology Journal*, 10(2), 237-247.

Howald, C., Tanzer, A., Chrast, J., Kokocinski, F., Derrien, T., Walters, N., Gonzalez, J. M., Frankish, A., Aken, B. L., Hourlier, T., Vogel, J. H., White, S., Searle, S., Harrow, J., Hubbard, T. J., Guigo, R., and Reymond A. (2012). Combining RT-PCR-seq and RNAseq to catalog all genic elements encoded in the human genome. *Genome Research*, 22(9), 1698-1710.

Huang, Y. Y., Matzke, A. J. M., and Matzke, M. (2013). Complete Sequence and Comparative Analysis of the Chloroplast Genome of Coconut Palm (Cocos nucifera). *Plos One*, 8(8), e74736.

Huang, Y. Y., Lee, C. P., Fu, J. L., Chang, B. C. H., Matzke, A. J. M., and Matzke, M. (2014). De novo transcriptome sequence assembly from coconut leaves and seeds with a focus on factors involved in RNA-directed DNA methylation. G3: *Genes Genomes Genetics*, 4(11), 2147-2157.

IPCC, (2014). Climate Change 2014: Impacts, Adaptation, and Vulnerability. Part A: Global and Sectoral Aspects. Contribution of Working Group II to the Fifth Assessment Report of the Intergovernmental Panel on Climate Change [Field, C. B., V. R. Barros, D. J. Dokken, K. J. Mach, M. D. Mastrandrea, T. E. Bilir, M. Chatterjee, K. L. Ebi, Y. O. Estrada, R. C. Genova, B. Girma, E. S. Kissel, A. N. Levy, S. MacCracken, P. R. Mastrandrea, and L. L. White (eds.)]. Cambridge University Press, Cambridge, United Kingdom and New York, NY, USA, pp. 1132.

Jacobsen, E., and Nataraja, K. N. (2008). Cisgenics-Facilitating the second green revolution in India by improved traditional plant breeding. *Current Science*, 94, 1365-1366.

Jayashree, R., Rekha, K., Venkatachalam, P., Uratsu, S. L., Dendekar, A. M., Jayasree, P. K., Kala, R. G., Priya, P., Sushamakumari, S., Sobha, S., Asokan, M. P., Sethuraj, M. R., and Thulaseedharan, A. (2003). Genetic transformation and regeneration of rubber tree (Hevea brasiliensis Muell. Arg) transgenic plants with a constitutive version of an antioxidative stress superoxide dismutase gene. *Plant Cell Report*, 22, 201-209.

Jeong, S., Ho, C., Gim, H., and Brown, M. (2011). Phenology shifts at start vs. end of growing season in temperate vegetation over the northern hemisphere for the period 1982-2008. *Global Change Biology*, 17, 2385-2399.

Jiang, C., Zheng, Q., Liu, Z., Xu, W., Liu, L., Zhao, G., and Long, X. (2012). Overexpression of Arabidopsis thaliana Na+/H+ antiporter gene enhanced salt resistance in transgenic poplar (Populus euramericana 'Neva'). *Trees*, 26, 685-694.

Kong, L., Anderson, J. M., and Ohm, H. W. (2005). Induction of wheat defense and stressrelated genes in response to Fusarium graminearum. *Genome*, 48(1), 29-40.

Krens, F. A., Schaart, J. G., van der Burgh, A. M., Tinnenbroek-Capel, I.E., Groenwold, R, Kodde, L. P., Broggini, G. A., Gessler, C., and Schouten, H. J. (2015). Cisgenic apple trees; development, characterization, and performance. *Frontiers in Plant Science*, 27(6), 286.

Krysan, P. J., Young, J. C., and Sussman, M. R. (1999). T-DNA as an insertional mutagen in Arabidopsis. *Plant Cell*, 11, 2283-2290.

Kuo, W. P., Kim, E. Y., Trimarchi, J., Jenssen, T. K., Vinterbo S. A., and Ohno-Machado, L. (2004). A primer on gene expression and microarrays for machine learning re searchers. *Journal of Biomedical Informatics*, 37(4), 293-303.

Lawson, S. S., and Michler, C. H. (2014). Overexpression of AtSTO1 leads to improved salt tolerance in Populus tremula x P. alba. *Transgenic Research*, 23, 817-826.

Leal G. A. Jr., Albuquerque, P. S., and Figueira, A. (2007). Genes differentially expressed in Theobroma cacao associated with resistance to witches' broom disease caused by Crinipellis perniciosa. *Molecular Plant Pathology*, 8(3), 279-292.

Lennon, G. G., and Lehrach, H. (1991). Hybridization analyses of arrayed cDNA libraries. *Trends in genetics*, 7, 314-317.

Li, D., Deng, Z., Chen, C., Xia, Z., Wu, M., He, P., and Chen, S. (2010). Identification and characterization of genes associated with tapping panel dryness from Hevea brasiliensis latex using suppression subtractive hybridization. *BMC Plant Biology*, 10, 140.

Li, Y., Wang, X., Chen, T., Yao, F., Li, C., Tang, Q., Sun, M., Sun, G., Hu, S., Yu, J., and Song, S. (2015). RNA-Seq based de novo transcriptome assembly and gene discovery of Cistanche deserticola fleshy stem. *Plos One*, 10(5), e0125722.

Liang, Y., Yuan, Y., Liu, T., Mao, W., Zheng Y., and Li, D. (2014). Identification and computational annotation of genes differentially expressed in pulp development of Cocos nucifera L. by suppression subtractive hybridization. *BMC Plant Biology*, 14, 205.

Liu, J. P., Xia, Z. Q., Tian, X. Y., and Li, Y. J. (2015). Transcriptome sequencing and analysis of rubber tree (Hevea brasiliensis Muell.) to discover putative genes associated with tapping panel dryness (TPD). *BMC Genomics*, 16, 398.

Liu, X. N., and Baird, W. V. (2003). Differential expression of genes regulated in response to drought or salinity stress in sunflower. *Crop Science*, 43, 678-687.

Lorenz, W. W., Alba, R., Yu, Y. S., Bordeaux, J., Simoes, M., and Dean, J. (2011). Microarray analysis and scale-free gene networks identify candidate regulators in drought-stressed roots of loblolly pine (P. taeda L.). *BMC Genomics*, 12(1), 264.

Ma, C., and Mitra, A. (2002). Intrinsic direct repeats generate consistent post-transcriptional gene silencing in tobacco. *Plant Journal*, 31, 37-49.

Maqbool, A., Zahur, M., Irfan, M., Younas, M., Barozai, K., *et al.* (2008). Identification and expression of six drought-responsive transcripts through differential display in desi cotton (Gossypium arboreum). *Molecular Biology*, 42, 492-498.

Margulies, M., Egholm, M., Altman, W., Attiya, S., Bader, J., Bemben, L., Berka, J., *et al.* (2005). Genome sequencing in micro-fabricated high-density picolitre reactors. *Nature*, 437, 376-380.

Milbau A., Graae B. J., Shevtsova, A., and Nijs, I. (2009). Effects of a warmer climate on seed germination in the subarctic. *Annals of Botany*, 104, 287-296.

Miller, J. K., and Barnes, W. M. (1986). Colony probing as an alternative to standard sequencing as a means of direct analysis of chromosomal DNA to determine the spectrum of single-base changes in regions of known sequence. Proceedings of the National Academy of Sciences, USA, 83, 1026-1030.

Mohanpuria, P., Kumar, V., Ahuja, P. S., and Yadav, S. K. (2011). Producing low-caffeine tea through post-transcriptional silencing of caffeine synthase mRNA. *Plant Molecular Biology*, 76(6), 523-534.

Mu, J., Peng, Y., Xi, X., Wu, X., Li, G., Niklas, K. J., and Sun, S. (2015). Artificial asymmetric warming reduces nectar yield in a Tibetan alpine species of Asteraceae. *Annals of Botany*, 116, 899-906.

Nethra, P., Nataraja, K. N., Rama, N., and Udayakumar, M. (2006). Standardizations of environmental condition for induction and retention of posttranscriptional gene silencing using tobacco rattle virus vector. *Current Science*, 90(3), 431-435.

Omidvar, B., Abdullah, S. N. A., Ebrahimi, M., Ho, C. L., and Mahmood, M. (2013). Gene expression of the oil palm transcription factor EgAP2-1 during fruit ripening and in response to ethylene and ABA treatments. *Biologia Plantarum*, 57(4), 646-654.

Paez, U. A. H., Romero, G. I. A., Restrepo, R. S., Gutiérrez, A. F. A., and Castano, M. D. (2015). Assembly and analysis of differential transcriptome responses of Hevea brasiliensis on interaction with Microcyclus ulei. *Plos One*, 10(8), e0134837.

Parinov, S., Sevugan, M., Ye, D. Yang, W. C., Kumaran, M., and Sundaresan, V. (1999). Analysis of flanking sequences from dissociation insertion lines: A database for reverse genetics in Arabidopsis. *Plant Cell*, 11, 2263-2270.

Parmesan, C., and Hanley, M. E. (2015). Plants and climate change: complexities and surprises. *Annals of Botany*, 116(6), 849-864.

Parvathi, M. S., and Nataraja, K. N. (2016). Emerging tools, concepts and ideas to track the modulator genes underlying plant drought adaptive traits: An overview, *Plant Signaling and Behavior*, 11(1), e1074370.

Perera, L., Russell, J.R., Provan, J., and Powell, W. (2000). Use of microsatellite DNA markers to investigate the level of genetic diversity and population genetic structure of coconut (*Cocos nucifera* L.). *Genome* 43, 15-21.

Perez-de-Castro, A. M., Vilanova, S., Canizares, J., Pascual, L., Blanca, J. M., Diez, M. J., Prohens. J., and Pico, B. Application of genomics tools in plant breeding. *Current Genomics*, 13, 179-195.

Prado, J. R., Segers, G., Voelker, T., Carson, D., Dobert, R., Phillips, J., Cook, K., Cornejo, C., Monken, J., Grapes, L., Reynolds, T., and Martino-Catt, S. (2014). Genetically engineered crops: From idea to product. *Annual Review of Plant Biology*, 65, 769-790.

Preethi, P., Divyalakshmi, B., Naganeeswaran, S., Hemalatha, N., and Rajesh, M. K. (2015). Construction of EST-SSR marker database in coconut. *International Journal of Innovative Research in Computer and Communication Engineering*, 3(7), 263-268.

Qin, B., Liu, X. H., Deng, Z., and Li, D. J. (2012). Identification of genes associated with tapping panel dryness in Hevea brasiliensis using oligonucleotide microarrays. *Chinese Journal of Tropical Crops*, 33(2), 296-301.

Rahman, A. Y. A., Usharraj, A., Misra, B., Thottathil, G., Jayasekaran, K., *et al.* (2013). Draft genome sequence of the rubber tree Hevea brasiliensis. *BMC Genomics*, 14, 75.

Robertson, D. (2004). VIGS vectors for gene silencing: many targets, many tools. *Annual Review of Plant Biology*, 55, 495-519.

Sanger, F., Nicklen, S., and Coulson, A. R. (1977). DNA sequencing with chain terminating inhibitors. *Proceedings of National Academy of Sciences*, USA, 74, 5463-5467.

Scaven, V., and Rafferty, N. E. (2013). Physiological effects of climate warming on flowering plants and insect pollinator and potential consequences for their interaction. *Current Zoology*, 59, 418-426.

Schena, M., Shalon, D., Davis, R. W., and Brown, P. O. (1995). Quantitative monitoring of gene expression patterns with a complementary DNA microarray. *Science*, 270, 567-570.

Settele, J., Spangenberga, J. H., Heongd, K. L., Burkharde, B., Bustamanteg, J. V., *et al.* (2015). Agricultural landscapes and ecosystem services in Southeast Asia the LEGATOProject. *Basic and Applied Ecology*, 16, 661-664.

Sham, A., Moustafa, K., Al-Ameri, S., Al-Azzawi, A., Iratni, R., and AbuQamar, S. (2015). Identification of Arabidopsis candidate genes in response to biotic and abiotic stresses using comparative microarrays. *Plos One*, 10(5), e0125666.

Singh, R., Ong-Abdullah, M., Low, E. T., Manaf, M. A., Rosli, R., *et al.* (2013). Oil palm genome sequence reveals divergence of inter-fertile species in Old and New worlds. *Nature*, 500, 335-339.

Slayback, D. A., Pinzon, J. E., Los, S. O., and Tucker, C. J. (2003). Northern hemisphere photosynthetic trends 1982-99. *Global Change Biology*, 9, 1-15.

Suwastika, I. N., Musliminb, Rifkaa, Aisyaha, N., Rahmansyaha, Mutmainaha, Ishizakic, Y., Basrid, Z., and Shiina, T. (2015). Genotyping based on SSR marker on local cacao (Theobroma Cacao L.) from Central Sulawesi. *Procedia Environmental Sciences*, 28, 8891.

Thakur, A. (2003). RNA interference revolution. *Electronic Journal of Biotechnology*, 6, 39-49.

Thumma, B. R., Sharma, N., and Southerton, S. G. (2012). Transcriptome sequencing of Eucalyptus camaldulensis seedlings subjected to water stress reveals functional single nucleotide polymorphisms and genes under selection. *BMC Genomics*, 13, 364.

Tissier, A. F., Marillonnet, S., Klimyuk, V., Patel, K., Torres, M. A., Murphy, G., and Jones, J. D. (1999). Multiple independent defective suppressor mutator transposon insertions in Arabidopsis: A tool for functional genomics. *Plant Cell*, 11, 1841-1852.

Townsend, J. A., Wright, D. A., Winfrey, R. J., Fu, F., Maeder, M. L., *et al.* (2009). High frequency modification of plant genes using engineered zinc finger nucleases. *Nature*, 459, 442-445.

Venkatachalam, P., Thulaseedharan, A., and Raghothama, K. G. (2009). Molecular identification and characterization of a gene associated with the onset of tapping panel dryness (TPD) syndrome in rubber tree (Hevea brasiliensis Muell.) by mRNA differential display. *Molecular Biotechnology*, 41, 42-52.

Waterhouse, P. M., Graham, M. W., and Wang, M. B. (1998). Virus resistance and gene silencing in plants can be induced by simultaneous expression of sense and antisense RNA. *Proceedings of National Academy of Sciences*, USA, 95, 13959-13964.

Wullschleger, S. D., and Difazio, S. P. (2003). Emerging use of gene expression microarrays in plant physiology. *Comparative and Functional Genomics*, 4, 216-224.

Xia, J., and Wan, S. (2013). Independent effects of warming and nitrogen addition on plant phenology in the Inner Mongolian steppe. *Annals of Botany*, 111(6), 1207-1217.

Xing, H. L., Dong, L., Wang, Z. P., Zhang, H. Y., Han, C. Y., Liu, B., Wang, X. C., and Chen, Q. J. (2014). A CRISPR/Cas9 toolkit for multiplex genome editing in plants. *BMC Plant Biology*, 14, 327.

Xiong, J. S., and Li, Y. (2015). Genome editing technologies and their potential application in horticultural crop breeding. *Horticulture Research*, 2, 15019.

Yang, Z., Peng, Z. S., Yang, H., Yang, J., Wei, S., and Cai, S. H. (2011). Suppression subtractive hybridization identified differentially expressed genes in pistil mutations in wheat. *Plant Molecular Biology Reports*, 29, 431-439.

Zentella, R., Yamauchi, D., and Ho, T. H. (2002). Molecular dissection of the gibberellin/abscisic acid signaling pathways by transiently expressed RNA interference in barley aleurone cells. *Plant Cell*, 14, 2289-2301.

Zhang, D. Y., Yang, H. L., Li, X. S., Li, X. Y., and Wang, Y. C. (2014). Overexpression of Tamarix albiflonum TaMnSOD increases drought tolerance in transgenic cotton. *Molecular Breeding*, 34(1), 1-11.

Zhang, G., Pan, L., Yin, Y., Liu,W., Huang, D., Zhang, T., Wang, L., Xin, C., Lin, Q., Sun, G., Abdullah, M. M., Zhang, X., Hu, S., Al-Mssallem, I. S., and Yu, J. (2012).

Large scale collection and annotation of gene models for date palm (*Phoenix dactylifera*, L.), *Plant Molecular Biology*, 79, 521-536.

Zhang, X., Xiong, H., Liu, A., Zhou, X., Peng, Y., Li, Z., Luo, G., Tian, X., and Chen X. (2014). Microarray data uncover the genome wide gene expression patterns in response to heat stress in rice post-meiosis panicle. *Journal of Plant Biology*, 57, 327-336.

2017, Impact of Climate Change on Plantation Crops *Pages* **209–218**
Editors: **K.B. Hebbar, S. Naresh Kumar & P. Chowdappa**
Published by: **ASTRAL INTERNATIONAL PVT. LTD., NEW DELHI**

Chapter 13

Farming System Approach to Reduce Impacts of Climatic Change

K.B. Hebbar, M. Arivalagan and Ravi Bhat

1. Introduction

Plantation crops mainly coconut, rubber, tea, coffee, oil palm, arecanut, cashew, cocoa are grown in ecologically sensitive areas such as coastal belts, hilly areas and areas with high rainfall and high humidity. Amongst these, coconut is the major crop in India grown in almost 2 m ha while others are grown in 0.5 m ha or less areas. To meet the projected demand of 22 billion nuts in 2025 from the present supply of around 15 billion nuts, need adaptation measures that are most likely to be effective in improving yields of coconut grown under drought, flood and high temperature conditions in future climates. Coconut is grown between 20° N and 20° S latitude. It can be grown even at 26° N latitude but the temperature is the main limitation. The optimum weather conditions for good growth and nut yield in coconut are well distributed annual rainfall between 130 and 230 cm, mean annual temperature of 27 °C, abundant sunlight ranging from 250 to 350 Wm^{-2} with at least 120 hours per month of sun shine period. Since, it is humid tropical crop it grows well above 60 per cent humidity (Child 1974, Murray 1977). The generally recommended levels of major nutrients like NPK is at 500 g N: 320 g P_2O_5: 1200 g K_2O per palm/year. The recommended irrigation levels are 200L/palm once in 4 days or @ 66 per cent Eo through drip irrigation (Rajgopal *et al.*, 1999). Any deviations from these optimal conditions cause the palms to experience the stress conditions.

Climate change will affect coconut plantation through higher temperatures, elevated CO_2 concentration, precipitation changes, increased weeds, pests, and disease pressure, and increased vulnerability of organic carbon pools. In order to predict the future coconut production a coconut simulation model Infocrop-COCONUT (CocoSim) was developed (Naresh Kumar *et al.*, 2008). Simulation analysis using the model indicates that under all storylines, coconut productivity is projected to go up by up to 10 per cent during 2020, up to 16 per cent in 2050 and up to 36 per cent in 2080 over current yields only due to climate change. However, in east coast yield is projected to decline by about 2 per cent in 2020, 8 per cent in 2050 and 31 per cent in 2080 scenario over current yields due to climate change. Yields are projected to go up in Kerala, Tamil Nadu, Karnataka, Maharastra while they are projected to decline in Andhra Pradesh, Odisha and Gujarat (Naresh Kumar and Aggarwal, 2009; 2013). Coconut has adaptive strategies to withstand or overcome the stress conditions at morphological, physiological, biochemical, anatomical and molecular levels (Kasturi bai *et al.*, 1997). In this chapter the response and adaptive strategies of coconut are discussed with respect to climatic factors like high CO_2 effect and the consequences of climate change like drought and high temperature.

2. Climate Change Impact

In open top chamber (OTC) experiments it was observed that coconut seedling growth was promoted with [ECO_2] while, elevated temperature [ET] 3°C above ambient reduced the growth (Naresh Kumar, 2007; Hebbar *et al.*, 2013b). 700 ppm [ECO_2] increased plant height, leaf area and biomass production of coconut seedlings by 18 per cent, 16 per cent and 15 per cent respectively, as compared to ambient 380 ppm. The higher root biomass accumulation indicated better CO_2 sequestration with [ECO_2]. Higher growth was due to both increased leaf area and photosynthesis. On the other hand, ET significantly reduced both photosynthesis and leaf area and thus the plant growth. In open top chamber (OTC) experiments it was further observed that the stimulatory effect of CO_2 under drought was less and it could increase the biomass by only 8 per cent at 700 ppm CO_2 (Hebbar *et al.*, 2013a). However, both under normal and water limited condition there was reduced stomatal conductance and transpiration with elevated CO_2. Thus, the water requirement to produce unit biomass in ECO_2 treatment is less. This indicated that, at the present level of moisture available coconut would produce more biomass under future climate. However, with the projected reduction in precipitation under future climate the biomass production and nut production may be reduced unless corrective measures are taken. Similar to the above ground biomass below ground biomass i.e, root biomass too increased with elevated CO_2. Thus, it is expected that there will be higher carbon sequestration under future climate which is an important option to mitigate the climate change effect.

2.1. Moisture Deficiency

More than 60 per cent of coconut cultivation is rainfed and over 50 to 60 per cent yield loss is due to drought stress. Extensive work has been carried out at CPCRI to characterize the drought prone areas, to assess the impact and identify the critical stages. Coconut drought is characterized in different agro-climatic zones *viz.*, Western ghats high rainfall zone (Kidu-Karnataka), Western coastal area - hot sub-humid-per-humid (Kasaragod - Kerala; Ratnagiri - Maharastra), hot semi arid

(Arisikere - Karnataka) and Eastern coastal plains- hot sub-humid (Veppankulum-Tamil Nadu; Ambajipeta- Andhra Pradesh). Weather data based characterization of drought and its intensity indicated variations in length and number of dry spells in each zone (Naresh Kumar *et al.*, 2007). Apart from this they also differed for rainfall, temperature regimes and light intensities, thus bringing about the different intensities of drought.

2.2 Response to Elevated CO_2 and High Temperature

In open top chamber (OTC) experiments it was observed that coconut seedling growth and biomass increased at elevated CO_2. At 550 and 700 ppm CO_2 the biomass increased by 8 and 25% respectively against ambient CO_2 concentration of 380 ppm (Hebbar *et al.*, 2013). High temperature on the other hand significantly reduced the biomass (Table 13.1). Elevated CO_2 to certain extent could offset the negative effect of temperature in coconut. Drought also significantly reduced the biomass production across all the treatments. The stimulatory effect of CO_2 under drought was less and it could increase the biomass by only 8% at 700 ppm CO_2. CO2 was closely associated with the photosynthesis (PN).

Table 13.1: Biomass Production of Coconut Seedling in OTC with Climate Change Variables ECO_2, ET and Drought

Climate variable *Volume (cm^3)* *Root length (cm)*	*Root wt (kg)*	*Stem wt (kg)*	*Leaf wt (kg)*	*Biomass (kg)*
Ambient 94 752	0.218	0.406	0.398	1.022
550 ppm CO2 98 875	0.247	0.446	0.433	1.126
700 ppm CO2 105 992	0.269	0.487	0.460	1.216
ET 110 638	0.195	0.346	0.370	0.911
ET+550 ppm CO2 116 736	0.218	0.380	0.390	0.988
Treatments				
Control 112 852	0.273	0.523	0.461	1.257
Drought 107 581	0.113	0.234	0.238	0.584
CD at 5%				
OTC 8 77	0.017	0.022	0.028	0.037
Treat 6.45 59.67	0.014	0.017 0	.022	0.029

Prolonged periods of temperatures above the maximum tolerable limit can hamper vegetative growth as well as reproductive development and consequently the yield. Pollen of crops, since they have to stay viable in the field after anthesis until pollination, are exposed to temperature and humidity fluctuations in the atmosphere for a longer period. Also, the pollen development and anthesis stages have been found to be highly sensitive to temperature changes. In coconut most of the genotypes tolerate temperature up to 30oC (Hebbar and Chaturvedi 2015; 2016). Beyond which pollen germination as well as pollen tube length of genotypes decreases. Genotypes among these groups viz. talls, dwarfs and hybrids too showed wide variability for pollen germination across temperatures. Tall variety WCT had high pollen germination at all the temperatures, followed by FMST. Hybrid MYD X WCT performed better than COD X WCT at high temperature. Amongst dwarfs COD was the best at all temperatures (Helen, 2016). So, studies on the effect of high temperature on pollen and the genotypic variation, if any, in the thermotolerance of pollen are important for identifying the ones that can survive in a warmer future. The results may be useful in future breeding programmes aimed at developing climate-smart varieties.

High temperatures can have both negative and positive impacts on growth and production in coconut. The negative impacts such as added heat stress, especially in areas at low to mid-latitudes already at risk today, but they also may lead to positive impacts in currently cold-limited high-latitude regions. Warming trends are noticed in most parts of the coconut growing areas of Karnataka, Kerala and Tamil Nadu. High temperature increases both photorespiration and the dark respiration and thus the total biomass production go down. Regression analysis indicated increase in T min increased the leaf emergence rate: increase in T max increased inflorescence emergence rate; pistillate flower production has curvilinear relationship with rainfall/month (150mm/month-opt), nut retention has curvilinear relationship with T max (32 oC-opt) and Tmin (20 oC-opt). Frequent but short periods of temperature below 15°C result in abnormalities of fruit such as bicarpelate nuts and lack of pollination under North Indian conditions.

Thus, climate change has far reaching implications for plant growth, production and food security, and approaches are required for adapting to new climates. One of the primary approaches broadly exist for adapting plantation crops to these conditions is devising new cropping systems and methods for managing crops in the field. These approaches include the specific strategies as discussed below.

3. Coconut Based Farming System

Coconut is mostly grown in coastal and hilly areas where the rainfall is very high and the soil is poor in nutrients. The soil is sandy or laterite which has very low water holding capacity. With the impending climate change projections of high temperature and reduced rainfall the coconut productivity in these soils may decline in the future climates. Studies conducted at CPCRI and elsewhere indicated that coconut based farming system approach is the best adaptation strategy to overcome the effect of climate change.

In a coconut based farming system, coconut trees are planted as a base crop and all other crops are intercropped using the vertical and horizontal spaces between coconut trees. Coconut is a tree which has no branches and grows straight vertically upwards providing more and more space under its canopy. Its leaves are such that it allows sun light to the crops grown under it. Because of these peculiar characteristics of this tree the coconut based farming system is quite different from other farming system based on other crops. Coconut based farming system is a combination of multiple cropping systems in vertical and horizontal dimensions.

Between two coconut trees, fruit trees such as lime, lemon, guava, pamegranate, custard apple, cocoa, nutmeg, clove, crop which are planted at 15 -20 ft distance. These are medium sized crops both in height as well as canopy and can easily fit in between two adjacent coconut trees. They can be planted simultaneously or after the coconuts are established. It takes 8 to 10 years for all the coconut trees to start yielding properly. Whereas a number of the above mentioned crops start yielding well from 3 -5 years after planting. However, their fruiting period will last only 15 -20 years. By the time the coconut will be in its peak yield stage and will be about 20 ft high. The intercrops may be replaced by any other crop and another cycle of medium sized intercrops can be established.

The space between the coconut tree and the first intercropping is about 15 -16 ft which is more than sufficient for a number of perennial, biennial and seasonal crops. For example within this space two banana plants can be planted. Between banana plants, shade loving crops such as pineapple, turmeric, ginger, yam, elephant foot, etc. are planted. If bananas are not planted then crops like tapioca for tuber as well as fodder (stem, leaves etc.) and fodder grasses can be planted. If fodder crops are planted between the intercrops herbivorous animal husbandry is included. The same area can be intercropped with vegetables, green crops such as maize, jowar, ragi, bajra, soyabean, peg ion pea, black gram, green gram, etc. Either these grains can be sold or may be fed to animals such as poultry, pigs, goats, sheep, rabbits etc. The dung and other waste from these animals will be brought back into the land as manures.

The advantages of cropping systems in terms of sustainability and ecology are

1) maximization of solar energy capture; 2) optimization of soil moisture use and retention capacities; 3) enhancement of soil fertility build up due to higher biomass generation over time; and 4) minimal soil erosion and nutrient losses largely attributed to effective and efficient crop canopies and root systems (Magat, 2007).

3.1. Soil and Water Conservation

The major problem in heavy rainfall areas is the soil erosion due to the direct impact of rainfall and the separation of soil aggregates. Increased ground cover due to adoption of cropping system can control soil erosion by 70- 90 per cent, compared to bare soil or un-cropped condition. The presence of undergrowth vegetation in coconut plantations minimizes soil and water loss through surface runoff as well as evaporation. Rain water infiltration and storage are the indirect benefits of ground cover. Any change in rainfall pattern due to climate change can thus be mitigated by adopting cropping system. The runoff in cropping system was reduced to 1.45 per cent from 14.85 per cent in coconut monocrop. Similarly the soil loss was reduced to 0.9 t/ha in cropping system from 3.12 t/ha in monocropping (Dhanapal *et al.*, 2002).

Nair and Balakrishna (1977) reported more equitable microclimate in side coconut + cocoa cropping system compared to monocropped stand. The crop mixes recorded lower mean maximum temperature, higher relative humidity and reduced evaporative demand. In another study, the soil temperature at 30 and 60 cm depths was 3 to 6°C lower and the variation in the mean monthly soil temperature was the

Figure 13.1: Soil and Water Conservation in Coconut Basins. (a) Mulching with coconut leaves, (b) Burial of coconut husk, (c) Burial of composted coir pith.

least in the coconut + cocoa mixed cropping system compared to the monoculture of coconut (Varghese *et al.,* 1978). Soil moisture conservation through leaf mulching, husk/coirpith burial in basin enhanced coconut yields in different agro-climatic regions (Figure 13.1) (Naresh Kumar *et al.,* 2006).

3.2. Moderating Temperature and Moisture Content in Soil

Coconut is being cultivated in tropical regions where temperature is high and in summer month's evaporation is also higher. Apart from moisture retention the organic materials can also bring down the soil temperature to the optimum level. It is well known phenomenon that using husk, leaves and other waste materials as a mulch in coconut garden will conserve the moisture as well as reduces the incident of sun rays reaching the earth surface. In coconut monocrop drip irrigation and mulching reduced the temperature to the extent of 4.3°C compared to unmulched rainfed crop at 15 cm depth in littoral sandy soil (Maheshwarappa *et al.,* 1998).

Adopting cropping system by growing intercrops in coconut garden has reduced the soil temperature as compared to coconut monocrop. Among intercrops, grass cultivation recorded lowest average temperature (26.5°C) followed by pineapple (27.1°C), vegetables (28.7°C) and monocrop has recorded highest average temperature (29.3°C). At the same time average maximum soil temperature in open place where no cropping is practiced was 36°C and minimum was 29.3°C (Subramanian *et al.,* 2013).The moisture content in soil is most important for growth and production of any crop. The technologies generated should also help in maintaining higher soil moisture in soil. The burial of organic materials and growing intercrops in coconut garden has helped in maintaining higher moisture level (5-7 per cent) as compared to control (3 per cent) during the crop growth period (Shivakumar, 2016). In another study, the available soil moisture under drip mulch was higher by 22.2 to 28.8 per cent compared to drip without mulch. In basin irrigation also, on fourth day after irrigation, the available soil moisture; stored in the mulched condition was 36.8 to 37.6 mm and it was 18.2 to 19.9 mm under unmulched condition, indicating the moisture depletion higher by 28.6 to 30.7 per cent compared to moisture present on first day after irrigation (Maheshwarappa *et al.,* 1998).

On the farm scale, shade can be essential for certain crops, particularly in the tropics. Cocoa and banana are important intercrops in coconut and arecanut garden have shady canopies they grow beneath. Trees in particular are also important for filtering air pollution and particulates, and help create a protected and nourishing microclimate in most places they are planted. Simulation analysis indicated that negative impacts of climate change can be overcome by adaptation strategies such as assured irrigation through drip system coupled with soil moisture conservation and by providing fertilizers/nutrients through organic and inorganic source in doses higher than those currently applied by the farmers. Such measures also maximize the positive impacts of climate change. Farmers who adopted soil moisture conservation practices or drip irrigation could reduce the drought impact on their plantations. In drought affected coconut gardens, farmers could grow short duration pulses, oil seeds and millets for their sustenance.

In Kerala, providing more fertilizers along with summer time irrigation and following soil moisture conservation practices could further improve the positive gains due to climate change by 7 to 21 per cent in different scenarios. In Karnataka, West Bengal, Gujarat, Maharashtra and Odisha assured irrigation and providing more fertilizers could not only off-set the negative impacts but could also result in higher yields.In North-Eastern States, providing summer irrigation and even low dose of fertilizers could further improve (in the range of 10-33 per cent) the positive impacts of climate change. Coconut plantations in islands, if managed scientifically by proper spacing, canopy management, summer irrigation and even with low dose of fertilizers the productivity could be enhanced to an extent of 2-25 per cent.

4. Conclusions

It is clear that climate change will have negative effects on coconut plantations due to increased temperature and water stress. However, coconut plantations can also be used to mitigate climate change which is an environment service by acting as C Sinks to absorb CO_2 from the atmosphere and control global warming while giving an additional income to the growers through C trading. In areas of poor performance of coconut yield, it is advisable to go for coconut based farming system due to a number of advantages like high productivity, maximum biomass utilization, increases in soil microbial population, and increases in soil water holding capacity which is helpful in plant growth and survival during low rain fall and drought condition. This is why we can call CBFS as permanent agriculture. Soil moisture conservation through leaf mulching, husk/coirpith burial in basin enhanced coconut yields in different agro-climatic regions (Naresh Kumar et al., 2006).

References

Bhaskara Rao, E.V.V.B., Pillai, P.V. and Mathew, J. (1991). Relative drought tolerance and productivity of released coconut hybrids. In: Silas EJ, Aravindhakshan M, Jose AI (ed.)Coconut Breeding and Management, KAU, Vellanikkara Thrissur, India.

Child, R. (1974). *Coconut 2nd ed*. Longman, London.

Coomans, P. (1975). Influence des facteurs climatiques sur les fluctuations saisonnieres et annuelles de la production du cocotier. *Oleagineux*, 30, 153-159

Dhanapal, R., Palaniswami, C., Mathew, A.C. and Manojkumar, C. (2002). Effect of Hybrid Grass (Co3) in Soil and Water Conservation in Western Ghat Region. pp. 43-46. In: *Proc. of National symposium on soil and water conservation measures and sustainable land use systems with special reference to the western ghats region* held at ICAR Research Complex for Goa, during 2000 November 16-17.

Hebbar, K B., Mathew, A C., Arivalagan M, Samsudeen K and Goerge V. Thomas. 2013. Value added products from Neera –Indian Coconut Journal, August, 2013. Pp28

Hebbar, K B., Sheena, T L., Shwetha Kumari, K., Padmanabhan, S., Balasimha, D., Mukesh Kumar. and George V. Thomas. (2013b). Response of coconut seedlings to elevated CO2 and high temperature in drought and high nutrient conditions. *Journal of Plantation Crops*, 41: 118.

Hebbar, K.B., Balasimha, D. and Thomas, G.V. (2013a). Plantation Crops Response to Climate Change: Coconut Perspective. In: Climate-Resilient Horticulture: Adaptation and Mitigation Strategies, DOI 10.1007/978-81-322-0974-4_16@ Springer India 2013, pp. 177-187.

Helen, R. (2016). In-vitro pollen germination technique to screen coconut genotypes for high temperature tolerance. Msc. thesis, ICAR-CPCRI, Kasaragod, 2016.

Kasturi Bai, K.V., Rajagopal, V., Balasimha, D. and Gopalasudaram, P. (1997). Water relations, gas exchange and dry matter production of coconut (*Cocos nucifera* L.) under irrigated and non-irrigated conditions. *Coconut Research and Development*, 13, 45-58.

Krishnakumar, K.N., Rao, G.S.L.H.V.P. and Gopakumar, C.S. (2008). Climate change at selected locations in the humid tropics. *Journal of Agrometeorology*,10, 59-64.

Magat, S.S. (2007). Coconut farming systems for social, ecological and economic benefits. pp. 53-87. In: Coconut for rural welfare (Eds. P.K. Thampan and K.I. Vasu). *Proceedings of the International Coconut Summit*, 2007, Kochi, India.

Maheswarappa, H.P., Gopalasundaram, P., Dhanapal, R., Subramanian, P. and Hegde, M.R. (1998). Influence of Irrigation and Mulching on Soil Moisture and Soil Temperature Under Coconut in Littoral Sandy Soil. *J.Plantn.Crops*, 26(1), 93-97.

Mathes, D.T. (1988). Influence of weather and climate on coconut yield. *CoconutBulletin* 5, 8-10.

Murray, D.V. (1977). Coconut palm. In. Alvim TA, Kozlowski TT (eds.) Ecophysiology of Tropical Crops. Academic Press, New York.

Nair, P. K. R. and Balakrishnan, T. K. (1977). Ecoclimate of a coconut+cocoa crop combination in the West Coast of India. *Agricultural Meteorology*, 18: 455-462.

Naresh Kumar, S. and Aggarwal, P. K. (2009). Impact of climate change on coconut plantations. In Global Climate Change and Indian Agriculture-case studies from ICAR Network Project (PK Aggarwal ed.), ICAR, New Delhi Pub., pp.24-27.

Naresh Kumar, S. and Aggarwal, P.K. (2013). Climate change and coconut plantations in India: Impacts and potential adaptation gains. *Agricultural Systems*, http://dx.doi.org/10.1016/j.agsy.2013.01.001.

Naresh Kumar, S., Rajagopal, V., Siju Thomas, T., Vinu K. Cherian, Ratheesh Narayanan M.K., Ananda, K.S., Nagawekar, D.D., Hanumanthappa M., Vincent S. and Srinivasulu, B. (2007). Variations in nut yield of coconut (*Cocos nucifera* L.) and dry spell in different agroclimatic zones of India. *Ind. J. Hort.* 64 (3): 309-313.

Naresh Kumar, S., V. Rajagopal, T. Siju Thomas and Vinu K. Cherian (2006). Effect of conserved soil moisture on the source-sink relationship in coconut (*Cocos nucifera* L.) under different agro-climatic conditions in India. *The Ind. J. Agril. Sci.* 76:5: 277-281.

Naresh Kumar, S. (2007). In: Climate change effects on growth and productivity of plantation crops with special reference to coconut and black pepper: Impact, adaptation and vulnerability and mitigation strategies'-ICAR Network Project Final Report (Naresh Kumar *et al.*, 2007), submitted to ICAR.

Naresh Kumar, S., Kasturi Bai, K.V., Rajagopal, V. and Aggarwal, P.K. (2008). Simulating coconut growth, development and yield with the Info Crop-coconut model. *Tree Physiology*, 28, 1049-58.

Naresh Kumar, S., Rajagopal, V., Siju Thomas, S., Vinu Cherian, K. M., Hanumanthappa, M., Anil Kumar, B., Srinivasulu, B.and Nagvekar, D. D. (2002). Identification and characterization of *in situ* drought tolerant coconut palms in farmers' fields in different agro-climatic zones. In: Sreedharan K, Vinod Kumar PK, Jayaram Basavaraj MC (eds.) *Proceedings of PLACROSYM XV*, Kerala.

Rajagopal, V. and Kasturi Bai, K.V. (1999). Water relations and screening for drought tolerance. In: Rajagopal V, Ramadasan A (eds), *Advances in Plant Physiology and Biochemistry of Coconut Palm*, Asian and Pacific Coconut Community. Jakarta.

Rajagopal, V., Kasturi Bai, K.V. and Voleti, S.R. (1990). Screening of coconut genotypes for drought tolerance. *Oleagineux*, 45, 215-223.

Rajagopal, V., Shivashankar, S. and Mathew, J. (1996). Impact of dry spells on the ontogeny of coconut fruits and its relation to yield. *Plant Rech Dévelopment*, 3,251-255.

Sage, R.F. and Kubien, D.S. (2007). The temperature response of C-3 and C-photosynthesis. *Plant, Cell and Environment*, 30, 1086-1106.

Shivakumar, S. N. (2016). Intercropping of flower crops in coconut with *in-situ* moisture conservation materials in littoral sandy soils of west coast. M.Sc. Thesis submitted to University of Horticultural Sciences. p. 124.

Subramanian, P., Dhanapal, R. and Harisha, C. B. (2013). Impact of soil conservation measures and inter cropping on soil temperature in coconut garden under coastal sandy soil. **In:** National Symposium on Climate change and Indian Agriculture: Slicing down the uncertainties. CRIDA, Hyderabad, 22-23rd January 2013.

Varghese, P.T., Nelliat, E.V. and Balakrishnan, T. K. (1978). Beneficial interactions of coconut-cacao combination. 383-392. *In: Proceedings of Ist Plantation Crops Symposium* (PLACROSYM I). Indian Society for Plantation Crops, Kasaragod, India.

2017, Impact of Climate Change on Plantation Crops *Pages* **219–235**
Editors: **K.B. Hebbar, S. Naresh Kumar & P. Chowdappa**
Published by: **ASTRAL INTERNATIONAL PVT. LTD., NEW DELHI**

Chapter 14

Climate Resilience Agriculture: Experiences from Kerala

G.S.L.H.V. Prasada Rao

1. Introduction

The State of Kerala under the Humid Tropics is one of the wettest places in the Humid Tropics where annual rainfall is of the order of 3000 mm, ranging from less than 1000 mm in the South to greater than 5000 mm in the North and High Ranges. About 68 per cent of the rainfall is obtained during southwest monsoon while 16 per cent in post monsoon and the rest from summer (14 per cent) and winter rainfall (2 per cent). Bi-model rainfall pattern is noticed towards southern districts due to influence of both southwest and northeast monsoons while uni-model rainfall pattern towards northern districts of Kerala. Coconut productivity is better towards south despite dreaded disease of root-wilt when compared to that of northern districts due to less dry spells during summer. Floods during monsoon adversely affect paddy production in the State while prolonged droughts during summer in the absence of post monsoon rainfall adversely affect plantation crops' production (not all plantations) to a considerable extent. The wetlands in Kerala are rich sources of water during summer and act as sink during monsoon season. Such wetlands are fast declining in Kerala and converted as garden lands. Decreasing wetlands might be one of the reasons for frequent floods and droughts in Kerala in recent years. The forest cover of Kerala also declined from 70 per cent to 24 per cent over a period of one- hundred- and - fifty years due to deforestation and forest fires. The years 1983, 2004 and 2013 experienced prolonged summer droughts in Kerala and many plantation crops' production was adversely affected and the State's economy was hit badly. The annual coconut production in Kerala during 2013-14 was only 5,200 million nuts as against 5,799 million nuts in 2012-13. The low coconut yield in 2013-14 could be attributed to the severe summer drought that occurred in 2012-13 across the coconut growing areas of the State. The onset of monsoon and monsoon behaviour appear to be erratic in recent years. Frequent failure or break in monsoon across the State lead to low water levels in major reservoirs and adversely affect hydro power generation. There was a deficit of 26 per cent on-an-average in monsoon rainfall during 2015 across Kerala while the deficit was high (39 per cent) in Wayanad District. Such a deficit rainfall is a very rare phenomenon across the State during the monsoon season. There is a significant change in cropping pattern too over a period of time. Rice and cashew area was declining while area under rubber and coconut increasing. Vanilla and cocoa were introduced and in the same fashion vanilla disappeared and area under cocoa declined. Increase in area under black pepper was noticed across the State and now declining. There was a time that oranges were plenty in Wayanad District and now disappeared almost. The paddy lands in Wayanad District are mostly converted to areca nut and banana gardens. As a whole, the index of food crops was declining while increasing non-food crops due to various socio-economic factors.

2. Climate Change/Variability over Kerala

The lull in monsoon over Kerala during 2002 followed by floods in October 2002 due to cyclonic storm over the Karnataka Coast devastated seasonal crops and plantations to a considerable extent. Kannur received 370 mm of rainfall on 14th October 2002, which was the highest and not received since 1924. Heavy monsoon rainfall in 2007 followed by unusual summer showers in March 2008 together adversely affected the paddy production of the State to a considerable extent during 2007-08. The State received record rainfall during post monsoon 2010. The widespread post monsoon rainfall during 2010 also devastated the paddy fields especially in upper Kuttanad and Kole lands of Thrissur. Low lands of southern districts were flooded due to the prolonged rainfall during November, 2010. In contrast, the prolonged dry spell during summer 2013 adversely affected the water resources and several plantation crops' production affected. Floods in August/ September 2014 across Wayanad District adversely affected paddy and arecanut fields. Yellowing in arecanut devastated large number of palms in Wayanad District in 2014. Are these frequent weather aberrations like monsoon uncertainties, floods and droughts, sunburns, atmospheric heat load during summer and UV radiation filtered to ground due to ozone depletion could be as a part of global warming and climate change? There were reports that the heat burst might be the reason for scorching leaves across the Coast of Kerala in June 2015. Sunburns and Heat bursts are new phenomena noticed in Kerala, the science behind is yet to be studied. The weather phenomenon like heat burst was noticed in Kerala when the entire south (except Kerala) was reeling under severe heat wave in June 2015. The climate change related issues over Kerala may be decline in rainfall, wetlands, land and ocean biodiversity, increase in temperature and sea level, floods and droughts, landslides, groundwater depletion and saline water intrusion, decline in forest area, frequent forest fires, unusual rains and hailstorms. As a whole, the State of Kerala was moving from wetness to dryness within the Humid Climate (B4 to B3). It is believed that almost all the crops are likely to be under threat due to climate change/variability in the ensuing decades. The threat from global warming and climate change could be seen in the form of decline in cropped area, production and productivity and quality of grains. The rate of increase in temperature is more across the High Ranges, followed by low lands due to increase in deforestation and sea surface temperature, respectively.

2.1. Onset of Monsoon

The monsoon directory of Kerala since last 145 years indicates that the onset of monsoon is on 1st June with ± 7 days, indicating that it varies from 25th May to 8th June in majority of the years. However, the earliest monsoon was recorded on 11th May in 1918 while belated monsoon on 18th June in 1972. The mean onset of monsoon during the tri-decade of 1901-30 was on 4th June while in other tri-decades it revolved around 1st June. The trend analysis since 1870 also indicated that the onset of monsoon is stable and it tends to be around 1st June (Figure 14.1). Though the trend analysis indicates that the onset of monsoon is likely to be on or before 1st June over Kerala, the fact file in recent years appears to be different since the onset of monsoon in 2011 was on 29th May and it was 5th June in 2012 and again on 1st June in 2013, 6th June in 2014 and 5th June in 2015. It indicates that inter-annual variations are expected within one standard deviation in majority of the years. It is also understood that the monsoon set may be early (before 25th May) or late (after 8th June) occasionally during which the monsoon rainfall is likely to be below normal or normal, indicating that the chances of excess rainfall in such years (early or late monsoon years) are likely to be less.

2.2. Rainfall

A decline in monsoon rainfall while increase in post monsoon is the trend across the State though cyclic trends of 40-60 years were noticed in annual and monsoon rainfall. Of course, the decline in annual rainfall is evident since last 50-60 years. It also reveals that the percentage contribution of rainfall in monsoon season to the annual rainfall was declining while increasing during the post monsoon season. However, increase in post monsoon rainfall may not compensate rainfall decrease in monsoon season. Interestingly, there was a shift in climate as a result of changes in thermal and moisture regimes over the State of Kerala. The State of Kerala as a whole was moving from wetness to dryness within the Humid Climate from B_4 to B_3 since last 100 to 150 years (Figure 14.2). One of the major factors could be due to alarming deforestation and forest fires that took place during the above said period across the Western Ghats within and outside the State of Kerala.

2.3. Temperature Projections

Warming Kerala is real as the trend in temperature was increasing significantly since 1980s in tune with the global warming. Within the State, the rate of increase in temperature was high across the Highranges, followed by the low lands while moderate increase along the midlands. It could be attributed to alarming deforestation across the Highranges and the effect of increase in sea surface temperature along the Coast. At the current rate of increase in temperature, it is projected that increase in maximum temperature is likely to be around 1.5°C by

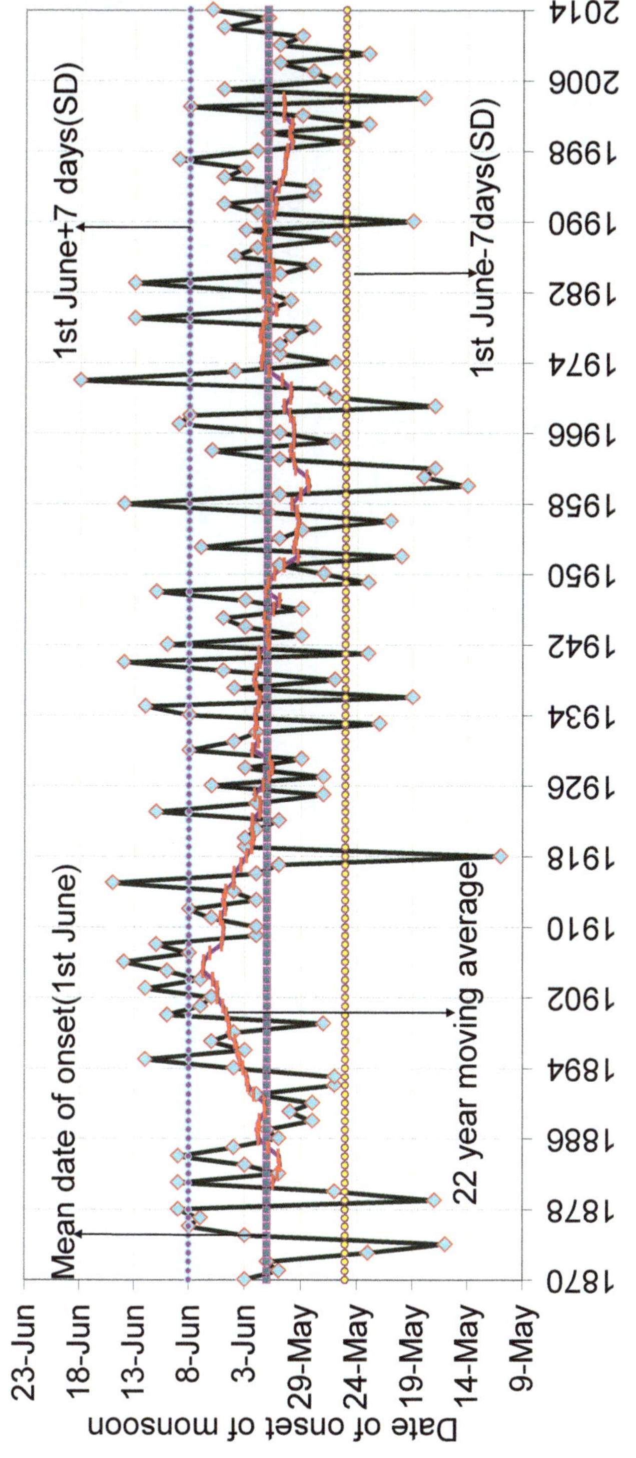

Figure 14.1: Onset of Monsoon over Kerala from 1870 to 2015.

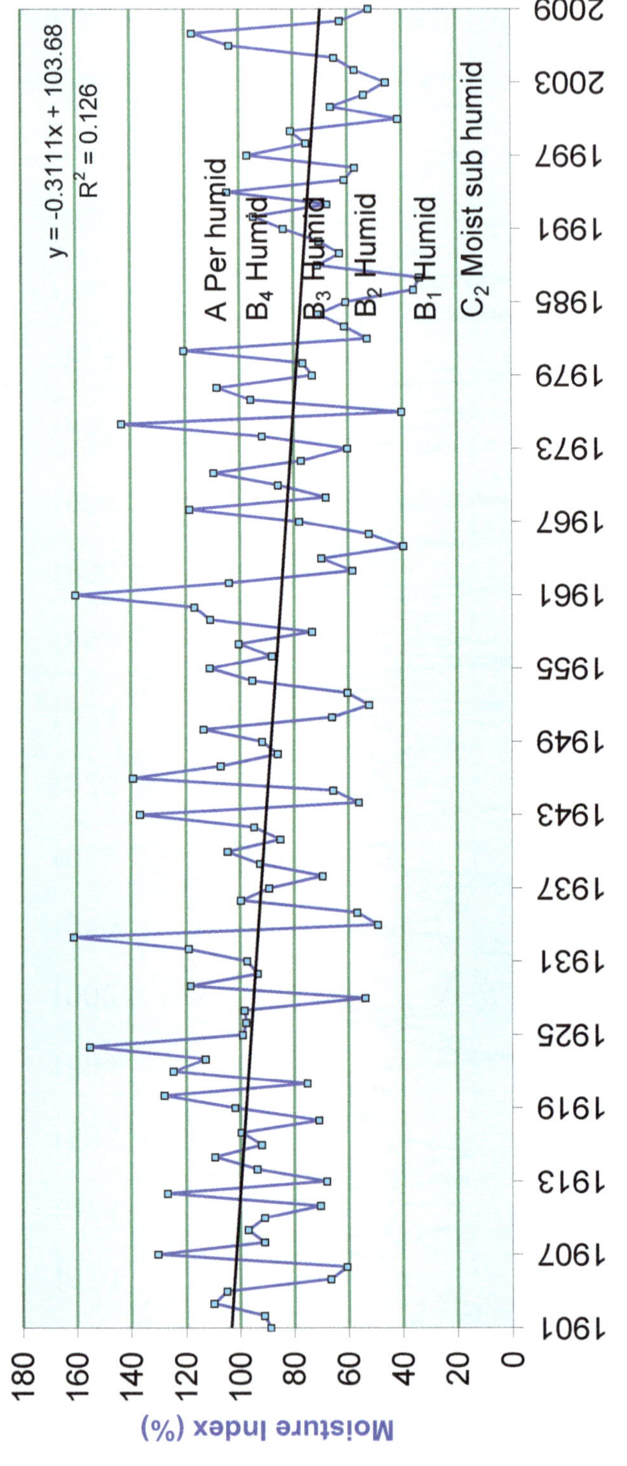

Figure 14.2: Climate Shifts over Kerala from 1901 to 2009.

2100 A.D. while 0.3-0.4°C in the case of minimum temperature. Increase in mean surface air temperature is likely to be less than 1°C by 2100 A.D. The decade 1981-90 was the warmest and driest decade in Kerala during which the plantation crops' production was adversely affected to a considerable extent. The year 1987 was the warmest, followed by 1983. Increase in night temperature is noticed during winter in recent years and flowering of fruit crops is adversely affected. It is more so in the case of mango.

2.4 Climate Change and Rice

Under the projected climate change scenario, monsoon uncertainties are expected in the State of Kerala on which rice production is dependant to a large extent during the first crop season. As already seen, prolonged monsoon coinciding with post monsoon rains and unusual summer rains in the form of floods is a threat to paddy production. There was a decline of 8.5 per cent on an average, ranging from 2 to 17 per cent depending upon the stage of crop coinciding with the occurrence of floods. Therefore, abrupt monsoon uncertainties are likely to influence the paddy production adversely in ensuing decades rather than the influence of climate change on long term basis. The rate of increase in mean temperature at the current level in ensuing decades is likely to be less than 1°C across the low lands of Kerala, where paddy is cultivated mostly. Increase in temperature of this nature during the first crop season is beneficial to paddy crop under better crop improvement (saline and flood tolerant varieties need to be introduced), crop management and crop protection measures. However, inter annual/seasonal variations were noticed in rice production and its productivity due to weather aberrations in the form of climate variability such as floods during the first crop season (Figure 14.3). Under such conditions, the rice yield is bound to decline. Therefore, the study indicates that the rice yields in Kerala are unlikely to decline due to long term climate change

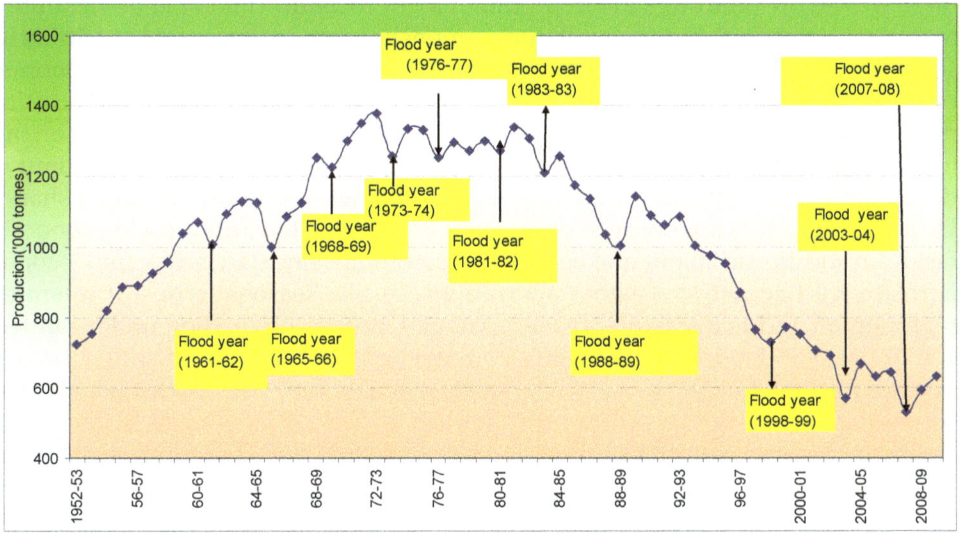

Figure 14.3: Impact of Heavy Rainfall on Rice Production in Kerala.

such as increase in temperature, but bound to decline to some extent through the abrupt short term changes as noticed in 2008, 2009 and 2010 in the form of monsoon uncertainties. Of course, there was a drastic decline in rice area and production since last 60 years and at the same time increase in rice productivity (1.37t/ha in 1960-61 to more than 2-2.5t/ha since last one decade) is noticed despite increase in atmospheric temperature. However, rice-milling percentage has come down significantly in recent years. It is attributed to global warming. The rate of increase in Indian foodgrains production was below the normal of 12.5 per cent since last 15 years due to frequent occurrence of weather abnormalities like floods, droughts, heat and cold waves. While area and production of paddy was declining in Kerala, the productivity was increasing during the study period. The decline in paddy production varied between 2 to 17.6 per cent due to heavy rainfall. On an average, the effect of heavy rainfall on paddy production was 8.5 per cent. In the case of paddy productivity also, the effect of heavy rainfall during *kharif* varied between 2.2 and 14.9 per cent. Erratic behaviour of monsoon, extended post monsoon rainfall with intermittent dryspell during post monsoon and unusual summer rains are detrimental to paddy production in the State as noticed in recent years. Under the projected climate change scenario, the frequency of such instances is likely to be more and hence it is a threat to paddy production in the State. In fact, the paddy productivity can be increased further at the current rate of increase in temperature during *kharif* through better crop improvement and management techniques, for which R and D initiatives need to be taken up on priority basis as a part of food security under the projected climate change scenario.

2.5. Climate Change and Coconut

Coconut productivity in Kerala is relatively low though the total coconut production of the State is high. With increase in area under coconut, Kerala contributes major share in terms of total coconut production of the country while there was no breakthrough in nut productivity of coconut. This could be attributed to several factors *viz.*, high rainfall during monsoon period, insignificant rainfall in several parts of Kerala during summer, nutrient status of soil, lack of agronomic practices and prevalence of root wilt disease in southern districts of Kerala. The price of coconut, copra and oil are also subjected to wide fluctuation depending upon the total coconut production of the State. At present, the price escalation of coconut is debatable as it is all the time high and coconut farmers' economy is in good shape. However, coconuts are not seen on coconut crown due to various socio-economic factors in addition to abiotic and biotic stress conditions since last two years. Coconut is relatively tolerant to weather aberrations, unlike seasonal crops. It tolerates high rainfall and temperature to some extent. Coconut production under rainfed conditions is influenced significantly by the length of dry spells at critical stages. The primordium initiation, ovary development and button-size nut development in coconut are sensitive to soil moisture availability. The crop performs better under irrigated conditions during summer in traditional and non-traditional areas.

2.5.1. Summer Drought and Coconut Yield

The coconut production of Kerala suffered to a great extent during 1983-84 due to unprecedented drought that occurred during summer 1983, followed by 2004 and 2013. The coconut production in Kerala during 2013-14 was declined by 10 per cent as per the recent survey reports. Almost all the coconut growing states experienced decline in coconut production in 2013-14. It was attributed to prolonged summer drought in 2012-13 across the coconut growing areas. The effect of high rainfall from June to September and prolonged dry spell during summer adversely affect the coconut production in the following year under the Kerala's situation. The decline in coconut productivity due to severe drought during summer could be seen in the following year under rainfed conditions. Similarly, summer showers influence coconut yield positively in the subsequent year. The effect of summer drought on monthly nut yield at various locations across Kerala indicated that the effect commences in the seventh, eighth or ninth month after the drought period is over by May or June, depending upon the receipt of pre-monsoon showers or onset of monsoon. Heavy monsoon rainfall with prolonged dry spell during summer influences the ontogeny of coconut and seasonality of nut production in coconut to a large extent. The decline in monthly nut yield due to drought continued for 12 months, reaching to its maximum decline in $12^{th}/13^{th}$ month after the drought period is over (Figure 14.4). The percentage decline in the subsequent year due to summer drought varies depending upon the yield group. In higher nut yield group (greater than 100 nuts per palm per annum), it revolves around 29-30 per cent while around 10 per cent in poor yielders (yield group of 40 and 60 nuts per palm per annum). It is intermediary (24.5 per cent) in group III, producing 60 to 80 nuts per palm per annum. On an average, the decline in nut yield due to drought in the subsequent year is 23.7 per cent and it varies based on the yield group. Good planting material, coconut hybrids, good agronomic practices, floor crops, intercropping and scarce water management practices and integrated pest management alleviate the ill effects of abiotic and biotic stresses to a large extent in coconut.

There was a marginal decline in coconut productivity during this tri-decade (5670 nuts/ha) when compared to previous tri-decade (5762 nuts/ha). It can be attributed to rise in temperature in tune to global warming. The negative trend in coconut productivity was much more (17.9 per cent) in the decade 1981-90, which was the warmest and driest decade. The decline in coconut productivity was maximum during 1983-84 (19.2 per cent), followed by 1986-87(6.2 per cent) and 4 per cent in 1987-88. All the above three low coconut productivity years fell in the decade 1981-90. Therefore, there is a threat to coconut productivity in the ensuing decades under the projected climate change scenario. Therefore, there is urgent need for pro-active measures to sustain coconut productivity in Kerala as a part of climate change adaptation.

2.6. Cashew

The cashew productivity over the decades has shown a steep decline. It is quite alarming that the cashew productivity declined from the ever highest value of 1600 Kg/ha during late 50s (1952-60) to the lowest productivity of 659 Kg/ha during

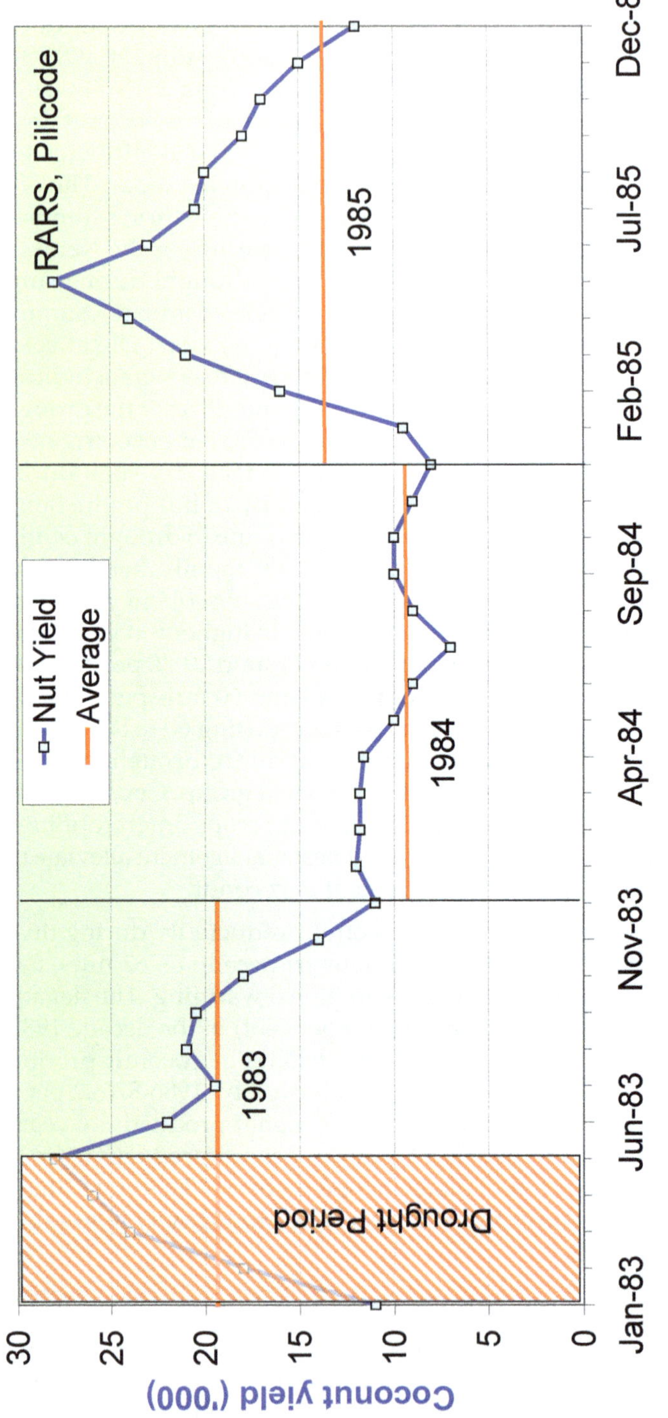

Figure 14.4: Effect of Drought during Summer 1983 on Coconut Yield at RARS, Pilicode.

1981-90. Slight improvement in cashew productivity was noticed during 1991-00 and 2001-10 (785 Kg/ha and 773 Kg/ha, respectively). The percentage decline in cashew productivity was 51 per cent over 1950s. In terms of tri-decadal area, an increase of 27 per cent was registered from 1951-80 (83.8 thousand ha) to 1981-10 (106.3 thousand ha). During the same period, tri-decadal production of raw cashew nuts has declined by 14 per cent from 1951-80 (89.6 thousand tonnes) to 1981-10 (977.1 thousand tonnes). The decline in tri-decadal productivity was 42 per cent during the period from 1951-80 (1200 Kg/ha) to 1981-10 (700 Kg/ha) (Figure 14.5). Farmers' reluctance to cultivate cashew crop due to poor returns and switching over to other profitable crops like rubber is one of the reasons for the decline in cashew area in the State. 1997-98 was a poor yielding year due to unfavourable weather conditions prevailed during the flowering phase of cashew. The yield recorded was only was only 56890 tonnes. In contrast, incidence of pest complex devastated the cashew plantation along the West Coast during 1998-99 though profuse flowering was noticed due to favourable weather. It was more so in Kannur and Kasaragod districts of Kerala, which account for more than 50 per cent of the total cashew production of the State. In 1999-2000 and 2000-01, a combination of adverse weather and incidence of pest complex adversely affected cashew production of Kerala. The percentage decline in cashew productivity was the highest (59 per cent) in 1981-90 due to the warmest and drought conditions in the form of decline in rainfall and increase in temperature. There was also a drastic decline in cashew yield during the recent years. Therefore, the decline in cashew productivity over a period of time can be attributed to climate variability and climate change.

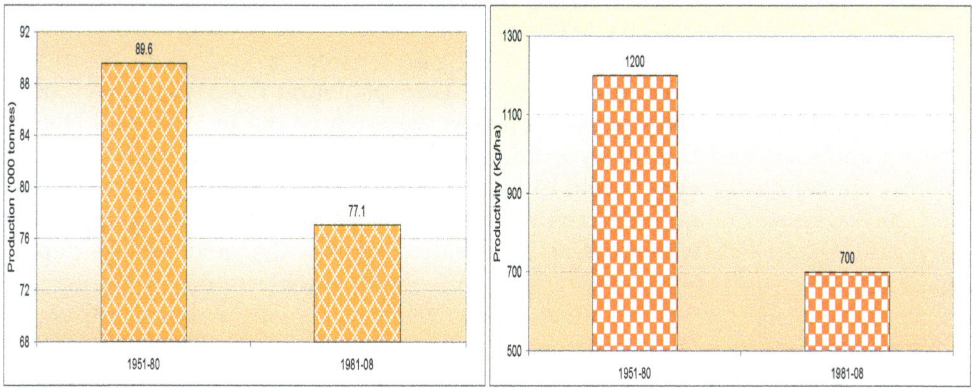

Figure 14.5: Tri-decadal Production and Productivity of Cashew in Kerala.

2.7. Coffee

The blossom and backing showers are very important in the case of coffee. Coffee yield is very poor in the absence of either of them. The growth of coffee appears to be affected when exposed, and hence it requires a certain degree of shade for better performance. Arabica is more sensitive to weather when compared to that of Robusta. Of course, the quality of Arabica appears to be better. Arabica prefers high altitudes when compared to that of Robusta. It is understood that crop in recent years matures early and berry weight declined. It is attributed to global warming and climate change. In Wayanad District, where coffee is predominantly grown, decline in rainfall and rise in temperature was noticed. Overall, the productivity of coffee in recent decades increased though inter-annual variation in coffee production is significant due to influence of blossom and backing showers. Sprinkler irrigation is practiced in the case of coffee to meet the blossom and backing showers in the absence of rains. The coffee growers are also of the opinion that drip irrigation during summer can be a substitute to sprinkler irrigation and beneficial to the crop.

Monsooned Malabar Coffee is a speciality coffee preferred by the people who like black coffee and it fetches high premium in the international market. If monsoon is weak during the monsoon season in which coffee beans are processed, the quality of coffee appears to be poor. The quality of 'Monsooned Malabar Coffee' was very poor for the first time due to bad monsoon in 2002. The quality of "Monsooned Malabar Coffee" is better if monsoon is heavy and continuous during July and August over the Malabar Region, where the coffee beans are processed. Continuous clouds with heavy rainfall and its better distribution, high atmospheric water vapour content, moderate surface air temperature and gentle sea breeze would lead to better moisture content in coffee beans if processed during the monsoon period. Probably, active monsoon period within the monsoon season may be more conducive for such processing of coffee beans. It is one of the factors leading to superior quality in addition to other genetical and environmental factors.

2.8. Black Pepper

Whenever summer showers are received, the yield of black pepper declines. Interestingly, if the same showers coincide with blossom and backing showers, coffee flourishes. Of course, mortality rate is high under poor management conditions in prolonged dry spell conditions in the case of young black pepper vines. It was the reason, why, black pepper gardens were wiped out in Wayanad District during summer 2004 due to hydrological drought.Another important aspect is that whenever black pepper yield is low the coffee yield is better. It is true in 2003 and 2008 when dried coffee beans were high the yield of dry black pepper was low. In 2006 and 2000 when dried coffee beans were low the yield of black pepper was high. Therefore, mixed cropping of coffee with black pepper is better against prolonged dry spell during summer rather than to allow as monocrop. Cashew, coffee and black pepper appear to be better across the mid-and-high lands of northern region of Kerala since the prolonged dry spell from November to May may not be detrimental for better yield in terms of quality and quantity depending upon the nature of the crop under rainfed conditions. Area under black pepper and arecanut is declining in recent years in Wayanad District. Heavy monsoon rains in 2014 across Wayanad District led to waterlogging and due to yellowing, many arecanut gardens are adversely affected. The wetlands in the District are converted into arecanut gardens and banana plantations.

2.9. Cocoa

The number of cocoa (cacao) pods is high during summer, but pod weight is low. A harvest of five pods during October/November may be equal to eight/ nine pods harvested in summer, showing superiority in cocoa pods harvested in

October/November. There is a lag period of 4 to 6 months between weather effects and cocoa yield. The maximum temperature from January to March influences cocoa yield to a considerable extent. A prolonged dry spell from November to May with high maximum temperature adversely affects the pod yield to the tune of 40 per cent depending upon the crop management. High rainfall versus cocoa yield showed inverse relationship, indicating that high rainfall during rainy season may not be conducive in the case of cocoa due to waterlogging and lack of soil aeration. Overall, it reveals that high maximum temperature during summer, followed by heavy rains during the Southwest monsoon may not be conducive for obtaining good yield in cocoa under the Humid Tropics. High rainfall has a malevolent effect in places where the crop is grown under waterlogged conditions, while summer rainfall has a benevolent effect on pod yield. High maximum temperatures (> 36°C) during February and March, 2004 resulted in low pod yield in cocoa over the central region of Kerala. Cocoa is grown as an intercrop in coconut or part of homestead garden. Both the crops cocoa and coconut respond identically with reference to weather and climate under the Humid Tropics. Of course, the phenology of both the crops is totally different. In case of cocoa, the difference in monthly yield is high during the monsoon season between good and bad yield years when compared to other seasons. In cocoa, mango and cashew the flowering period is very much influenced by the geographical co-ordinates such as latitude, longitude and altitude of the crop growing regions and appears to follow the famous Hopikins Bio-climatic Law. It states that "A biotic event in North America will, in general, show a lag of four days for each one degree of latitude, five degree of longitude and 400 feet of Altitude, northward, eastward and upward in spring and early summer".

2.10. Cardamom

A sharp decline in area under cardamom was noticed across the cardamom tract of the Western Ghats in recent years. However, the production and productivity of small cardamom were increasing. The important physiological stages like panicle initiation and subsequent growth (forwarding) depend on receipt of showers from January to May. The failure of showers during this period results in poor growth and crop yield. The unprecedented drought in summer 1983 adversely affected the growth and yield of cardamom in Coorg district, and the same trend prevailed in other cardamom-growing tracts of South India. The drought that prevailed during 1982–83 resulted in as high as 50 per cent mortality in some of the cardamom estates. In the Palani hills of Tamil Nadu, both panicle and flower production are seen in October and November, during which northeast monsoon rains are received. It indicates that both panicle and flower production in cardamom are commensurate with the rainfall pattern of the cardamom tract. There is a close relationship between cardamom production and rainfall distribution during summer. Summer rains during March, 2008 have benefited the cardamom crop of 2008–2009 up to 20–30 per cent. Therefore, summer rains have significant positive influence on cardamom. There was a strong negative relationship between soil moisture stress and cardamom yield. Climate risk is more in the case of cardamom in Karnataka region, followed by Ambalavayal and neighbouring areas in northeast of Kerala while climate risk is less in Thandikudi and Idukki regions in south of cardamom tract.

2.11. Rubber

Area, production and productivity of rubber have been increasing over a period of time. The rate of increase in rubber productivity was affected during 1959-63 (-24.3 per cent), followed by 1979-83 (-8.3 per cent), 1984-88 (2.2 per cent), 1999-03 (13.5 per cent) and 2004-08 (18.9 per cent). Despite increase in temperature, aridity index, number of droughts, decline in rainfall and moisture index (index of climate shifts) the rubber productivity was in increasing trend. However, decline in rate of increase in rubber productivity was noticed. At the same time, the prolonged dryspells during summer due to absence of rainfall during northeast monsoon and failure of summer showers is likely to affect rubber productivity adversely as seen in recent years. Abrupt changes in weather may affect the productivity of the crop as noticed in 2007-08, during which the yield declined (1471 Kg/ha) due to continuous rainfall and more number of rainy days when compared to that of previous year 2006-07 (1554 Kg/ha). Climate change in the form of climate variability, especially unpredictable, irregular and deficient rainfall pattern adversely affects the growth and productivity of natural rubber. Changes in rainfall pattern triggers fungal infection of leaves, which adversely affects productivity. It indicated that the long term increase in temperature is likely to affect rubber yield adversely and the rate of increase is likely to decline.

3. Climate Resilience Agriculture

The State Action Plan on Climate Change (SAPCC) indicated that Alappuzha, Palakkad, Wayanad and Idukki are the 'Climate Change Hot spots' in Kerala. Idukki, Wayanad, Thiruvananthapuram and Kannur are highly vulnerable to Climate Change. Kerala is predominantly a plantation state, falls under the Tropical Humid Climate, moving from wetness to dryness within the Humid Climate ($B_{4\ to}$ B_3 as per the Thorthwaite's climate classification). Monsoon rainfall across the State was declining since last 60 years (cyclic trend of 40-60 years is also noticed with annual and monsoon rainfall) in temperature is evident. Of course, rate of increase in temperature was alarming across the Highranges (where cardamom, coffee and tea are grown) due to deforestation. It is also true to some extent along the Coast (low land) due to increase in sea surface air temperature. The decade 1981-90 was the driest and warmest decade and warming Kerala is true since last three decades. 1987 was the warmest year, followed by 1983. The decadal and tri-decadal changes in climate and crop productivity were taken into account to explain the influence of climate change on crops with reference to Kerala.

Rice productivity was increasing while not so in cashew and coconut and to some extent in the case of rate of increase in rubber too. In contrast, the crop productivity was increasing in cocoa, cardamom, coffee and black pepper and not so in the case of tea. At the same time, all the plantation crops showed low productivity or low rate of increase in productivity during the warmest and driest decade 1981-90. Therefore, the adverse affect of climate variability on plantation crops is evident and at the same time the effects of climate change cannot be quantified though decline in crop productivity could be seen in coconut, cashew, tea and rubber. However, crop growth simulation models indicate that crop yields including cropped area are likely to decline in the ensuing decades under the projected global warming and climate change scenario. Climate change adaptation strategies and awareness are particularly important for managing the risks posed by climate variability and change, not only on plantation crops but also across all the society linked sectors that are sensitive to weather and climate. Various agroclimatic techniques have been used to effectively manage some of the risks posed by climate change to plantation crop productivity. Expansion and further development of such techniques will be vital for the continued sustenance of plantation crops' production in the Humid Tropics. As pointed out by Prof. M.S. Swaminathan "India's strength lies in its ability to manage monsoons" instead of saying" Indian agriculture is a gamble of the monsoon".

Climate smart agriculture, climate resilience agriculture, precision farming, hi-tech farming, conservation agriculture, traditional agriculture, protected cultivation, mixed farming and integrated farming are some of the technologies/ tools are coined to minimize the ill effects of climate change on crops. These technologies may be sound in the case of field crops/seasonal crops, but not in the case of plantation crops since the plantation crops are in the field round the life time, having economic yield age of 20 to 60 years depending upon the type of crop. In the case of coconut depending upon the type of cultivar or hybrid, the first flowering commences between fourth and eighth year after planting under better management conditions and coconut yield is stabilized by its 15/20th year and yields economically up to eighty and hundred years. Therefore, integrated farming as done traditionally or crop combinations in the case of coffee and black pepper or multi-tier cropping systems may be strategic approach to withstand against the ill effects of climate change/variability rather than looking for mono-crops. The farm-level weather insurance may be another tool which should be seriously examined for the benefit of the farmers in the event of crop losses due to weather extremes. Unlike seasonal crops, the effects of climate change may be standing many more subsequent years depending upon the phenology of the crop. It is also observed that crop mix/integrated farming may withstand relatively better against the ill effects of climate variability/change as seen in the case of coffee and black pepper. When black pepper performs better coffee suffers and vice-versa. Recent findings indicate that the climate change affects not only crop output but also commodity quality and price. The current hike in coconut price could be attributed to low coconut production in southern states since last three years. One of the factors for overall price escalation in recent years could be attributed to global warming and climate change. Therefore, it is high time that the government agencies and policy makers to be proactive with short term and long term strategies to mitigate the ill effects of weather related disasters with the people's participation and minimize the losses against the weather related disasters. Need of the hour is expect the unexpected guests in the form of weather related disasters under the projected climate change scenario and tune to the system to minimize the losses with well preparedness at various levels. Keeping the above in view, the National Action Plan on Climate Change (NAPCC) identified 'Sustainable Agriculture' is one of the thrust areas to tackle the climate change related issues.

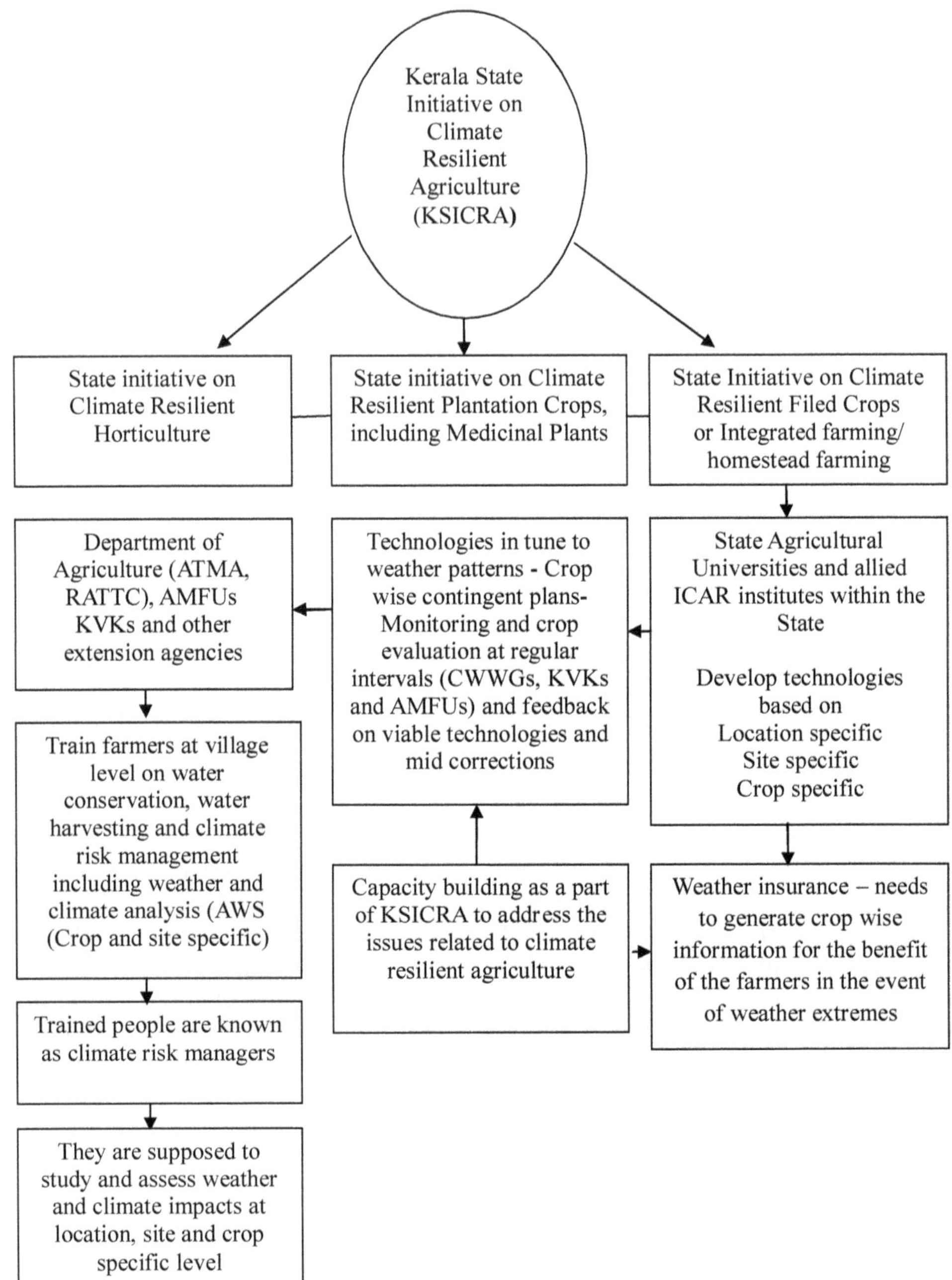

Figure 14.6: Concept of SICRA.

The mission intends making Agriculture more resistant to climate change by identifying and developing new varieties of crops that are thermal resistance and capable of withstanding extreme weather. These issues are being tackled as a part of National Initiative on Climate Resilient Agriculture (NICRA). When it comes to the State of Kerala such initiatives need to be taken up on priority basis in developing technologies against the weather extremes such as saline/flood resistance and drought resistance/tolerance rather than thermal resistance, for which a definite State Action Plan on Climate Change (SAPCC) in Agriculture Sector need to be highlighted and tackled with concerted efforts. To tackle the climate change related issues in Agriculture, a definite policy framework must be chalked out and related institutes should look into. To me it appears, we need have an action plan similar to that of NICRA at the State level with technical people in the form of 'State Initiative on Climate Resilient Agriculture'. It can be known as SICRA at the State level similar to that of NICRA at the national level. It may be related to plantation crops or horticulture or medicinal plants or seasonal or as a whole integrated/ homestead faming (Figure 14.6). Such system should be made operational and in vogue at the earliest possible time since climate change/variability is a big threat in plantation sector.

References

Martinez, R., Hemming, D., Malone, L., Bermudez, N., Cockfield, G., Mushtaq, S., Diongue, A. Hansen, J., Hildebrand, A., Ingram, K., Jakeman, G., Kadi, M., McGregor, G.R. Rao, P., Pulwarty, R., Ndiaye, O., Srinivasan, G., Eh.R., Seck, White, Nand Zougmore, R. (2012). Improving Climate Risk Management at Local Level – Techniques, Case Studies, Good Practices and Guidelines for World Meteorological Organization Members-Chapter 21 (477-532): In Risk Management-Current Issues and Challenges (Ed: Nerija Banaitiena). In Tech Publishers. Rijeka, Croatia, 584p.

Nybe, E.V. and Rao, G.S.L.H.V.P. (2005). Drought management in plantation crops. Kerala Agricultural University. Thrissur, Kerala, India. 176p.

Rao, G.S.L.H.V. P. (2011). Climate change adaptation strategies in agriculture and allied sectors. Scientific Publishers (Jodhpur), India, 336p.

Rao, G.S.L.H.V.P. (2015). Agricultural Meteorology (Third printing). PHI Learning Private Limited, New Delhi, India. 352p. 9.6 Climatic normals for crop production.

Rao, G.S.L.H.V.P. (2015). Climate Change and Agriculture - Chapter-1. Climate Resilient Crops for the Future. New India Publishing Agency. New Delhi, India. 450p.

Rao, G.S.L.H.V.P. (2015). Climate Change and Horticulture-Chapter-5. Basics of Horticulture (Ed. K.V.Peter). 2nd Revised and Enlarged Edition. New India Publishing Agency. New Delhi, India. 704p.

Rao, G.S.L.H.V.P. and Gopakumar, C.S. (2011). Climate change and plantation crops in Kerala. Kerala Agricultural University. Thrissur, Kerala, India. 255p.

Rao, G.S.L.H.V.P. and Krishnakumar, K.N. (2011). Climate change and coconut productivity in Kerala. Kerala Agricultural University. Thrissur, Kerala, India. 219p.

Rao, G.S.L.H.V.P. and Manikandan, N. (2009). Climate Variability and small cardamom across the Western Ghats. Kerala Agricultural University. Thrissur, Kerala, India. 117p.

Rao, G.S.L.H.V.P. and Manikandan, N. (2009). Crop weather relationships of cocoa in the Humid Tropics. Kerala Agricultural University. Thrissur, Kerala, India. 117p.

Rao, G.S.L.H.V.P. and Nair, R.R. (1988). Agrometeorology of plantation crops. Kerala Agricultural University. Thrissur, Kerala, India. 187p.

Rao, G.S.L.H.V.P., Rao, A.V.R.K., Krishnakumar, K.N. and Gopakumar, C.S. (2009). Climate variability and food security. Kerala Agricultural University. Thrissur, Kerala, India. 117p.

Rao, G.S.L.H.V.P., Rao, G.G.S.N. and Rao. V.U.M. (2010). Climate change and agriculture over India. PHI Learning Private Limited, New Delhi, India. 320p.

2017, Impact of Climate Change on Plantation Crops *Pages* 237–245
Editors: K.B. Hebbar, S. Naresh Kumar & P. Chowdappa
Published by: ASTRAL INTERNATIONAL PVT. LTD., NEW DELHI

Chapter 15

Climate Change in Plantation Crops: A Socio-Economic Reflection on Vulnerability, Risks and Way Forward

S. Jayashekar

1. Introduction

Climate change is one of the manifestations of the environmental change. It has gained global attention since 1990s when the first assessment report of the Intergovernmental Panel on Climate Change (IPCC), which is the major scientific body associated with climate change at international level, was published in 1990 and when United Nations Conference on Environment and Development (UNCED) developed United Nations Framework Convention on Climate Change (UNFCCC) in 1992. In 2001, the IPCC defined climate change as "any change in climate over time, whether due to natural variability or as a result of human activity". According to the report from IPCC, there is convincing evidence that climate change is occurring and that it poses important global risks (IPCC, 2007). Agriculture is one sector of the economy that is exposed directly and considerably affected by climate and its changes. The impacts of climate change are likely to be highly spatially variable. The report of fourth assessment report of IPCC in the year 2007 showed that there would be significant impact on the agricultural productivity and potential for food production in different regions of the world as a result of the projected changes in the future climate.

Coconut, the small holders' plantation crop is cultivated in 18 states in India and support the livelihood of over 10 million people. The coconut growing regions in the country are scattered to different agro climatic zones, which are characterized by variable rainfall and temperature. Estimated trends of projections of climate models show that along with inherent climate variability, some of these regions are vulnerable to current and future climate change. Attributable to the strong relationship between the weather and plantation crops' output, climate change will significantly impact crop yields and farm income. The climate change is characterized by extreme weather events such as unprecedented drought, flood, sea level rise, cyclones etc. and coconut is known for its higher vulnerability to these kinds of extremities.This paper outlines different conceptual and methodological approaches to impacts, vulnerability and adaptation assessment in climate change research. Further the paper attempts throw light on issues pertinent to plantation sector with special reference to coconut sector.

2. Assessment of Vulnerability to Climate Change

Numerous definitions to vulnerability can be found in climate change literature. Some of them are discussed below to have an understanding on different dimensions of climate change vulnerability. So far there is no consensus in their application to vulnerability studies. We can find many studies using different vulnerability assessment framework based on the definitions and conceptual approaches developed by different organisations. IPCC (2001) definition for vulnerability is "the degree to which the system is susceptible to, or unable to cope with, adverse effects of stresses including climate variability and extremes. Vulnerability is a function of the character, magnitude, and rate of change in stresses to which a system is exposed, its sensitivity, and its ability to adaptation or adaptive capacity".

The natural disaster literature defines vulnerability as "the characteristics of a person or group and their situation that influence their capacity to anticipate, cope with, resist and recover from the impact of a natural hazard" (Paavola, 2008). Moser (1998) argues vulnerability is closely linked to asset ownership. The more assets people have, the less vulnerable they are, and the greater the erosion of people's assets, the greater their insecurity. Vulnerability also varies across geographical scales, temporal scales and must be addressed within complex and uncertain conditions and hence calls for interdisciplinary and multiple expertise (TERI, 2005). Chaudhuri, Jalan and Suryahani (2002) pointed out that vulnerability is the *ex ante* risk that households will be poor, if not so currently, and if they are currently poor, the risk that they will remain poor. Poverty is therefore an *ex post* measure of wellbeing, and vulnerability an *ex-ante*. In a vulnerability framework, the poor are considered active agents, and interventions can build on those strengths (Baschieri, A, 2009).

According to numerous authors who worked on vulnerability, poverty and its causes such as inequality and marginalisation are among the most important determinants of vulnerability (Adger and Kelly, 1999; Ribot, 1996 In Olmos, 2001). However, some authors argue that both the wealthy and the poor could be adversely affected by the impact of extreme weather events (O' Brien and Leichenko, 2000 In Olmos, 2001).From the above discussions on vulnerability, it can be viewed as "a function of exposure, sensitivity and adaptive capacity". The term "exposure" addresses the incidence of climate change impacts, the term "sensitivity" addresses the capacity of actors to be affected by climate change impacts and the term "adaptive capacity" or "resilience" addresses ability of actors to shield themselves and to recover from adverse climate change impacts. Vulnerability thus comes from social and physical factors, among physical factors concentrates on exposure and sensitivity and social factors accounts for assets, distribution of income, class etc. This means environmental and social systems together construct the vulnerability of households in a context dependent way (Paavola, 2008).

3. Conceptual Approaches to Vulnerability Assessment

3.1. Socio Economic Approach

Adger, 1999 described that the socioeconomic approach is mainly concerned with the social, economic, and political aspects of society and it focuses on the assessment of the socioeconomic and political status of individuals or social groups. The main limitation of the socioeconomic approach is that it focuses only on variations within society or social groups and overlooks the environment-based intensities, frequencies, and probabilities of environmental shocks, such as drought and flood (Deressa *et al*, 2008).

3.2. Biophysical or Impact Assessment Approach

The biophysical, or impact assessment, approach is mainly concerned with the physical impact of climate change on different attributes, such as yield and income (Fussel and Klein 2006). Fussel (2007) pointed that this is a risk-hazard approach and Kelly and Adger (2000) considered it as an end-point analysis. One major limitation of this approach is it only focuses on the physical damages due to climate change variables (Deressa *et al*, 2008).

3.3. Integrated Approach

According to Fussel (2007) the integrated assessment approach combines both the socioeconomic and the biophysical attributes in vulnerability analysis. He argued that this approach to vulnerability analysis conceptualizes vulnerability as a function of adaptive capacity, sensitivity, and exposure to climate change. According to Fussel and Klein, 2006 the risk-hazard framework (biophysical approach) corresponds most closely to sensitivity in the IPCC terminology. Adaptive capacity is largely consistent with the socioeconomic approach (Fussel, 2007). In the IPCC framework, exposure has an external dimension, whereas both sensitivity and adaptive capacity have internal dimension, which is implicitly assumed in the integrated vulnerability assessment framework (Fussel, 2007 in Deressa *et al*, 2008).

Even though the integrated assessment approach corrects the weaknesses of the other approaches, it also has its limitations. The main one is that there is no standard method for combining the biophysical and socioeconomic indicators. The relative importance of different variables used in this approach has not been taken into account and thus need much care in using this approach. The other weakness of this approach is that it does not account for the dynamism in vulnerability. The dynamism underlying in the process of adaptation involves continual change of strategies to take advantage of opportunities and which is missing in this approach (Deressa *et al*, 2008).

4. Adaptive Capacity and Vulnerability to Climate Change

Adaptive capacity is defined by IPCC (2001) as the "potential, capability, or ability of a system to adapt to climate change stimuli or their effects of impacts." One of the major focuses of our study is to characterize adaptive capacity of the selected regions to understand the factors influencing the adaptive capacity of that region. The Table 15.1 outlines the situations of climate risk based on the degree of impacts and adaptive capacity of a region or group or community. The quadrants represent artificial boundaries of our knowledge of anticipated impacts of climate change and capacity of livelihoods or regions to adapt to climate change impacts.

Table 15.1: Climate Change Adaptation Matrix: Common Situations of Vulnerability and Clusters of Climate Risk

Impacts	*Adaptive Capacity*	
	Low	*High*
High	Vulnerable communities	Development opportunities
Low	Residual risks	Sustainability

Source: Tol *et al.*, 2004.

Presumably, adaptive capacity varies widely but, to date, it has been difficult to empirically measure it and establish the relative importance of the factors identified by the IPCC (Brooks *et al.* (2005), In Alberini *et al.*, 2005). Brooks *et al.* (2005) argue that because adaptation does not occur instantaneously, the relationship between adaptive capacity and vulnerability depends crucially on the timescales and hazards with which we are concerned. The vulnerability, or potential vulnerability, of a system to climate change that is associated with anticipated hazards in the medium- to long-term will depend on that system's ability to adapt appropriately in anticipation of those hazards. However, vulnerability to hazards associated with climate variability that may occur in the immediate future will be related to a system's existing short-term coping capacity rather than its ability to pursue long-term adaptation strategies. We need hazard- and context-specific, determinants of vulnerability as vulnerability will exhibit substantial sub-national geographical and social differentiation at regional levels (Brooks *et al.*, 2005).Downing (2003) argue that adaptive capacity is seen as beyond wealth, which includes technology, institutions, social networks etc. which includes the kinds of processes underlying sustainable livelihoods.

5. Climate Change and Plantations

The study by Sivakumar *etal* (2005) showed that warming trend in India is about 0.57°C per 100 years. Analysis of rainfall data for India highlights the increase in the frequency of severe rainstorms over the last fifty years. Kumar and Parikh,

2001 showed that even after accounting for farm level adaptation, a 2 °C rise in mean temperature and a 7 per cent increase in mean precipitation will reduce net revenues by 8.4 per cent in India. These research results show that people are and will be with the challenge to address new threats beyond their experience or capacity to cope.Research study conducted at CPCRI showed that the elevated CO_2 and temperature influence entire physiology and biochemistry of coconut, arecanut and cocoa plants. Varietal differences in response were also observed in this study. Based on the simulation studies, it was reported that the coconut yields are likely to increase in Kerala, parts of Tamil Nadu and parts of Karnataka due to climate change. However, yield is projected to reduce in Andhra Pradesh (mostly irrigated area) and rain fed areas of other states. This study also indicated that climate change will alter the quality of copra and coconut oil during storage and the shelf life of copra and coconut oil is likely to reduce if the current storage practices are continued. Vulnerability to climate change comes from social and physical factors, among physical factors concentrates on exposure and sensitivity and social factors accounts for assets, distribution of income, class etc. (Paavola, 2008). The impacts and vulnerability are integrated with adaptation responses of farmers at their farm or household level, which may reduce the vulnerable status by affecting the livelihoods of farmers who are engaged in growing plantation crops.

The field studies conducted at CPCRI indicated that soil moisture conservation is one of the potential and important adaptation strategies to reduce the climate change impacts particularly in water scarce/limited conditions. Efficiency of adaptation strategies depend heavily on the accuracy and reliability of information collected. Extensive data collection on crop characteristics, soil and other natural resources in a large geographical platform is very time consuming and almost impossible using the traditional approaches. Indian Council of Agricultural Research (ICAR) and Indian Space Research Organization (ISRO) jointly conducted first multi spectral air borne study for identification of root wilt disease in coconut in 1969. Later on country level studies related to application of remote sensing technologies were initiated after the launch of IRS-1A. Crop acreage estimation, Yield and production estimation, disease stress identification, etc are some of the areas wherein remote sensing and advances in GIS technologies were utilized. Application of remote sensing is widely employed in natural resources management and precision farming. Advanced GIS tools compliment remote sensing in compilation, analysis and presentation, which in turn can be utilized for quick action plans, especially in disaster management. Adaptation of efficient strategies on cropping systems, microbial dynamics, crop and resource management is therefore a plausible option for increasing the resilience of farmers whose livelihood is based from growing plantation crops. Thus, there is strong need to assess and analyse the existing adaptation strategies followed in the plantation crops sector. Further, developing innovative climate resilient adaptation strategies in a multidisciplinary system approach will help to strengthen the coconut economy of India by alleviating the associated climate change risks in the sector.

6. The Research Gaps and Way Forward

Climate change phenomenon has positive or adverse effects on biophysical and socio economic aspects. These changes vary from region to region depending on their biophysical characteristics, development status etc. Therefore, these factors have much to do with impacts and vulnerability to climate change and climate change response options in that region. Thus, climate change impacts and vulnerability of a region need to be understood thoroughly to plan appropriate adaptation strategies for climate change resilience in plantation crops. The impacts of climate change on coconut based cropping systems have emerged as a critical concern in recent years as these are strictly suffering already from constrained use of natural resources, particularly water. The cropping system changes in these areas have much to do with the natural resource use, income, and living standards of the people. The management aspects in coconut based cropping systems related to plant health, soil and water management techniques are to be evaluated and modified in order to fit in the new climate change scenarios with higher efficiency in the pursuit of sustainable production.

Climate change and its variability significantly affect the crop-weather relationship and yield potential of the crop, physiology of which has to be understood wherein advanced crop modeling studies can be employed. Climate change and associated changes in structural and microbial dynamics of soil has a key role in crop growth and hence in nutrient management strategies. Weather parameters play a major role in determining the occurrence and severity of pests and diseases in coconut. Hence an efficient forewarning system incorporating the weather associated risks and uncertainties needs to be developed to improve the preparedness to handle biotic stress. Information on natural resources, socio economic variables, plant characteristics and their temporal changes in association with climate change is vital in planning climate adaptive strategies. Remote sensing and GIS tools are sophisticated and effective in collection analysis of extensive spatial and geographical data. Disaster management is complete only when the research findings and thus developed strategies reach the beneficiaries. A major link in this process is the effective collaboration of research organizations and implementing agencies.

It is imperative to assess the climate change impacts and vulnerability of plantation crop sector in the present scenario of unprecedented climate fluctuations. In this regardit is pertinent to evolve location specific climate resilient adaptation measures in plantation crops by conducting strategic research on cropping systems, associated microbial dynamism, plant health management and management of natural resources. Mapping of biophysical and socio- economic vulnerability to climate change in the plantation crops sector of the selected regions and analysis of past and present adaptation practices to evaluate its resilience to climate change are important aspects. The satellite data assimilation techniques to monitor crop systems and hydrological state over coconut based plantations incorporating GIS based thematic maps can create a big impact. It is also desirable to strengthen the functional linkages with various research and development institutions including Disaster Management Cells of different states.

7. Conclusions

Most of the regions in developing countries, particularly semi arid tropics, are vulnerable to the current climate change and climatic shocks in future. Therefore, understanding the nature and degree of vulnerability is the initial concern of any climate change study which helps to build adaptation strategies along with understanding the ground level adoption of adaptation strategies, to cope better with the climate change and variability in future.

Unlike in the case of annuals, it is not possible to make sudden changes in cropping systems with perennials and hence it is required to evolve appropriate adaptation measures in a time bound manner for sustained production and risk reduction. This in turn requires strategic research on cropping systems, associated microbial dynamism, plant health management and management of natural resources. For evolving robust technology products, functional linkages with concerned institutions need to be strengthened. For effective management of resources in the event of natural disasters, a pre-requisite is its impact and vulnerability assessment, appropriate protocols with remote sensing techniques and GIS has to be evolved in collaboration with front line research organizations. Further, liaison with State Disaster Management (SDM) Cells is required to deliver the technology products evolved for the benefit of farming communities in the occurrence of any disaster.

References

Adger, N. and Kelly, M. (1999). Social Vulnerability to Climate Change and the Architecture of Entitlements, In Olmos, S. (2001), Vulnerability and Adaptation to Climate Change: Concepts, Issues, Assessment Methods, Climate Change Knowledge Network, Foundation Paper,www.cckn.net.

Adger, W. N. (1999). Social vulnerability to Climate Change and Extremes in Coastal Vietnam. *World Development*, 27(2), 249–269.

Baschieri, A, (2009). Poverty and vulnerability: a static vs dynamic assessment of a Population subjected to climate change shock in Sub-Saharan Africa. XXVI IUSSP International Population Conference, 2009(forthcoming), Session: 1203 Interrelations between population and climate change, Morocco.

Brooks, N., Adger, W.N. and Kelly P.M. (2005). The determinants of vulnerability and adaptive capacity at the national level and the implications for adaptation. *Global Environmental Change*, 15, 151–163.

Chaudhuri, S., Jalan J., Suryahadi A. (2002). "Assessing Household Vulnerability to poverty from Cross-Sectional Data: A Methodology and Estimates from Indonesia". Columbia University, Department of Economics, Discussion Paper Series 0102-52.

Deressa, T., Hassan, R. M. and Ringler, C. (2008). Measuring Ethiopian Farmers' Vulnerability to Climate Change Across Regional States, IFPRI Discussion Paper 00806. International food policy research institute(IFPRI) Washington.

Deressa, T.T., Hassan, R.M., Ringler, C., Alemu, T. and Yusuf, M. (2009). Determinants of farmers' choice of adaptation methods to climate change in the Nile Basin of Ethiopia. *Global Environ. Change*, doi,10.1016/j.gloenvcha.2009.01.002.

Downing, T.E. (2003). Lessons from famine early warning and food security for understanding adaptation to climate change: toward a vulnerability/adaptation science, In (Eds)Joel, B. Smith, Richard, J.T Klein and Huq, S, Climate change, adaptive capacity and development, chapter 5, pp. 71- 100 imperial college press.

Fussel, H. (2007). Vulnerability: A Generally Applicable Conceptual framework for Climate Change Research. *Global Environmental Change*, 17(2), 155–167.

Fussel, H.-M., and R. J. T. Klein. (2006). Climate change vulnerability assessments: An evolution of conceptual thinking. *Climatic Change* 75(3), 301–329.

IPCC, (2007). Climate Change 2007: The Physical Science Basis. Contribution of Working Group I to the Fourth Assessment Report of the Intergovernmental Panel on climate Change, Solomon, S., D. Qin, M. Manning, Z. Chen, M. Marquis, K.B. Averyt, M. Tignor and H.L. Miller (eds.). Cambridge: Cambridge University Press.

IPCC, (2001). Climate change, 2001: Impacts, Adaptation and Vulnerability. Contribution of working group II to the third assessment report of the IPCC. Geneva: UNEP/WMO.

Kelly, P. M., and W. N. Adger. (2000). Theory and practice in assessing vulnerability to climate change and facilitation adaptation. *Climatic Change*,47(4), 925–1352.

Kumar, K.S.Kavi, and Jyoti Parikh (2001). Socio-economic Impacts of Climate Change on Indian Agriculture, *International Review of Environmental Strategies*, 2(2), pp. 277-293.

O' Brien, K. and Leichenko, R. (2000). Double Exposure: Assessing the Impacts of Climate Change within the Context of Globalisation, In: Olmos, S. (2001), Vulnerability and Adaptation to Climate Change: Concepts, Issues, Assessment Methods, Climate Change Knowledge Network, Foundation Paper, www.cckn.net.

Olmos, S. (2001). Vulnerability and Adaptation to Climate Change: Concepts, Issues, Assessment Methods, Climate Change Knowledge Network, Foundation Paper, www.cckn.net.

Paavola, J. (2008). Livelihoods, Vulnerability and Adaptation to Climate Change in Morogoro, Tanzania, Environmental science and policy, II,642-654.

Ribot, J., Magalhaes, A., and Panagides, S. (eds). (1996). Climate Variability, Climate Change and Social Vulnerability in the Semi-arid Tropics, In Kao, C-W(2009). A White Paper on the Impacts of Climate Change on Drylands, Seminar on Climate Change, Department of Geography, University of Florida, USA.

Sivakumar M. V. K., DasH. P., Brunini O. (2005). Impacts of present and future climate variability and change on agriculture and forestry in the arid and semi-arid tropics. *Clim. Change*, 70,31–72.

TERI. (2005). Adaptation to Climate Change in the Context of Sustainable Development, Background paper, The Energy and Resources Institute (TERI), New Delhi, India.

Tol, R.S.J., Downing, T.E., Kuik,O.J. and Smith, J.B. (2004). Distributional aspects ofclimate change impacts, Global Environmental Change,14, pp. 259–272.

Index

A

Abscisic acid (ABA) 171, 178

Agricultural management 56

Agro-climatic requirements 128

Agronomic adaptation 10

Amino acids 25, 175

Areca catechu 61, 71, 73

Arecanut 6, 7, 12, 13, 24, 61-73, 83-85, 158, 159, 166

Atmospheric carbon dioxide 105

B

Biochemical response 45

Biofuel 36, 101

Biotechnological tools 190, 192

Biotic stresses 53, 55, 140, 226

Black pepper 7, 75-79, 229

C

C_3 plants 17, 105, 118, 120

Canopy temperature 176

Carbon dioxide 22, 87, 88, 96, 105, 113-121, 131-133, 141, 157, 164, 182, 183, 186

Carbon-isotope discrimination (CID) 174

Carbon sequestration 38, 55, 69, 70, 96, 113, 151, 157, 158, 165

Carbon stock 159-162

Cardamom 7, 75, 81, 84, 230

Cashew 6, 9, 87-99, 159, 165, 226-229

CDNA's 193

CH_4 1, 88, 94

Chlorophyll florescence 172

Chlorophyll Index 173

Chloroplasts 50

Cisgenesis 198

Clean development mechanism (CDM) 113, 116, 166

Climate change 2-4, 15, 27, 41, 58, 61, 72, 80-92, 102, 106, 118-122, 143, 150, 168, 182, 209, 217, 232-244

Climate change hot spots 231

Climate proofing 39

Climate resilience agriculture 219, 232

Climate resilient varieties 39, 51

Climate smart agriculture 232

Climate warming 149, 150
Climatic geography 61
Clustered regularly interspaced short palindromic repeat (CRISPR) 199
CO_2 1-3, 5, 15, 17-25, 40-51, 56-88, 94, 110, 113, 123, 131, 134, 141, 161-171, 174, 183, 187, 209-216, 241
Cocoa 6, 8, 24, 61-72, 83, 84, 158-166, 195, 215, 229, 230
COCOA Genome database 190
Coconut 6, 12, 15, 18-32, 159, 169, 180, 183, 195, 202, 209, 215-219, 225-227, 237
Coconut based farming system 212
Coffee 6, 9, 159, 164, 167, 195, 228, 229
Colletotrichum gleosporoides 91
Composting 115
Conservation agriculture 232
Contingency plan 95
Crop diversification 35
Crop germplasm 52
Crop insurance 10, 35, 96
Crop management 10, 32, 33, 54, 81, 95, 116, 125, 224, 230
Crop models 54
Cropped area 221, 232
Crop varieties 54, 56, 112, 125, 140
Crop yield 134, 136, 137, 179

D

Deforestation 1, 15, 82, 113, 219, 221, 231
DNA arrays 192, 201
DNA technology 197
Drip irrigation 30, 34, 35, 38, 54, 69, 70, 72, 92, 99, 209, 215, 229
Drip system 54, 215
Drought 3-6, 16-18, 26, 39, 82, 87, 94, 95, 103, 106, 111, 219, 220, 225, 231
Drought intensity index (DII) 53, 179
Drought stress 11, 17, 50, 57, 171-180, 185, 187, 200, 210
Drought tolerant coconut palms 34, 58, 218
DsRNA 196, 197

E

ECO2 51, 210, 211
Elaeis guineensis 120, 122, 190
Elettaria cardamomum 75, 84, 85
Elevated temperature (ET) 51
El Nino 27, 107-110, 117, 118
El Niño Southern Oscillation (ENSO) 134
Empty fruit bunches (EFB) 115
Environmental factors 17, 41, 71, 124, 176, 229
Environmental modelling 4, 18
Environmental rating 89

F

Farming system 209, 212
Food and Agriculture Organization (FAO) 169
Fossil fuels 2, 3, 15, 113
Free atmospheric carbon dioxide enrichment (FACE) 19
Free atmospheric temperature elevation (FATE) 21
Fresh fruit bunches (FFB) 109

G

Gene expression 192, 195
Genetic adaptation 9
Genetic improvement 34
Geographic information system (GIS) 138
Geometric mean (GM) 53
Geometric mean productivity (GMP) 52, 179
Germination 23, 25, 97, 178
Gleosporium mangiferae 91

Global agriculture production 169
Global climate models 3
Glyricidia 94
Green credits 39
Greenhouse gases (GHGs) 1-3, 10, 15, 36, 88, 92, 94, 107, 114
Green manuring 94
Green technologies 39
Gross domestic product (GDP) 45

H

H_2O 1, 116
Hairpin RNA (hpRNA) 196, 197
Harvest index 35, 164, 186
Hevea brasiliensis 11, 145, 152-, 166, 190, 202-206
High density multi species cropping system (HDMSCS) 162
Hi-tech farming 232
Hybridization method 192
Hydrogen peroxide 175

I

Indian Space Research Organization (ISRO) 241
InfoCrop-COCONUT model 21, 27
Infra-Red Thermography 176
In situ 10, 34, 58, 68, 93, 218
Integrated approach 239
Integrated farming 68, 232
Intercropping 72, 94, 218
Inter-Governmental Panel on Climate Change (IPCC) 2, 11, 40, 102, 189, 237
Intragenesis 198, 201
Irrigated conditions 8, 58, 60, 63, 101, 108, 113, 121, 176, 186, 217, 225

L

Land ecological suitability evaluation (LESE) 138
Leaf night respiration 170
Leaf water relation 173

M

Marker assisted selection 9, 59
Mean annual temperature 1, 45, 47, 89, 209
Mean productivity (MP) 52, 179
Microarrays 192
Moisture content 215
Moisture deficit 45, 170
Molecular breeding 192
Mulching 93, 214, 217

N

National action plan on climate change (NAPCC) 232
National initiative on climate resilient agriculture (NICRA) 234
Natural resource management 35
Next generation sequencing (NGS) 193
Normalized difference vegetation index (NDVI) 177
Novel Genes 193

O

Oil palm 101, 105, 109, 110, 113-115, 119, 121, 158, 159, 165, 205
Omics 199
Open top chamber (OTC) 19, 51, 210, 211
Osmotic adjustment 50, 175

P

Palm oil mill effluent (POME) 115
PCR (RT-PCR) 194
Pest and disease management 96
Pest proliferation 61
Pests and diseases 107
Phenological development 61
Phenology 61, 116, 202
Phenotyping 117, 169, 181, 186

Phenotyping tools 169
Photosynthesis 17, 23, 39, 48-53, 56, 66, 71, 96, 103-106, 116, 117, 119, 122, 132, 157, 170-177, 181-188, 210, 211, 218
Physiological mechanism 175
Physiological traits 175, 186
Phytophthora capsici 79, 83
Phytophthora spp. 148
Piper nigrum 7, 13, 75, 83-86
Plantation belt 62
Plantation crops 1, 5, 6, 11, 15, 19, 21, 31, 36, 42-45, 60, 64, 70-75, 84-87, 99, 101, 121-123, 145, 152, 157-159, 166-169, 183, 188, 189, 195, 198, 209, 216-219, 237
Plantation crops systems 36
Plant physiology 190, 197, 206
Precipitation 16, 64, 103
Precision farming 232, 241
Proline 25, 104, 175, 176, 184
Protected cultivation 232
Proteomics 190

R

Rainfall 3-9, 16, 17, 19, 25, 27, 45-48, 53, 55, 61-70, 75-90, 94-96, 102, 107, 110, 112, 123, 129, 130, 134-138, 145-154, 189, 209-213, 219-240
Rainfall deficit management 95
Rainfed conditions 88, 108, 225, 226, 229
Reforestation projects 157
Relative humidity 6, 7, 61-67, 77, 80, 89-91, 130, 148, 176, 213
Rice productivity 231
RNA interference (RNAi) 196
RNA sequencing (RNA-Seq) 193
Rubber 6, 9, 11, 12, 85, 121, 145-155, 159, 163, 166, 167, 185, 195, 231
Rubber plantations 151

S

Sesbania 94
Simple sequence repeat (SSR) 192
Single nucleotide polymorphism 192
Site specific nutrient management (SSNM) 94
Small cardamom 81
Socio economic approach 239
Soil and land management 112
Soil and water conservation 93, 99, 213-216
Soil conservation 54, 97, 98
Soil moisture conservation 30-34, 54, 215, 216, 241
Soil organic matter 151
Soybean 17, 18, 182
State action plan on climate change (SAPCC) 231, 234
Statistical analysis 129
Stomatal conductance 22, 47, 48, 56, 104-106, 120, 174-176, 210
Stress tolerance 9-11, 52, 56, 175, 179-183, 187, 200
Stress tolerance index (STI) 52, 179
Super oxide dismutase (SOD) 50
Sustainable agriculture 143, 232
Synthetic rubbers 152

T

TaMnSOD (ROS scavenger) 197
T-DNA insertional mutagenesis 195
Tea 6, 8, 96, 123-128, 131, 138, 141, 158, 164, 168, 195
Tea mosquito bug (TMB) 96
Temperature 16, 21-26, 40-43, 47, 51, 56, 79, 102, 117, 129, 135, 137, 149, 152, 170, 176, 183, 211, 215, 217, 221
Theobroma cacao 61, 62, 71-74, 167, 191, 194, 200, 203
Tolerance (TOL) 52, 179

Traditional agriculture 232
Trait specific gene discovery 190
Transcriptionactivator like effector nucleases (TALENS) 199
Transcriptomic analysis 194
Transcriptomics 190, 194
Transpiration 44, 46, 47, 48, 51, 67, 102-106, 110, 174-176, 210

U

United Nations Conference on Environment and Development (UNCED) 237
United Nations Framework Convention on Climate Change (UNFCCC) 2, 15, 237

V

Value addition 10, 35
Viral diseases 79

W

Water conservation 93, 99, 213, 214, 216
Water consumption 45
Water harvesting 35
Water management 54, 99, 114
Water use efficiency 50, 112
Weather forewarning 35

Z

Zea mays L. 183, 187, 191
Zero burning 114, 115
Zinc finger nucleases (ZFN) 199

Colour Plates

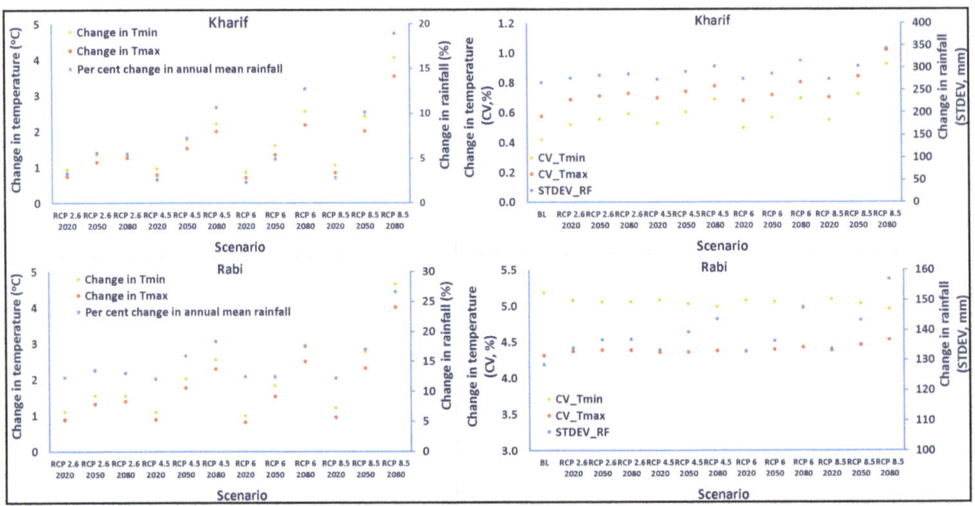

Figure 1.1: Projected Climatic Variability in *Kharif* and *Rabi* Seasons in India. p. 4

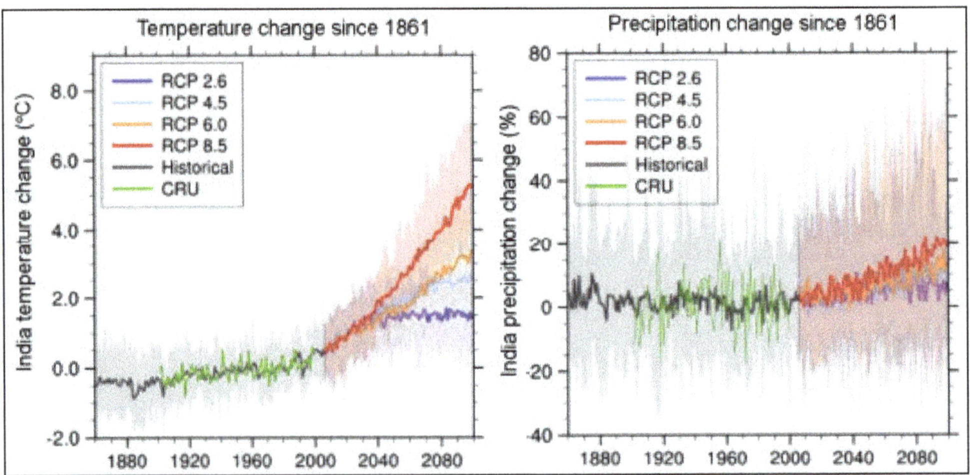

Figure 2.1: Projected Change in Mean Annual Temperature and Precipitation Over India in different Climate Scenarios (*Source*: Chaturvedi *et al.,* 2012). p.16

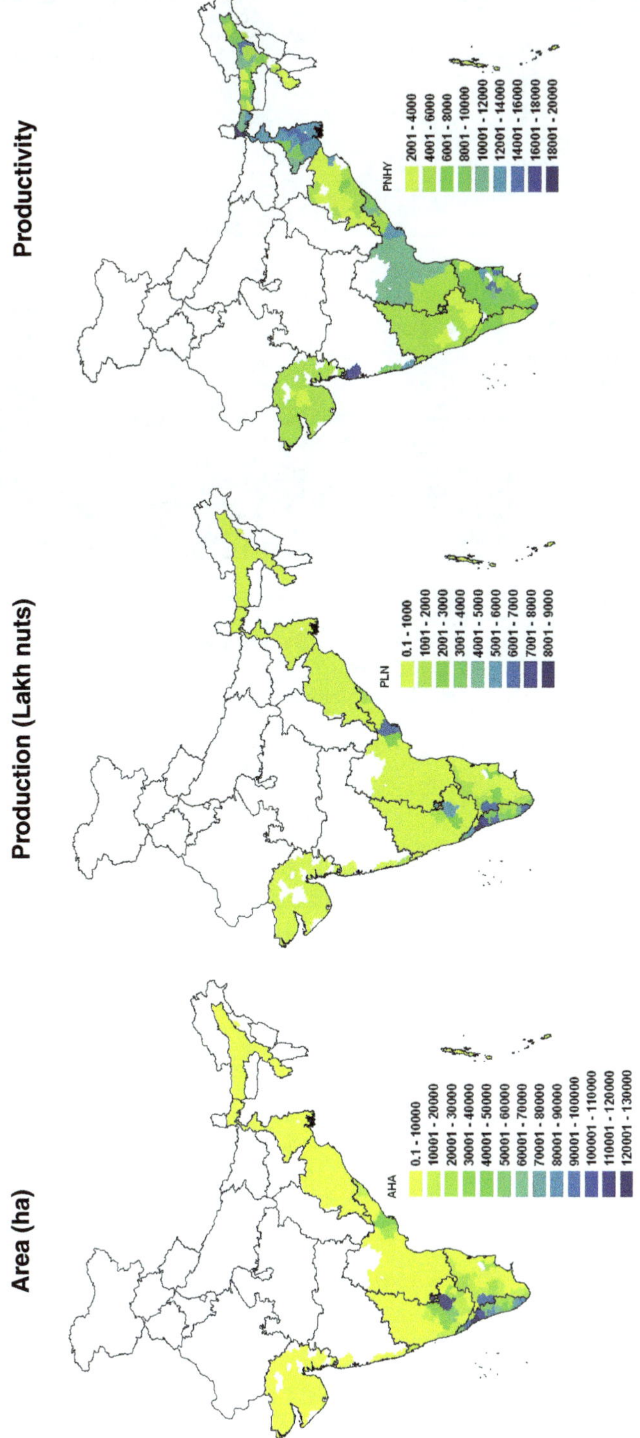

Figure 2.2: Spatial variation in area production and productivity of coconut in India p.20

Figure 2.3: Open Top Chamber Facility for Studying the Effect of Elevated CO_2 and Temperature at CPCRI, Kasaragod, India. p.22

Figure 2.4: Response of Coconut Seedling to Elevated CO_2; (a) Control (b) 550 ppm CO_2 and (c) 700 ppm CO_2. p.22

Figure 2.5: Response of Coconut Seedling to High Temperature; (a) Control (b) Ambient +3°C (c) 550 ppm CO_2 + Ambient +3°C. p. 23

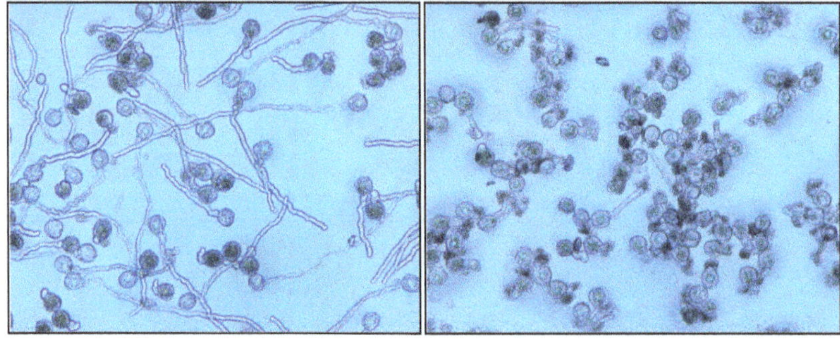

Figure 2.6: Pollen Germination of WCT at 25 (Left) and 40°C (Right). p. 23

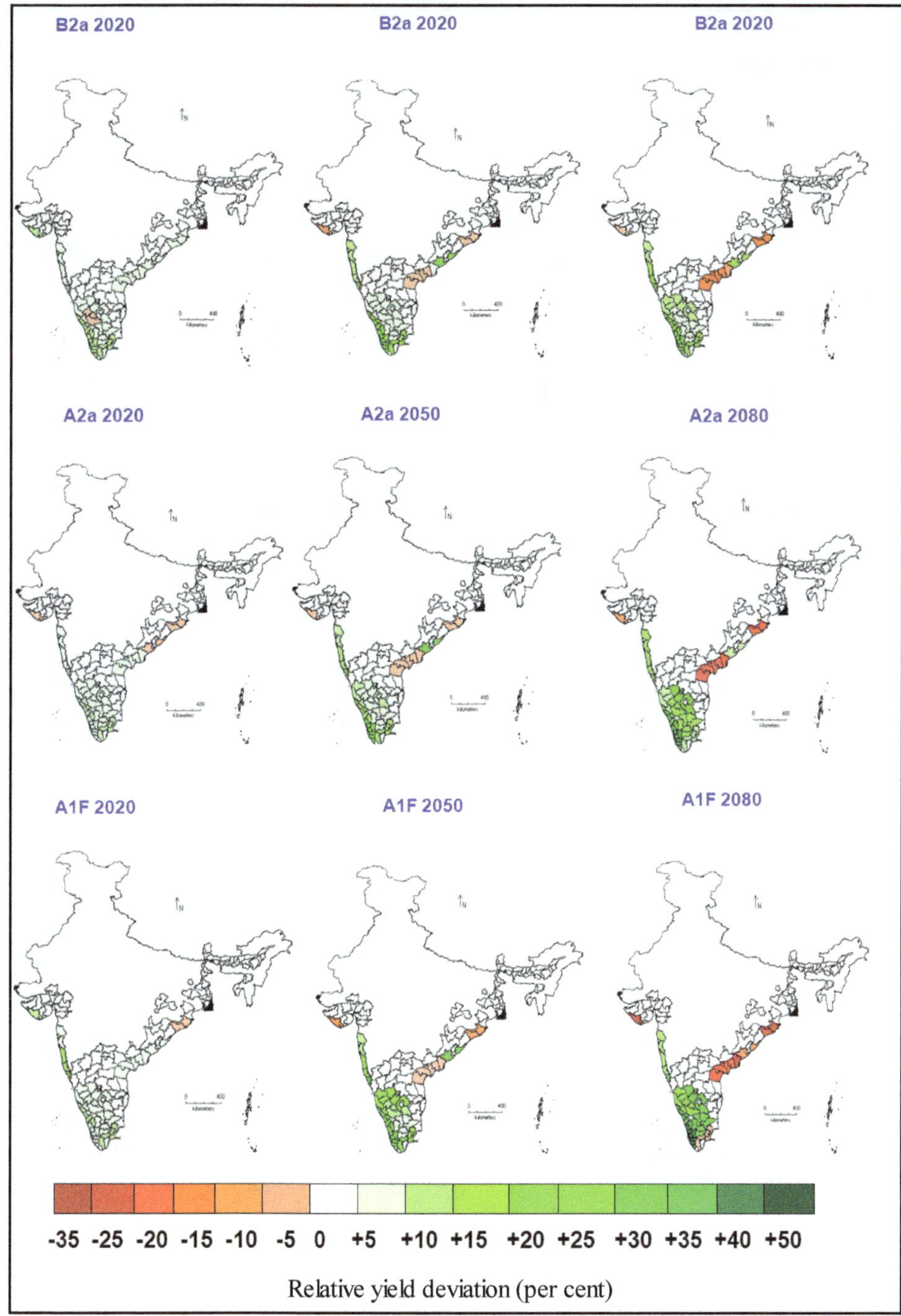

Figure 2.9: Impact of Climate Change on Coconut Productivity in 2020, 2050 and 2080 Climate Scenarios Based on HadCM3 Scenarios. The yield change is relative to mean productivity of coconut for five years (2000-2005) period (*Source*: Naresh Kumar, 2010). p.28

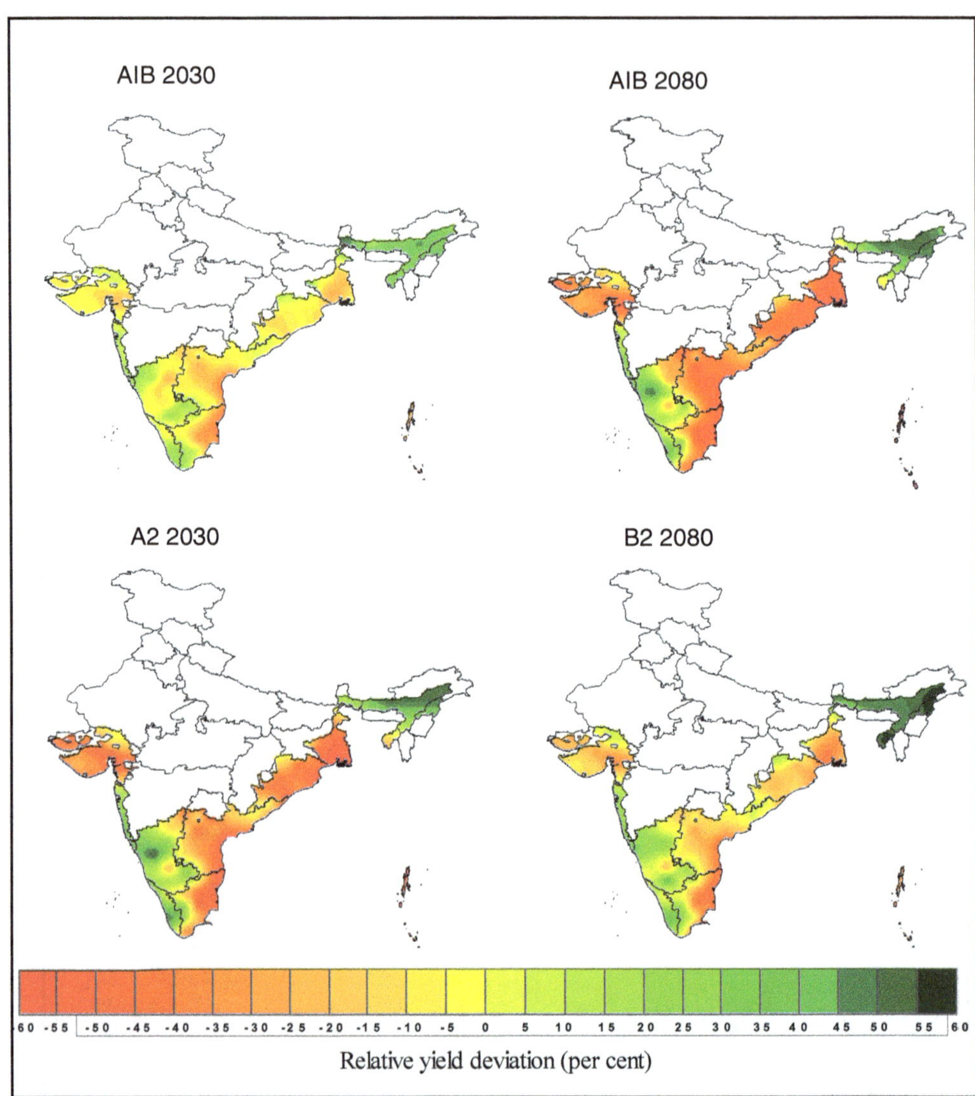

Figure 2.10: Spatial Variation in Impact of Climate Change on Coconut Productivity in India Based on High Resolution PRECIS A1B 2030 and 2080; A2-2080 and B2-2080 scenarios. The yield change is relative to mean productivity of coconut for five years (2000-2005) period (*Source*: Naresh Kumar and Aggarwal, 2013). p.29

Figure 2.14: Effect of Drip Irrigation on Coconut Plantations during Consecutive Drought Years (1998-2002) in Pollachi Area of Tamil Nadu. Plantation in the left side is with drip irrigation. p.34

Figure 2.15: *In situ* Drought Tolerant Palms Identified in Farmers' Fields in different Agro-climatic Zones and being Used in Population Improvement Studies (*Source*: Naresh Kumar, 2004). p.35

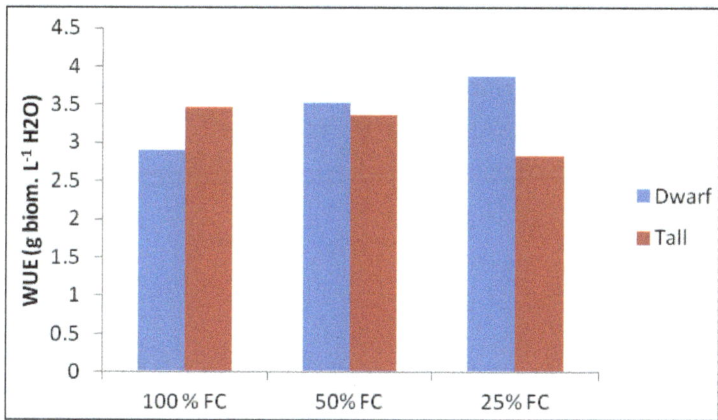

Figure 3.1: WUE of Tall and Dwarf Genotypes at different Moisture Regimes (CD at 5 per cent : 0.8). p.47

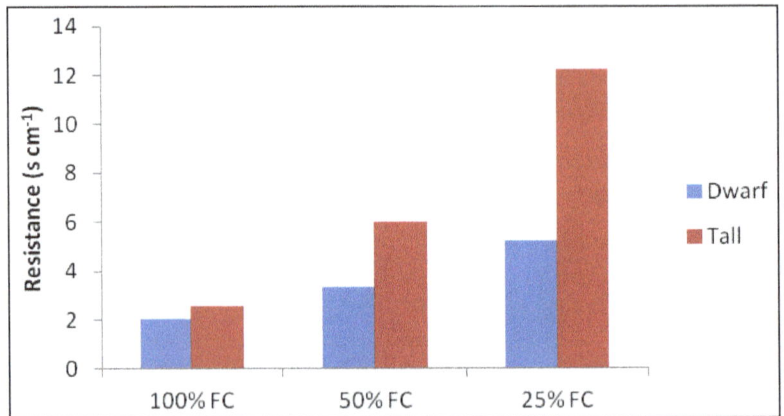

Figure 3.2: Stomatal Resistance of Tall and Dwarf Genotypes at different Moisture Regimes (CD at 5 per cent : 2.1). p.48

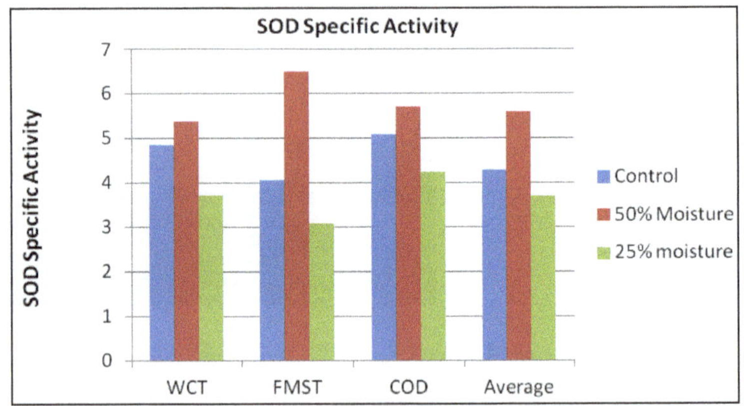

Figure 3.3: SOD Specific Activity under Induced Drought Stress. p.50

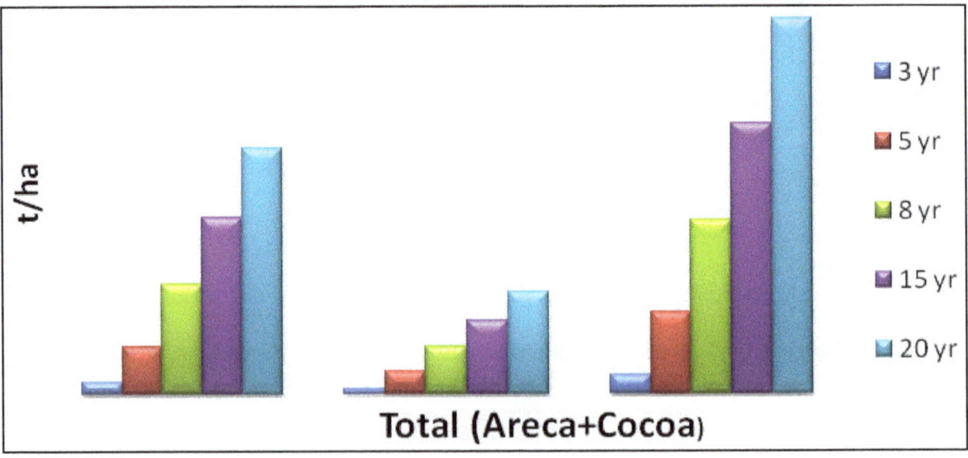

Figure 4.1: Carbon Sequestration in Areca and Cocoa System. p.70

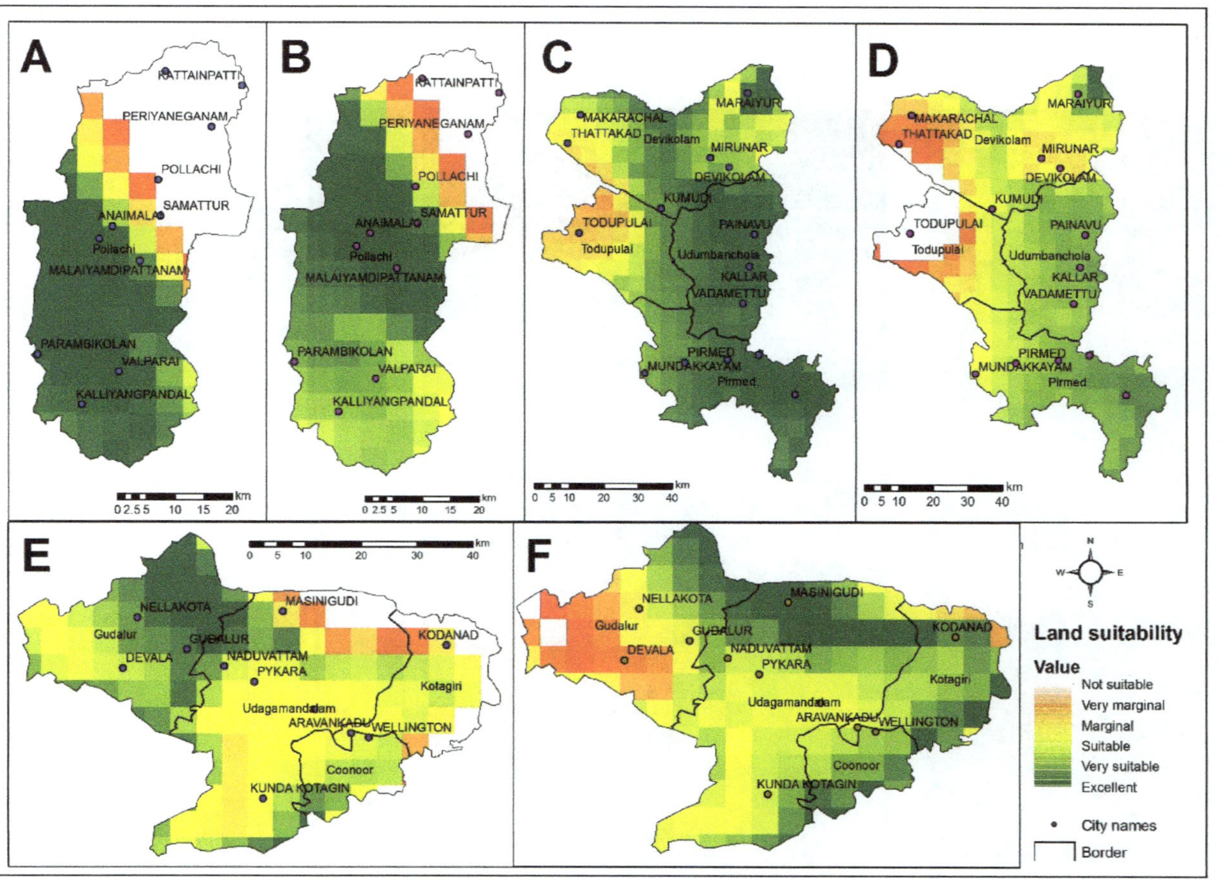

Figure 8.11: A Land Suitability Evaluation (LESE) Map of Tea Production Regions of South India. Land suitability under baseline (a, c, d); future scenario (b, d, f); a and b: the Anamalais; c and d Munnar and Vandiperiyar; e and F: the Nilgiris. p.139

Figure 13.1: Soil and Water Conservation in Coconut Basins. (a) Mulching with coconut leaves, (b) Burial of coconut husk, (c) Burial of composted coir pith. p.214

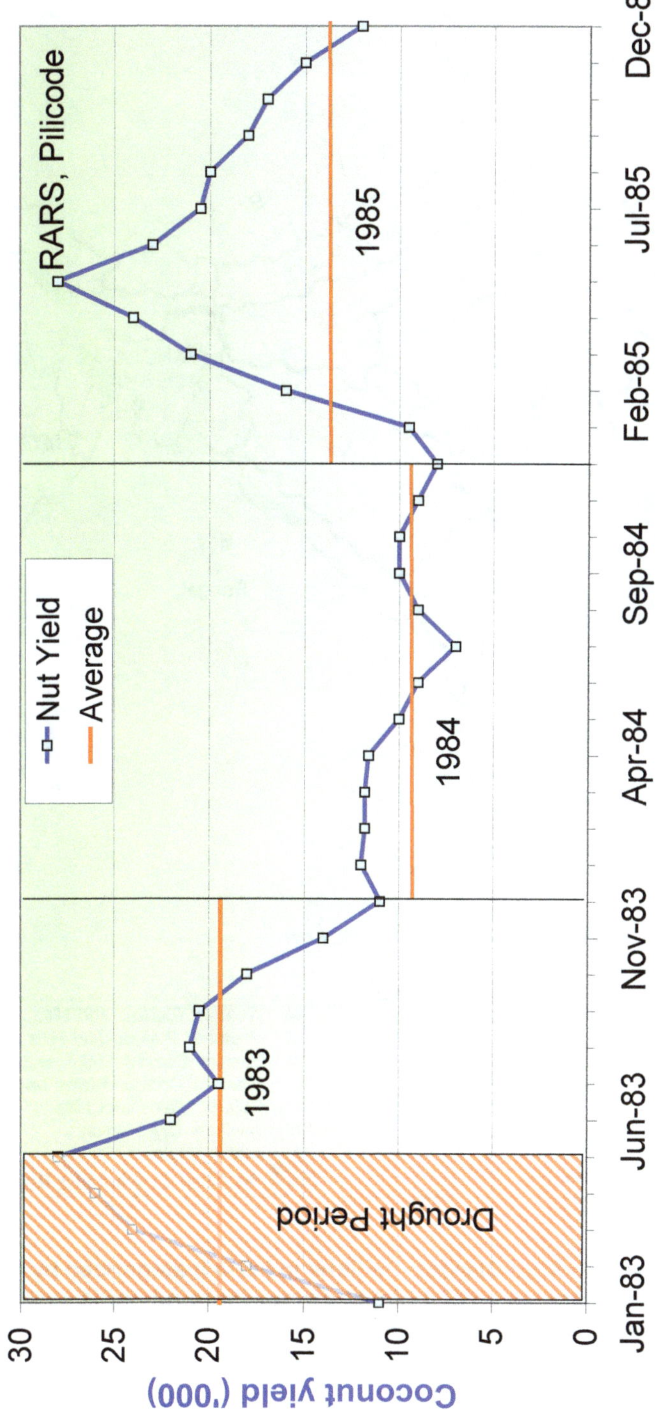

Figure 14.4: Effect of Drought during Summer 1983 on Coconut Yield at RARS, Pilicode. p.227

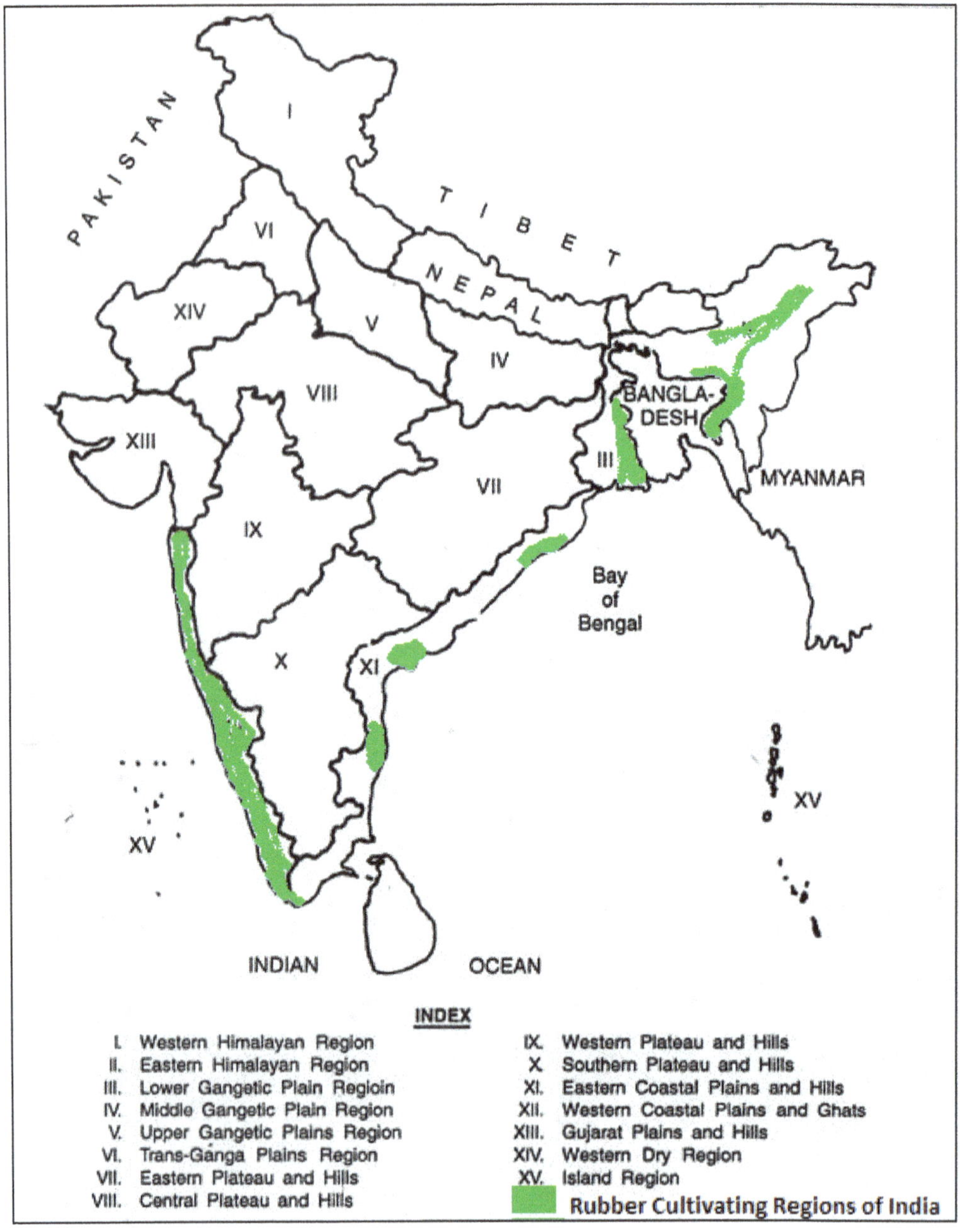

Figure 9.1: Agro Climatic Regions in India (categorised by the Planning Commission of India) showing where rubber is cultivated. Hatched areas only indicative and not drawn to scale. p.146

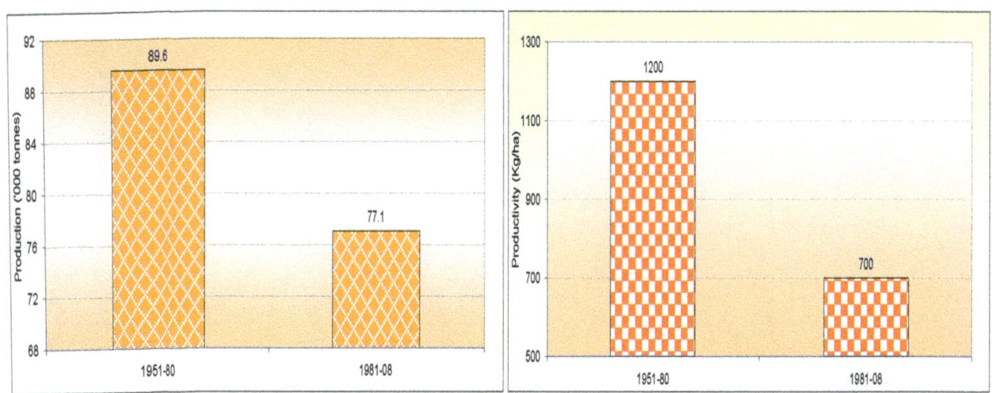

Figure 14.5: Tri-decadal Production and Productivity of Cashew in Kerala. p.228

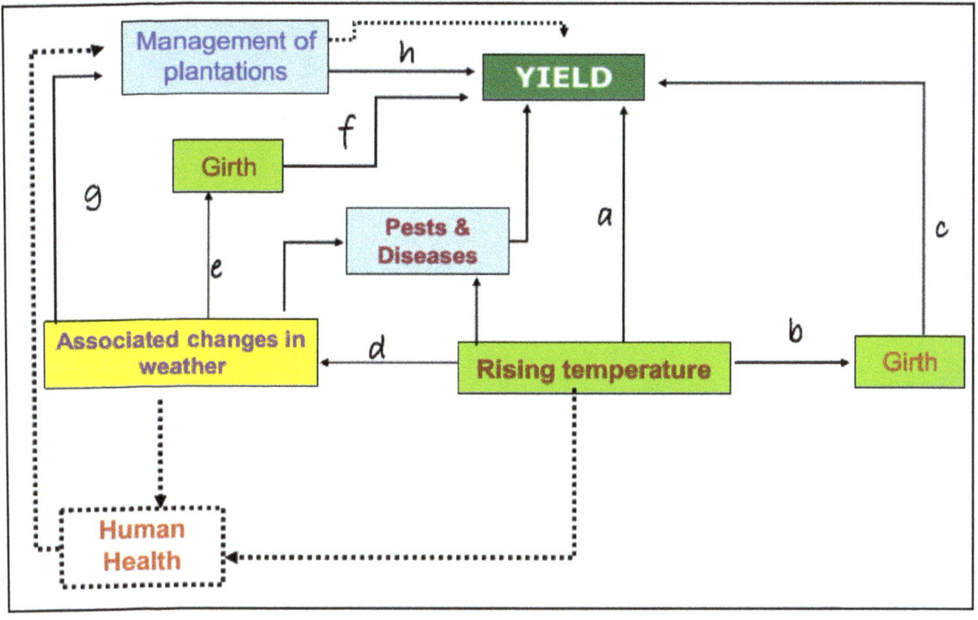

Figure 9.3: Schematic Chart Showing the Direct and Indirect Effects of Climate Warming on Growth and Yield of Rubber (after Satheesh and Jacob, 2011). p.150

www.ingramcontent.com/pod-product-compliance
Ingram Content Group UK Ltd.
Pitfield, Milton Keynes, MK11 3LW, UK
UKHW021010290726
14059UKWH00001BA/54

9 789386 071712